Optical Fiber Technology and Applications

Recent advances

IOP Series in Emerging Technologies in Optics and Photonics

Series Editor

R Barry Johnson, a Senior Research Professor at Alabama A&M University, has been involved for over 50 years in lens design, optical systems design, electro-optical systems engineering, and photonics. He has been a faculty member at three academic institutions engaged in optics education and research, employed by a number of companies, and provided consulting services.

Dr Johnson is an IOP Fellow, SPIE Fellow and Life Member, OSA Fellow, and was the 1987 President of SPIE. He serves on the editorial board of *Infrared Physics & Technology* and *Advances in Optical Technologies.* Dr Johnson has been awarded many patents, has published numerous papers and several books and book chapters, and was awarded the 2012 OSA/SPIE Joseph W Goodman Book Writing Award for Lens Design Fundamentals (second edition). He is a perennial co-chair of the annual SPIE Current Developments in Lens Design and Optical Engineering Conference.

Foreword

Until the 1960s the field of optics was primarily concentrated in the classical areas of photography, cameras, binoculars, telescopes, spectrometers, colorimeters, radiometers, etc. In the late 1960s optics began to blossom with the advent of new types of infrared detectors, liquid crystal displays (LCDs), light emitting diodes (LEDs), charge coupled devices (CCDs), lasers, holography, fiber optics, new optical materials, advances in optical and mechanical fabrication, new optical design programs, and many more technologies. With the development of the LED, LCD, CCD, and other electro-optical devices, the term 'photonics' came into vogue in the 1980s to describe the science of using light in the development of new technologies and the performance of a myriad of applications. Today optics and photonics are truly pervasive throughout society and new technologies are continuing to emerge. The objective of this series is to provide students, researchers, and those who enjoy self-education with a wide-ranging collection of books that each focus on a relevant topic in the technologies and applications of optics and photonics. These books will provide knowledge to prepare the reader to be better able to participate in these exciting areas now and in the future. The title of this series is *Emerging Technologies in Optics and Photonics* where 'emerging' is taken to mean 'coming into existence', 'coming into maturity', and 'coming into prominence'. IOP Publishing and I hope that you find this series of significant value to you and your career.

Optical Fiber Technology and Applications

Recent advances

Edited by
Mário F S Ferreira
Department of Physics, University of Aveiro, Aveiro, Portugal

Mukul Chandra Paul
CSIR–Central Glass & Ceramic Research Institute, Fibre Optics and Photonics Division, Govt. of India, Kolkata, India

IOP Publishing, Bristol, UK

ISBN 978-0-7503-3243-9 (ebook)
ISBN 978-0-7503-3241-5 (print)
ISBN 978-0-7503-3244-6 (myPrint)
ISBN 978-0-7503-3242-2 (mobi)

DOI 10.1088/978-0-7503-3243-9

Version: 20210801

IOP ebooks

British Library Cataloguing-in-Publication Data: A catalogue record for this book is available from the British Library.

Published by IOP Publishing, wholly owned by The Institute of Physics, London

IOP Publishing, Temple Circus, Temple Way, Bristol, BS1 6HG, UK

US Office: IOP Publishing, Inc., 190 North Independence Mall West, Suite 601, Philadelphia, PA 19106, USA

'Dream, dream, dream. Dreams transform into thoughts and thoughts result in action.'

'To succeed in your mission, you must have single-minded devotion to your goal.'

'Creativity is seeing the same thing but thinking differently.'

'Failure will never overtake me if my determination to succeed is strong enough.'

'All of us do not have equal talent. But, all of us have an equal opportunity to develop our talents.'

—Dr APJ Abdul Kalam

Contents

Preface xiii

Acknowledgments xvii

Editor biographies xviii

Contributors xx

List of abbreviations xxiii

1 A new class of specialty optical fibers based on a novel material composition of the doping host for the study of optical amplification 1-1

Mukul Chandra Paul, Alexander V Kir'yanov, Yuri Barmenkov, José Duarte, Joaquim P Leitão, Mário F Ferreira, Ana M Rocha, Randall Youngman, Mrinmay Pal, Anirban Dhar, Shyamal Das and Debjit Dutta

1.1 Introduction 1-1

1.2 Erbium-doped specialty optical fiber based on alumina–germania–calcia–yttria–chromium–silica glass 1-4

1.2.1 Fabrication and material properties 1-4

1.3 Optical properties 1-6

1.3.1 Linear absorption 1-6

1.3.2 Nonlinear absorption 1-7

1.3.3 Emission spectra and lifetimes 1-8

1.4 Gain 1-13

1.5 An erbium-doped specialty optical fiber based on phase-separated alumina–silica glass 1-15

1.5.1 The basic principle for the phase separation phenomenon of optical fiber preforms 1-15

1.5.2 Sample preparation 1-17

1.5.3 Material and optical characterization 1-19

1.6 Results and discussion 1-20

1.7 An erbium-doped specialty optical fiber based on zirconia–yttria–alumina–barium silica glass 1-30

1.7.1 Fiber fabrication and characterization 1-30

1.7.2 Optical amplification 1-34

1.8 Conclusions 1-36

Acknowledgements 1-37

References 1-38

2 A wideband optical amplifier with a hafnia–bismuth–erbium co-doped fiber as the active medium 2-1
Alabbas A Al-Azzawi, Mukul Chandra Paul and Sulaiman Wadi Harun

2.1 Introduction 2-1
2.2 The fabrication and characteristics of HB-EDF 2-2
2.3 The wideband HB-EDFA using a parallel structure 2-3
2.3.1 A parallel HB-EDFA with a total erbium fiber length of 2 m 2-4
2.3.2 A parallel HB-EDFA with a total erbium fiber length of 1.72 m 2-6
2.4 Performance comparison of the proposed parallel HB-EDFAs 2-7
2.5 A wideband HB-EDFA using a series configuration structure 2-7
2.5.1 Forward pumping 2-10
2.5.2 Backward pumping 2-11
2.6 Comparison of the performance of forward and backward pumping 2-13
2.7 The effect of total HB-EDF length on the backward pumped amplifier 2-13
2.8 Conclusions 2-15
References 2-17

3 Photon noise in a continuous-wave ytterbium-doped fiber laser 3-1
Yuri Barmenkov, Alexander V Kir'yanov, Pablo Muniz-Cánovas, José L Cruz and Miguel V Andrés

3.1 Introduction 3-1
3.2 The ytterbium-doped fiber laser: structure and basic characteristics 3-3
3.2.1 Ytterbium-doped fiber laser arrangement 3-3
3.2.2 Basic characteristics of the ytterbium-doped fiber laser 3-4
3.3 Coherence properties 3-5
3.3.1 Autocorrelation function of the laser signal 3-5
3.3.2 Repeatability of the noise pattern 3-8
3.4 Laser photon statistics 3-8
3.4.1 Theoretical background 3-9
3.5 Experimental results 3-11
3.6 The effect of photon statistics on optical spectrum broadening 3-13
3.7 Laser RF spectra 3-14
3.8 Conclusions 3-16
Acknowledgements 3-16
References 3-16

4 Development of borosilicate optical fibres for particle ionizing radiation dosimetry 4-1

H T Zubair, Amal Alqahtani, S F Abdul Sani, Hairul A Abdul-Rashid and D A Bradley

4.1 Introduction 4-1

4.2 The synthesis and physical characteristics of borosilicate media 4-4

4.2.1 The melt quenching technique 4-4

4.2.2 The vapour deposition technique 4-6

4.2.3 Heavy metal coating 4-7

4.3 Outlook for borosilicates for radiation dosimetry 4-8

4.4 Neutron detection with borosilicate glass 4-11

4.4.1 Motivation 4-11

4.4.2 The case for borosilicate glass as a potential neutron detection media 4-11

4.4.3 Photoneutron cross-section 4-12

4.4.4 Photoneutron production 4-13

4.4.5 Neutron detection factors 4-13

4.4.6 Materials and methods 4-13

4.5 Results and discussion 4-14

4.5.1 Activation products in borosilicate glass 4-18

4.6 A germano-borosilicate flat fibre dosimeter 4-18

4.6.1 Experimental procedures 4-20

4.6.2 Experimental studies 4-20

4.7 Conclusions 4-25

References 4-25

5 The ionizing radiation effect of erbium, ytterbium, or bismuth doped/co-doped optical fibres 5-1

Yanhua Luo, Desheng Fan, Mingjie Ding, Qiancheng Zhao, Binbin Yan, Jianxiang Wen and Gang-Ding Peng

5.1 Introduction 5-1

5.2 Ionizing radiation effects on Er/Yb co-doped optical fibre 5-3

5.2.1 Radiation-induced photodarkening 5-3

5.2.2 A reversible thermal effect in irradiated EYDF 5-4

5.3 Ionizing radiation effects in Bi-doped optical fibres 5-4

5.3.1 Activation of BAC from inactive bismuth 5-5

5.3.2 Factors that influence the radiation effect 5-7

5.3.3 Thermally induced effects in irradiated BEDF 5-8

5.4 Radiation hardening 5-10
5.4.1 Combined thermal treatment of silica fibre cladding materials 5-11
5.4.2 H_2 loading of BEDF 5-11
5.5 Applications 5-12
5.5.1 A radiation-independent fibre-optic temperature sensor 5-12
5.6 Conclusion 5-13
Funding 5-13
Acknowledgement 5-14
References 5-14

6 Optical fiber sensors 6-1
Daniele Tosi

6.1 Introduction 6-1
6.2 Overview of fiber optic sensor techniques 6-2
6.3 Intensity-based sensors 6-4
6.3.1 Fiber bundles 6-5
6.3.2 Bent fiber sensors 6-6
6.4 Fiber Bragg grating sensors 6-7
6.4.1 FBG arrays 6-8
6.4.2 Chirped FBG 6-9
6.4.3 Tilted FBG 6-10
6.5 Fabry–Perot interferometers 6-10
6.6 Distributed sensors 6-12
6.6.1 Distributed temperature sensors 6-14
6.6.2 Optical backscatter reflectometry 6-16
6.7 Biosensors 6-18
6.8 Conclusions 6-20
References 6-20

7 Recent advances and perspectives on pulsed fiber lasers 7-1
Qian Li, Xuanyi Liu, Renlai Zhou, M S Aruna Gandhi and H Y Fu

7.1 *Q*-switched fiber lasers 7-2
7.2 Gain-switched fiber lasers 7-4
7.3 Mode-locked fiber lasers 7-6
7.3.1 Pulse picking in mode-locked fiber lasers 7-6
7.3.2 Pulse shaping in mode-locked fiber lasers 7-10
7.4 Chirped pulse fiber laser amplifier 7-12
7.4.1 Nonlinear chirped pulse amplification 7-12

7.4.2 Pre-chirp managed amplification 7-13
7.4.3 Self-similar amplification and pre-chirp managed self-similar amplification 7-13
7.4.4 Parabolic pre-shaping amplification 7-14
7.4.5 Gain-narrowing compensation 7-14
7.5 NALM fiber lasers 7-15
7.6 Conclusions 7-19
References 7-19

8 Nonlinear fiber optics **8-1**
Mário F S Ferreira
8.1 Introduction 8-1
8.2 Fiber dispersion 8-3
8.3 The Kerr effect 8-4
8.4 The nonlinear Schrödinger equation 8-5
8.5 Nonlinear effects in optical fibers 8-6
8.5.1 Self-phase modulation 8-6
8.5.2 Cross-phase modulation 8-8
8.5.3 Four-wave mixing 8-9
8.5.4 Stimulated Raman scattering 8-11
8.5.5 Stimulated Brillouin scattering 8-13
8.6 Optical solitons 8-15
8.7 Highly nonlinear fibers 8-18
8.7.1 Highly nonlinear silica fibers 8-18
8.7.2 Microstructured optical fibers 8-21
8.7.3 Non-silica fibers 8-23
8.8 Supercontinuum generation 8-25
Acknowledgements 8-31
References 8-31

9 Pulsating dissipative solitons in optical fiber systems **9-1**
Sofia C V Latas and Mário F S Ferreira
9.1 Introduction 9-1
9.2 The cubic-quintic complex Ginzburg–Landau equation 9-3
9.3 The variational approach to CGLE solitons 9-6
9.3.1 Sech-type ansatz 9-7
9.3.2 Gaussian-type ansatz 9-8
9.4 Higher-order effects 9-9

9.5 Plain pulsating solitons 9-11
9.6 Creeping solitons 9-13
9.7 Erupting solitons 9-16
9.8 Conclusions 9-20
Acknowledgments 9-21
References 9-21

10 Optical fiber networks for 5G environments **10-1**
Francesco Matera

10.1 Introduction 10-1
10.2 A short overview of 5G networks 10-3
10.3 From radio to optics 10-6
10.3.1 Radio spectrum management 10-6
10.3.2 Wireless network architectures 10-7
10.3.3 The xHaul segment: backhauling and fronthauling 10-8
10.4 Management of 5G networks 10-11
10.4.1 The network partition for 5G: slicing 10-11
10.4.2 SDN for 5G: from the centralized up to the distributed model 10-12
10.5 Further support of optical technologies for 5G 10-15
10.6 Conclusions 10-17
References 10-17

Preface

Optical fiber based photonic devices currently play an important role in the life of modern humans, from high speed broadband Internet to medical surgery. Indeed, such devices have been adopted across the entire spectrum of scientific, industrial, and commercial applications. Specialty optical fibers with either special waveguide structures or novel material compositions have become the heart of all-fiber based advanced photonic devices and components. Some of the important milestones in the development of specialty optical fibers along with their universal photonic devices occurred at the beginning of the twenty-first century with the invention of high power optical amplifiers for cable-TV, fiber to the home, light detection and ranging, free space communication systems, high power fiber lasers for high precision metal cutting, welding, and military applications, broadband supercontinuum sources for optical coherence tomography applications, along with the invention of a novel fabrication technique called the stack-and-draw process. The topic of this book, *Optical Fiber Technology and Applications: Recent Advances*, is a multidisciplinary field covering the fundamental physics and chemistry related to material science and optoelectronics.

This book is a compilation of works presenting recent developments and practical applications in optical fiber technology. It contains ten chapters from various research organizations that represent global research on various topics such as the new material composition based erbium-doped optical fiber (EDF) for broadband communication systems, highlighting alumina–germania–calcia–yttria–chromium glass, phase-separated alumina–silica glass, and zirconia–yttria–alumina–barium silica glass, hafnia–bismuth erbium co-doped fibers, photon noise in continuous-wave ytterbium-doped fiber lasers, borosilicate optical fibers for particle ionizing radiation dosimetry scattering, the ionizing radiation effect of erbium, ytterbium, or bismuth-doped/co-doped optical fibers, optical fiber sensors, recent advances and perspectives on pulsed fiber lasers, nonlinear fiber optics, pulsating dissipative solitons in optical fiber systems, and optical networks for 5G. This book presents a comprehensive account of the recent advances and research in optical fiber technology with a focus on new material composition based active fibers along with their applications as broadband optical amplifiers, fiber lasers, radiation sensors, etc.

Chapter 1 describes the fabrication of erbium-doped specialty optical fibers based on alumina–germania–calcia–yttria–chromium glass, phase-separated alumina–silica glass, and zirconia–yttria–alumina–barium silica glass by a modified chemical vapor deposition (MCVD) process in combination with a solution doping technique. This novel Cr,Er AGCYS fiber demonstrates, in the near-infrared (NIR) range, all optical properties inherent to doping with Er^{3+} ions, well-known for EDFs, as well as a set of advantages stemming from its co-doping with Cr^{3+}/Cr^{4+} ions. This chapter also highlights a set of characteristics that affect NIR (>1 μm) applications, namely absorption and fluorescence spectra, fluorescence decay, resonant-absorption bleaching, and gain spectra. Finally, the chapter discusses the special features of such erbium-doped specialty optical fibers for use as optical fiber amplifiers in broadband communication systems compared to standard EDFs.

Chapter 2 discusses the optical amplification properties of hafnia–bismuth erbium co-doped fiber as an active medium. The proposed amplifier is investigated by utilizing two short lengths of HBEDF to fulfill amplification in the C and L telecommunication bands where the wideband HB-EDFA is achieved using two stages in both series and parallel structures in conjunction with a double-pass scheme. Using such a novel glass composition based optical fiber, both parallel and backward pumping series HB-EDFAs show wideband operation between 1520 nm to 1610 nm.

Chapter 3 focuses on photon noise in a continuous-wave ytterbium-doped fiber laser. In this chapter, the noise properties of a continuous-wave ytterbium-doped fiber laser with moderate optical power are discussed. The laser is implemented in a conventional Fabry–Perot cavity configuration with two fiber Bragg gratings as reflectors. The properties of laser noise are addressed in terms of optical and radio frequency (RF) spectra, a probability density function for photons, an autocorrelation function for oscilloscope traces of laser signal, and the distributions of the noise pulses' power. It is demonstrated that the laser noise causes broadening of the laser line through the effect of self-phase modulation.

Chapter 4 presents the ability of borosilicate media in optical fibers to act as a dosimeter of energetic radiation, with neutron sensing being used as a particular exemplar of the versatility of the medium, the medium also being shown to offer linear sensitivity over a range of extremely high electron doses, in the kGy to several hundred kGy range. This range covers radiation sterilization and radiation processing requirements. The borosilicates provide the capability to confirm the doses that are given to such products, and in principle can also be implemented at very low cost, also challenging the capability of the existing dosimetric technology applied within this range. Considering recent investigations on silica-based thermoluminescent dosimeters, the optical fiber structure doped with boron appears to have dosimetry capabilities for both particle and non-particle ionizing radiation detection. The fabrication and study of novel borosilicate optical fiber structures is an interesting avenue that is yet to be explored, promising both scientific and commercial value.

Chapter 5 describes the ionizing radiation effect of erbium-, ytterbium-, or bismuth-doped/co-doped optical fibers. Here the radiation-induced photo-darkening in erbium/ytterbium co-doped optical fiber (EYDF) and the thermal effect in these irradiated EYDFs are studied, which evidently display a reversible thermal effect. Then, the radiation-induced activation of the bismuth active center (BAC) from inactive bismuth is assessed through the analysis of the absorption, emission, and lifetime of bismuth-doped optical fiber (BDF) and bismuth/erbium co-doped optical fiber (BEDF). The influential factors, such as Yb co-doping and radiation type, upon the radiation effect are also inspected. In this chapter thermal treatment is implemented to improve the performance of BEDF after gamma/electron irradiation. To improve the radiation resistance of the fibers, the authors also attempt to investigate radiation hardening by combining thermal treatment and hydrogen loading. Finally, a radiation-independent temperature sensor based on the fluorescence intensity ratio (FIR) of EYDF is demonstrated, indicating great potential for the field of sensing in a radiation environment.

Chapter 6 provides an overview of the main optical fiber sensor architectures and applications, defining the basic principles and simple methods of detection up to the most advanced distributed sensors and biosensors. Optical fibers are compact, immune to electromagnetic fields, and can implement several high-performance sensing systems that outperform many electronic or mechanical counterparts. The key feature of optical fiber sensors is the possibility to stack several sensors within the same fibers, using advanced multiplexing methods that can detect hundreds of sensors up to distributed sensors with their hundreds of thousands of sensing points.

Chapter 7 focuses on the progress of *Q*-switched, gain-switched, and mode-locked fiber lasers. *Q*-switched fiber lasers are promising sources for obtaining high-energy pulses, which can reach the order of millijoules. On the other hand, gain-switched fiber lasers can deliver ultrashort pulses in the mid-infrared wavelength, which is of great significance for applications in the long wavelength region. However, the applications of *Q*-switched and gain-switched fiber lasers are limited due to their wide pulse width (at the nanosecond level). By employing mode-locking techniques, the performance of fiber lasers can be fully optimized to the unprecedented levels. Mode-locked fiber lasers exhibit state-of-the-art performance (femtosecond pulse width, high peak power, low noise), which expands their application range significantly. Recent advances in pulse amplification technology and nonlinear amplifying loop mirror (NALM) fiber lasers are reviewed.

Chapter 8 provides an overview of the main nonlinear effects occurring in both conventional and highly nonlinear fibers. Such nonlinear effects impose different limitations on communications systems, but they also offer a variety of possibilities to perform diverse all-optical signal processing functions. Different types of highly nonlinear fibers have been developed for this purpose, using silica or other glasses with a greater intrinsic material nonlinearity. New phenomena are observed in these types of fibers. Among them, special attention is paid to supercontinuum generation, which can find applications in diverse areas, such as bio-medical optics, optical frequency metrology, spectroscopy, optical telecommunications, etc.

Chapter 9 introduces the concept of dissipative solitons, focusing on fiber systems which can be described by the cubic-quintic complex Ginzburg–Landau equation (CGLE). In particular, the CGLE can describe all nonlinear phenomena occurring in mode-locked fiber lasers and long-haul soliton transmission lines. Dissipative solitons have many unique properties which differ from those of their conservative counterparts and which make them similar to living things. Both the soliton perturbation theory and the variational approach are discussed in order to find approximate pulse solutions for the CGLE. However, to fully explore the CGLE enormous numerical simulations must be carried out. Both stationary and pulsating soliton solutions of the CGLE can be found in this way. The dynamics of different pulsating solitons, namely plain pulsating, creeping, and erupting solitons, is described. The effect that some higher-order effects, namely the third-order dispersion, intrapulse Raman scattering, and self-steepening, can have on these pulses is also discussed.

Chapter 10 provides an overview of the evolution of optical networks to support the fifth-generation (5G) environment. 5G will produce a profound transformation of the entire telecommunications network, namely in terms of interoperability between heterogeneous wireless networks and wired networks from the access segment to the core. 5G will lead to a new range of services that will also affect sectors that until now had not made use of the potential offered by the ICT mode, such as the automotive, energy, and health sectors, etc. In addition to increasing the transmission capacity of terminals, 5G will allow very low latencies, a much higher degree of mobility, and above all it will subvert the classic network paradigms by introducing new traffic management methods.

In summary, this book, *Optical Fiber Technology and Applications: Recent Advances*, discusses some of the recent topics that have had a tremendous impact on the growth of science and technology during the last few years. The purpose of this book is to document the leading-edge work in the field of new material composition and waveguide design based specialty optical fibers and their photonic devices through a collection of research performed by some prominent researchers in the field of optical fiber technology. This book will be devoted solely to the most recent achievements in the development of different varieties of specialty optical fibers and will serve as an universal resource for future development. This book describes the recent developing field of optical fiber technology and applications. The collected topics in the field of optical fiber technology and applications will play an important role in providing students, researchers, and technology managers with valuable, timely, and unbiased information on the subject. We hope that this book will be useful for students, researchers, and professionals who work in the area of fiber optics. This book will also serve as an interesting and valuable reference that will affect, stimulate, and promote further advances in the field of optical fiber technology. This body of results will be useful to stimulate future research efforts to reach a deeper understanding of optical fiber materials for optical fiber technology for the development of fiber lasers, optical amplifiers, and fiber sensors.

Acknowledgments

We are extremely satisfied with the result achieved through the collective efforts of our wonderful team of authors, who are leading-edge scientists in the field of optical fiber technology and applications, and have gathered and shared their knowledge to create the final form of this book. First, we wish to thank the authors of all the chapters in this book for their enthusiasm and effort as well as the many reviewers around the world who, despite their many other commitments, provided high-quality reviews of the chapters. This book address all the branches of optical fiber technology, covering fiber lasers, amplifiers, sensors, etc, and all the chapters presented in this book offer insight into the most interesting and recently developed aspects in the field of optical fiber technology.

We must thank Elsevier, Springer, OSA, and IEEE for allowing us to reproduce the diagrams/images published in various journals in this book. Thanks are due to Ashley Gasque, Senior Commissioning Editor, Optical Sciences, and Robbie Trevelyan, ebooks Editorial Assistant, at IOP Publishing for their prompt help, boundless energy, and excellent organizational skills in helping us to meet the deadlines as well as their stimulating encouragement which helped us to serve as Editors of this book.

Editor biographies

Mário F S Ferreira

Mário F S Ferreira graduated in physics from the University of Porto, Portugal, and received his PhD degree in physics in 1992 from the University of Aveiro, Portugal, where he is now a Professor in the Physics Department. Between 1990 and 1991 he was at the University of Essex, UK, performing experimental work on external cavity semiconductor lasers and nonlinear optical fiber amplifiers. His research interests have concerned the modeling and characterization of multi-section semiconductor lasers for coherent systems, quantum well lasers, optical fiber amplifiers and lasers, soliton propagation, polarization, and nonlinear effects in optical fibers. He is the leader of the Optics and Optoelectronics Group of the I3N—Institute of Nanostructures, Nanomodelling and Nanofabrication. He has written about 400 scientific journal and conference publications, and several books, namely *Optics and Photonics* (Lidel, 2003, in Portuguese), *Topics of Mathematical Physics* (Editora Ciência Moderna, 2018, in Portuguese), *Optical Fibers: Technology, Communications and Recent Advances* (Editor, NOVA Science Publishers, 2017), *Advances in Optoelectronic Technology and Industry Development* (CRC Press, 2019), *Nonlinear Effects in Optical Fibers* (Wiley, OSA, 2011), and *Optical Signal Processing in Highly Nonlinear Fibers* (CRC Press, 2020). He was the Guest Editor of four Special Issues of *Fiber and Integrated Optics* (Taylor and Francis): 'Fiber Optics in Portugal' (2005), 'Nonlinear Fiber Optics' (2015), 'Optical Fiber Sources and Amplifiers' (2020), and 'Quantum Communications' (2020), a joint Special Issue of *Optics Express* and *Applied Optics* (OSA) on 'Optical Sensors and Sensing' (2019), and a Special Issue of *Fibers* (MDPI) on 'Optical Fiber Communications' (2020).

He is a Senior Member and a Traveling Lecturer of both the Optical Society of America (OSA) and SPIE—The International Society for Optical Engineering. He has served in various committees of OSA and SPIE, as well as in the organizing and scientific committees of various international conferences. He also serves as an Associate Editor and as an Advisor Board Member of several international journals in the field of optics and photonics.

Mukul Chandra Paul

Mukul Chandra Paul received his MSc in 1989 from Burdwan University and his PhD from the Jadavpur University of Kolkata (India) in 2003. His research interests include the development of specialty optical fibers for fiber lasers, amplifiers, broadband sources, and sensors. Currently, Dr Paul is a Senior Principal Scientist at CSIR–Central Glass and Ceramic Research Institute (CSIR–CGCRI), India, where he leads a research group focused on specialty optical fiber fabrication technology towards applications as fiber lasers, optical fiber amplifiers, broadband sources, and FBG writing. His significant contributions/achievements in R&D have involved the development of various novel glass composition based specialty optical fibers co-doped with Er_2O_3, Yb_2O_3, PbO, CaO, Sc_2O_3, Y_2O_3, Li_2O, ZrO_2, Cr_2O_3, HfO_2, Al_2O_3, GeO_2, P_2O_5, Ce_2O_3, Al_2O_3, and Bi_2O_3, etc, as well as fiber based photonics devices and components using such specialty optical fibers, with the active participation of Indian photonics companies. Dr Paul has received many awards, including the BOYSCAST Fellowship from DST (2003), Ministry of Science, Technology and Innovation (MOSTI) award by the Federal Government Administrative Centre, Malaysia (2010), IEEE Photonic Distinguished Lecture Award by the Multimedia University, Malaysia (2011), CSIR Technology Award (2012), DST-UKIERI Award (2013), CSIR Technology Award for Innovation (2015), Indo-Russian (RFBR) Collaborative Research Grant Award (2016), Indo-Portugal (FCT) Collaborative Research Grant Award (2017), Senior Visiting Scientist award by the National Taiwan University of Technology (NTUT), Taiwan etc. He has strong collaborative research links with different countries such as the UK, Russia, Portugal, Mexico, Malaysia, Vietnam, etc.

He is a Senior Member of OSA and IEEE and a Life Member of the Materials Research Society of India (MRSI) and the Indian Ceramic Society (ICS). He has also served as a Guest Editor for several special issues of *Optical Materials Express* (OMEX), current nano-science journals, the *IEEE Photonics Journal*, etc. He holds seven US patents, four Indian patents, has published seven book chapters, edited two books on fiber lasers, and has authored over 300 SCI papers in peer-reviewed journals and conferences.

Contributors

S F Abdul Sani
Department of Physics, Faculty of Science, University of Malaya, 50603, Kuala Lumpur, Malaysia

Hairul A Abdul-Rashid
Fibre Optics Research Centre, Faculty of Engineering, Multimedia University, Jalan Multimedia 63100, Cyberjaya, Malaysia

Alabbas A Al-Azzawi
Department of Electrical Engineering, University of Malaya, 50603 Kuala Lumpur, Malaysia
Department of Computer Techniques Engineering, Al-Esraa University College, Baghdad, Iraq

Amal Alqahtani
College of Medicine, University of Imam Abdulrahman Bin Faisal, Dammam, Saudi Arabia

Miguel V Andrés
Departamento de Física Aplicada, Instituto de Ciencia de Materiales, Universidad de Valencia, 46100 Valencia, Spain

M S Aruna Gandhi
School of Electronic and Computer Engineering, Peking University, Shenzhen 518005, China

Yuri Barmenkov
Centro de Investigaciones en Optica, Leon 37150, Mexico

D A Bradley
Sunway University, Centre for Biomedical Physics, Jalan Universiti, 47500, Subang Jaya, Malaysia
Department of Physics, University of Surrey, Guildford, GU2 7XH, United Kingdom

José L Cruz
Departamento de Física Aplicada, Instituto de Ciencia de Materiales, Universidad de Valencia, 46100 Valencia, Spain

Shyamal Das
CSIR–Central Glass and Ceramic Research Institute, Jadavpur, Kolkata-700 032, India

Anirban Dhar
CSIR–Central Glass and Ceramic Research Institute, Jadavpur, Kolkata-700 032, India

Mingjie Ding
Photonics & Optical Communications, School of Electrical Engineering, University of New South Wales, Sydney 2052, NSW, Australia

José Duarte
I3N–Institute of Nanostructures, Nanomodelling and Nanofabrication, Department of Physics, University of Aveiro, 3810-193 Aveiro, Portugal

Debjit Dutta
CSIR–Central Glass and Ceramic Research Institute, Jadavpur, Kolkata-700 032, India

Desheng Fan
Photonics & Optical Communications, School of Electrical Engineering, University of New South Wales, Sydney 2052, NSW, Australia

Mário F Ferreira
I3N-Institute of Nanostructures, Nanomodelling and Nanofabrication, Department of Physics, University of Aveiro, 3810-193 Aveiro, Portugal

H Y Fu
Shenzhen International Graduate School, Tsinghua University, Shenzhen, China

Sulaiman Wadi Harun
Department of Electrical Engineering, University of Malaya, 50603 Kuala Lumpur, Malaysia
Department of Engineering, Faculty of Vocational Studies, Airlangga University, Surabaya, Indonesia

Alexander V Kir'yanov
National University of Science and Technology 'MISIS', Moscow, Russia
Centro de Investigaciones en Optica, Leon 37150, Mexico

Sofia C V Latas
I3N-Institute of Nanostructures, Nanomodelling and Nanofabrication, Department of Physics, University of Aveiro, 3810-193 Aveiro, Portugal

Joaquim P Leitão
I3N-Institute of Nanostructures, Nanomodelling and Nanofabrication, Department of Physics, University of Aveiro, 3810-193 Aveiro, Portugal

Qian Li
School of Electronic and Computer Engineering, Peking University, Shenzhen, China

Xuanyi Liu
School of Electronic and Computer Engineering, Peking University, Shenzhen, China

Yanhua Luo
Photonics & Optical Communications, School of Electrical Engineering, University of New South Wales, Sydney 2052, NSW, Australia

Francesco Matera
Fondazione Ugo Bordoni, Roma, Italy

Pablo Muniz-Cánovas
Centro de Investigaciones en Óptica, A.C., Loma del Bosque 115, Col. Lomas del Campestre, 37150, León, Gto., Mexico

Mrinmay Pal
CSIR–Central Glass and Ceramic Research Institute, Jadavpur, Kolkata-700 032, India

Mukul Chandra Paul
CSIR–Central Glass and Ceramic Research Institute, Jadavpur, Kolkata-700 032, India

Gang-Ding Peng
Photonics & Optical Communications, School of Electrical Engineering, University of New South Wales, Sydney 2052, NSW, Australia

Ana M Rocha
Instituto de Telecomunicações, Campus Santiago, 3810-193 Aveiro, Portugal

Daniele Tosi
Nazarbayev University, School of Engineering, 010000 Nur-Sultan, Kazakhstan
National Laboratory Astana, Laboratory of Biosensors and Bioinstruments, 010000 Nur-Sultan, Kazakhstan

Jianxiang Wen
Key Laboratory of Specialty Fiber Optics and Optical Access Networks, Joint International Research Laboratory of Specialty Fiber Optics and Advanced Communication, Shanghai Institute for Advanced Communication and Data Science, Shanghai University, Shanghai 200444, China

Binbin Yan
State Key Laboratory of Information Photonics and Optical Communications, Beijing University of Posts and Telecommunications, Beijing 100876, China

Randall Youngman
Science and Technology Division, Corning Incorporated, Corning, NY 14830, USA

Qiancheng Zhao
Photonics & Optical Communications, School of Electrical Engineering, University of New South Wales, Sydney 2052, NSW, Australia

Renlai Zhou
School of Electronic and Computer Engineering, Peking University, Shenzhen, China

H T Zubair
Fibre Optics Research Centre, Faculty of Engineering, Multimedia University, Jalan Multimedia 63100, Cyberjaya, Malaysia
Lumisyns Sdn. Bhd., 47600, Subang Jaya, Selangor, Malaysia

List of abbreviations

5G	fifth-generation
AGCYS	alumina–germania–calcia–yttria–silica
ASE	amplified spontaneous emission
BBU	base band unit
CDF	chromium doped fiber
Cr,Er AGCYS	Cr, Er co-doped alumina–germania–calcia–yttria–silica
CrDF	chromium doped fiber
CGLE	complex Ginzburg–Landau equation
CMS	cladding mode stripper
cP	centipoise
CP	composite pulse
CPA	chirped pulse amplification
CW	continuous wave
DCF	double clad fiber
DEF	dose enhancement factors
DFB	distributed feedback
DND	direct nano-particle deposition
DSF	dispersion-shifted fiber
DWDM	dense wavelength division multiplexing
EDF	erbium-doped fiber
EDFA	erbium-doped fiber amplifier
EDFL	erbium-doped fiber laser
EDX	energy dispersive x-ray
eMBMS	evolved multimedia broadcast multicast service
EFPI	extrinsic Fabry–Perot interferometer
EPMA	electron probe microanalysis
ESA	excited state absorption
EXAFS	extended x-ray absorption fine structure
FBG	fiber Bragg grating
FEGSEM	field emission gun scanning electron microscope
FPI	Fabry–Perot interferometer
FTTB	fiber to the building
FTTH	fiber to the home
FWHM	full width half maximum
FWM	four-wave mixing
GDR	giant dipole resonances
GPON	gigabit passive optical networks
GVD	group-velocity dispersion
Gy	gray (unit)
HBEDF	hafnium–bismuth–erbium-doped fiber
HF	hydrogen fluoride
HNLF	highly nonlinear fiber
HR	high reflection
HTEM	high resolution transmission electron microscopy
IA	induced absorption
IFPI	intrinsic Fabry–Perot interferometer
IoT	Internet of things

IRS	intrapulse Raman scattering
LD	laser diode
LET	linear energy transfer
LIDAR	light detection and ranging
LR	low reflection
MCVD	modified chemical vapor deposition
MFD	mode field diameter
MI	modulation instability
MLC	multi-leaf collimators
MMF	multimode fiber
MOF	microstructured optical fiber
MPLS	multi-protocol label switching
NA	numerical aperture
NALM	nonlinear amplifying loop mirror
NBO	non-bridging oxygen
NBOHC	non-bridging oxygen hole center
NF	noise figure
NFV	network function virtualization
NIR	near-infrared
NLSE	nonlinear Schrödinger equation
NMR	nuclear magnetic resonance
NOLM	nonlinear optical loop mirror
MPLS	multi-protocol label switching
NP	noise pulsing
NPR	nonlinear polarization rotation
NSR	nonsoliton radiation
OBR	optical backscatter reflectometry
OFDR	optical frequency-domain reflectometry
OPC	optical phase conjugation
OFS	optical fiber sensors
OPM	optical power meter
OSA	optical spectrum analyzer
OSL	optically stimulated luminescence
OTDR	optical time-domain reflectometry
PCE	power conversion efficiency
PCF	photonic crystal fiber
PCMA	pre-chirp managed amplification
PD	photodiode
PDF	probability density function
PM	polarization-maintaining
POA	programmable optical attenuator
POF	polymer optical fiber
QCW	quasi-continuous wave
QKD	quantum key distribution
QoS	quality of service
RAN	radio access network
RE	rare earths
RI	refractive index
RIP	refractive index profile
RoF	radio over fiber

RRH	remote radio head
RT	round trip
RX	receiving fiber
SBS	stimulated Brillouin scattering
SC	supercontinuum
SD	solution doping
SDM	space division multiplexing
SDN	software defined networks
SE	spontaneous emission
SEM	scanning electron microscopy
SESAM	semiconductor saturable absorber mirror
SHB	spatial hole burning
SMF	single-mode fiber
SON	self-organizing network
Si-ODC	silicon oxygen defect center
SPM	self-phase modulation
SPR	surface plasmon resonance
SRS	stimulated Raman scattering
SSFS	soliton self-frequency shift
SST	self-steepening
SWCNT	single-walled carbon nanotube
TDM	time division multiplexing
TEM	transmission electron microscopy
TOD	third-order dispersion
TLD	thermoluminescent dosimeter
TLS	tunable laser source
TX	transmitting fiber
WDM	wavelength division multiplexing
XPM	cross-phase modulation
XRD	x-ray diffraction
ZDW	zero dispersion wavelength
Zr–Y–Al–Ba	zirconia–yttria–alumina–barium

Chapter 1

A new class of specialty optical fibers based on a novel material composition of the doping host for the study of optical amplification

Mukul Chandra Paul, Alexander V Kir'yanov, Yuri Barmenkov, José Duarte, Joaquim P Leitão, Mário F Ferreira, Ana M Rocha, Randall Youngman, Mrinmay Pal, Anirban Dhar, Shyamal Das and Debjit Dutta

Currently, specialty optical fibers developed through modification of the composition of the doping host of active rare-earth (RE) elements play an important role in the field of photonics for use as fiber lasers, optical fiber amplifiers, and broadband sources. The topics we discuss in this chapter include the fabrication of erbium-doped specialty optical fibers based on alumina–germania–calcia–yttria–chromium glass, phase-separated alumina–silica glass, and zirconia–yttria–alumina–barium silica glass using a modified chemical vapor deposition (MCVD) process in combination with a solution doping (SD) technique. Here we report their detailed material and optical characterization along with their spectroscopic and amplification properties. Finally, we discuss the special features of such erbium-doped specialty optical fibers as optical fiber amplifiers in broadband communication systems compared to standard erbium-doped fibers.

1.1 Introduction

Among the different types of fiber amplifiers, erbium-doped fiber amplifiers (EDFAs) are finding applications in most current communication systems [1, 2]. EDFAs present a low noise figure [3], the absence of crosstalk [4], high gain, high saturation output power, polarization independent gain [1–5], and low insertion loss. EDFAs are also used in wavelength-division multiplexing due to their large gain bandwidth which helps to amplify several channels simultaneously. The non-uniform gain spectrum of EDFAs

doi:10.1088/978-0-7503-3243-9ch1

causes problems when multiple EDFAs are cascaded in a system. The amount of data transmission is limited in such a system by this characteristic. This limitation may be overcome by employing gain equalizing filters, long-period fiber gratings, or chirped fiber Bragg gratings [6–8]. However, the use of such schemes increases the system's overall cost and complexity.

Extensive research has been done on EDFAs with different glass hosts and co-dopant materials such as silica, fluorozirconate, chalcogenides, bismuth, aluminum, phosphorous, and lanthanum. [9–11]. The results show that these materials have a significant impact on the overall performance and applicability of the optical amplifier. With increasing erbium concentration, the materials suffer concentration quenching [12] and cluster formation [13].

For telecoms, laser, medical, and military applications, researchers around the world are also searching for new silica based optical fibers with pronounced fluorescence and amplifying capacity in the near-infrared (NIR) region beyond 1 μm. However, a very limited number of 'active' fibers are currently known as confirmed candidates for this spectral domain. In fact, a spectral gap between the well-developed ytterbium (Yb)-doped fibers and erbium-doped fibers (EDFs) with operation ranges of 1.0–1.15 μm and 1.5–1.65 μm, respectively, still remains, at least for high-power laser/amplifying applications. This unfavorable situation remains unresolved despite such candidates as bismuth- (Bi) [14–16] and chromium (Cr)-doped silica fibers [17, 18], which possess fluorescence bands in the NIR, which have been proposed and partly investigated during the last two decades. However, both bismuth and chromium have some drawbacks, such as the absence of a reliable understanding of the type of active centers responsible for NIR emission/amplification, low concentrations of active dopants allowing efficient operation (in the case of Bi-doped fibers), and technological difficulties arising when attempting the stabilization of a proper valence of Cr ions (usually, Cr^{3+}/Cr^{4+}), along with the inherent problems of room-temperature operation (in the case of Cr-doped fibers). Furthermore, the method of co-doping with different active ions has attracted a great deal of attention in recent times [19]. In particular, co-doping of EDFs with other rare-earth ions such as Yb and thulium has been under extensive investigation. Yb/Er co-doping is known to provide very effective amplification using corresponding double-clad fibers for high-power applications for 1.5–1.6 μm. Meanwhile, the co-doping of Bi/Er [20–25] has only been tackled recently, which focused on the realization of ultra-broadband luminescence and extended gain in the NIR. Co-doping EDFs with Cr ions is unknown for optical silica based fibers, mostly because of the above-mentioned problem of Cr ion stabilization in the desired (Cr^{3+}/Cr^{4+}) valence (although such co-doping has been studied for bulk materials; see e.g. [26]). However, our recent studies in the field [27, 28] revealed some promise, provided by suitable engineering of the fiber lattice structure, for stabilizing Cr in these valent states and the functioning of Cr/Er co-doped fibers in the NIR.

The development of new fiber materials is one of the approaches for achieving a flat gain with a wide amplification bandwidth [29, 30] to avoid expensive components. IsoGain^{TM} erbium doped fiber (EDF) has been introduced to obtain flat gain for use in a high-channel count dense wavelength-division multiplexing (DWDM) system [31]. In order to achieve uniform signal gain one common approach is Al^{3+} co-doping with Er^{3+} in EDFs [32–37]. Tailoring of the doped host based on material characterization of the core-glass matrix within the optical fiber core region for incorporation of Er^{3+} ions is a crucial factor for the improvement of the spectroscopic properties of the fibers to obtain an intrinsic flat gain of EDFA. Interestingly, when Er_2O_3 is doped into an alumina–silica based modified phase-separated glass host, it produces efficient spectral broadening and thereafter it is possible to obtain a large gain bandwidth with improved flatness of the gain spectrum. Alterations in the Er^{3+} local environment, symmetry, and ligand field strength are the key factors to achieve the spectral broadening of the Er^{3+} emission. Since the host glass environment affects the spectroscopy of Er^{3+} ions, a number of fiber host materials, including silica, fluorozirconate glasses, and chalcogenides, have been investigated in an attempt to identify a host that will provide a larger gain bandwidth for the $^4I_{13/2}\rightarrow{}^4I_{15/2}$ transition of erbium [38–40]. Although fluoride and telluride hosts doped with Er^{3+} offer high quantum efficiencies for 1.5 μm transitions, they are not suitable for practical device applications due to their poor mechanical strength compared to their pure silica glass counterpart. It is extremely difficult to fabricate low-loss fluoride glass fibers as they are prone to crystallization and suffer from poor chemical durability. The normal alumina–silica glass based EDF under high Al co-doping shows better flatness compared to other silica EDFs such as germanium–silica and phosphosilicate [32]. Accordingly, the fabrication of high Al-doped silica based optical fibers has gained a lot of attention in recent years and various techniques such as the MCVD-SD technique, direct nano-particle deposition (DND), the sol–gel method etc, have been used. However, fabrication of phase-separated core glass attracts special attention as it can create more non-uniformity surrounding the incorporated Er ions and leads to enhanced stark splitting. Although there are many reports related to the fabrication of phase-separated RE-doped glasses doped with Er, Yb, Tm, etc [41–44], here our goal is to optimize the associated fabrication process parameters, including the heat treatment of the fabricated preform through different types of material characterization, which is not well reported in the literature, in particular for high alumino-silicate core-glass based fibers. Some reports are available on alumino-phosphosilicate based glass fibers fabricated through an MCVD-SD technique followed by heat treatment [45, 46], but not in pure alumino-silicate based glass fibers doped with Er ions.

In this chapter we focus on three different varieties of erbium-doped specialty optical fibers based on alumina–germania–calcia–yttria–chromium–silica glass, phase-separated alumina–silica glass, and zirconia–yttria–alumina–barium silica glass made using an MCVD process in combination with an SD technique. First we present the details of fabrication, the material characterization, and the analysis of the basic optical properties, including the absorption, fluorescence, and gain properties of Cr and Er co-doped alumina–germania–calcia–yttria–silica (AGCYS) based fiber, a novel type of active optical fiber for NIR applications.

Second, in order to obtain enhanced flatness around ±0.60 dB over the C-band region, Er^{3+} is incorporated into a modified alumina–silica glass host of suitable composition which forms the phase-separated region within the glass matrix under optimized annealing conditions. A detailed investigation was performed to optimize the different process parameters associated with the fabrication of Er_2O_3-doped alumina–silica based phase-separated EDFs along with the material characterization to correlate the nature and performance of the fiber obtained from the phase-separated glass. The spectroscopic properties and the nature of the gain spectra of the EDFs fabricated from both phase-separated and non-phase-separated alumino-silicate based Er_2O_3-doped preforms are investigated and explained on the basis of the achieved experimental results.

Third, we have developed nano-engineered Er^{3+}-doped zirconia–yttria–alumina–barium silica glass based optical fibers by increasing the doping levels of Er_2O_3 along with ZrO_2 and a minor amount of BaO into a silica glass based host. Incorporation of a larger content of Zr^{4+} into the yttria–alumina–silica glass host reduces the cluster formation of Er^{3+} in silica which improves the luminescence property [47]. On the other hand, the replacement of Al_2O_3 by a glass modifier ZrO_2 helps to increase the number of non-bridging oxygens through breaking of the silica network structure. Such an Er_2O_3-doped modified silica glass host shows much broader emission spectra compared to silica glass above the 1580 nm wavelength due to its larger emission cross-section [48]. Here the main purpose for incorporation of BaO, as for other co-dopants Al and P, is to change the surrounding environment of the RE ions [49, 50], to increase the solubility of the REs, as well as to increase the glass forming zone for Er_2O_3 [51, 52] through minimization of the phase separation. Here, incorporation of BaO as a glass modifier alters the glass structure along with the local electrostatic field closer to the RE ion because of the ionic radius mismatch in the glass matrix. Thus the glass symmetry as well as covalence of RE ions will be different from that of silica glass. Such glass modifications strongly affect the various luminescence transitions along with up-conversion, as shown by Courrola and Kassab *et al* [49]. The inclusion of Y_2O_3 in the host matrix serves the additional purpose of slowing down or eliminating changes in the ZrO_2 crystal structure. This is a crucial factor in the fabrication process [53]. The novelty of such an Er_2O_3-doped glass host is achieving a broadband optical amplification from 1530 to 1600 nm with better gain flatness as well as a lower noise figure.

1.2 Erbium-doped specialty optical fiber based on alumina–germania–calcia–yttria–chromium–silica glass

1.2.1 Fabrication and material properties

The fabrication of the Cr,Er co-doped optical preform based on the AGCYS glass was achieved through an MCVD process [54] in combination with an SD technique. The deposition of the porous SiO_2–GeO_2 layer was performed at a high temperature (1350 °C ± 10 °C) into a silica tube with 20 mm/17 mm outer/inner diameters, followed by soaking of the porous layer for incorporation in the SD of different precursors (co-dopants) in a suitable composition from a water-based solution. The deposition temperature during MCVD was monitored using an IR pyrometer, moved synchronously with an oxy-hydrogen burner along the tube in a forward direction. A photograph of the deposition process is provided in figure 1.1(a).

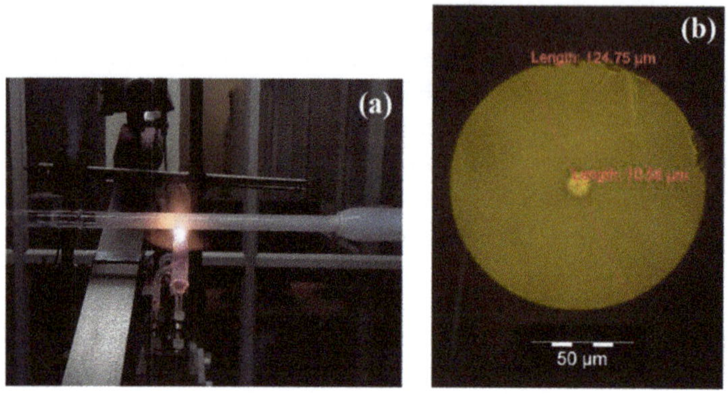

Figure 1.1. (a) Photograph of the deposition of the porous SiO_2–GeO_2 layer onto the preform during MCVD. (b) Cross-sectional view of the Cr,Er AGCYS fiber. Copyright (2019) IEEE. Reprinted, with permission, from [56] (CC BY 4.0).

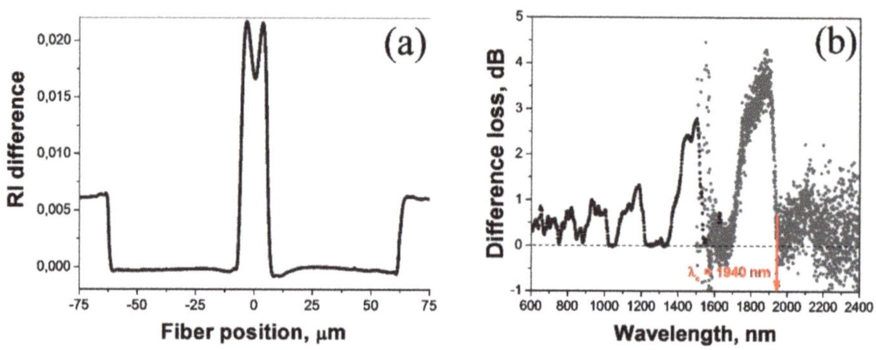

Figure 1.2. (a) RIP of the Cr,Er AGCYS fiber and (b) its loss difference spectrum, measured in straight and coiled positions (the arrow shows the spectral position of the cut-off wavelength, ~1940 nm). Copyright (2019) IEEE. Reprinted, with permission, from [56] (CC BY 4.0).

The co-dopants (Al, Er, Cr, Ca, and Y) were introduced into the porous frit employing the procedure described in [27, 55]. The doping levels of different co-dopants inside the core were optimized through variation of the solution strength of the dopant precursors comprising $ErCl_3.6H_2O$ (Alfa Aesar), $AlCl_3.6H_2O$ (Alfa Aesar), $CrCl_3.6H_2O$ (Alfa Aesar), $CaCl_2.6H_2O$ (Alfa Aesar), and $Y(NO_3)_3.xH_2O$ (Alfa Aesar), dissolved into doubly distilled water. The porous layer, soaked for a period of one hour, was air dried and remounted on the glass working lathe followed by subsequent processing to obtain the final preform. The fiber was drawn from the preform using a fiber drawing tower with on-line resin coating. A cross-sectional image of the fabricated Cr,Er AGCYS fiber, recorded using an Olympus microscope, is shown in figure 1.1(b); its outer (cladding) and inner (core) diameters were found to be ~125.0 and ~10.5 μm, respectively.

The fiber's refractive-index profile (RIP), measured using a fiber analyzer (NR-9200 EXFO), is shown in figure 1.2(a). The average numerical aperture (NA)

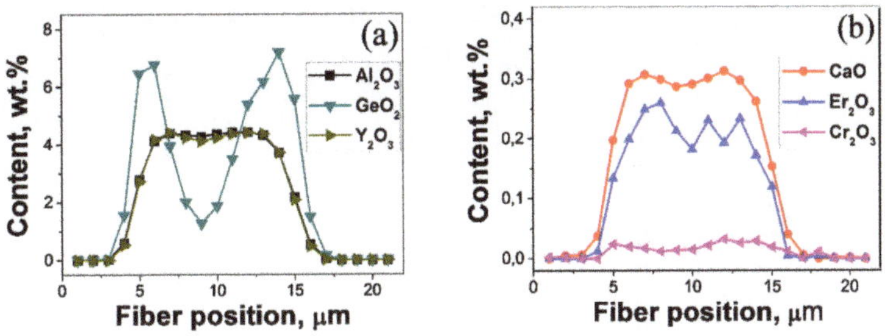

Figure 1.3. Radial distributions of the co-dopants: (a) Al_2O_3, GeO_2, and Y_2O_3; (b) CaO, Er_2O_3, and Cr_2O_3within the core area of the Cr,Er AGCYS fiber. Copyright (2019) IEEE. Reprinted, with permission, from [56] (CC BY 4.0).

of the Cr,Er AGCYS fiber was found to be ~0.23 (in the step-index approximation). Its cut-off wavelength (λ_c) was obtained using a fiber-adapted white-light source (Yokogawa AQ4305) and a couple of optical spectrum analysers with optical bands of 400–1650 nm (OSA1: ANDO 6315A) and 1200–2400 nm (OSA2: Yokogawa AQ6370B) employing standard transmission measurements with the fiber samples placed in straight and coiled positions. The result (the difference loss) is shown in figure 1.2(b). As can be seen, within our main spectral range of interest (1.1–1.7 μm), the Cr,Er AGCYS fiber demonstrates a slightly multimode regime of propagation, with $\lambda_c \sim 1.94$ μm.

The elemental composition of the Cr,Er AGCYS fiber was measured using electron-probe microanalysis (EPMA). The co-dopants' distribution profiles across the fiber's core area are shown in figure 1.3; their maximal doping levels were found to be 7.1 wt.% (GeO_2), 4.3 wt.% (Al_2O_3), 4.3 wt.% (Y_2O_3), 0.025 wt.% (Cr_2O_3), 0.23 wt.% (Er_2O_3), and 3.1 wt.% (CaO). Co-doping with Er and Cr was intended to obtain emission in the NIR (beyond ~1 μm; see the spectral analysis in section 1.3), Al and Ge were introduced to manage the RI profile of the fiber (see figure 1.2(a)), the addition of Y was chosen to enhance the fluorescent abilities of the optically active elements (Er and Cr), while Ca was chosen to facilitate the stabilization of Cr ions in the tetrahedral (Cr^{4+}) state, responsible for fluorescence beyond ~1.2 μm. Also note that, according to the glass technology principles, with increasing content of a structural modifier such as Y, the interstitial gap in a glass network can be increased via creating a large amount of non-bridging oxygens. This facilitates the dissolving of other atoms of the lattice into the silica based glass and hence reduces their clustering.

1.3 Optical properties

1.3.1 Linear absorption

The loss spectrum of the Cr,Er AGCYS fiber was measured using the cutback method. White light from the fiber-adapted source was launched into the BDF samples through splices and its transmitted part was recorded by the spectrum analyzers OSA1/OSA2. The measured transmission spectrum was re-calculated into the loss (absorption) spectrum, $\alpha_0(\lambda)$, presented in figure 1.4.

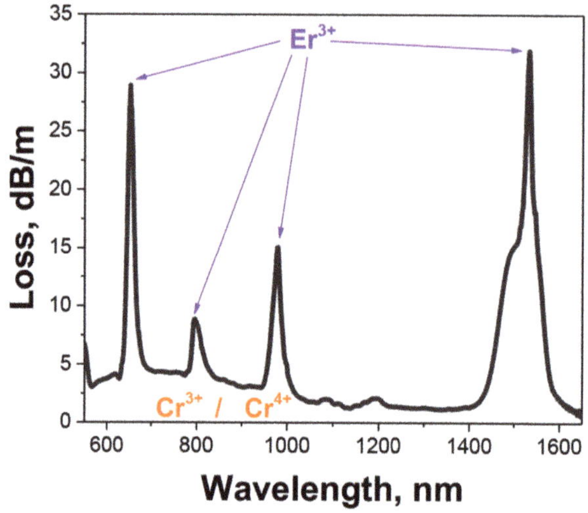

Figure 1.4. Absorption spectrum of the Cr,Er AGCYS fiber, where the spectral positions of the Er^{3+} and Cr^{3+}/Cr^{4+} bands are highlighted. Copyright (2019) IEEE. Reprinted, with permission, from [56] (CC BY 4.0).

As seen from the figure, this spectrum demonstrates the superposition of the spectral signatures of Er^{3+} ions (with the characteristic absorption peaks marked by violet arrows) and Cr^{3+}/Cr^{4+} ions, seen as a broad absorption band where the contribution of Cr^{3+} ions dominates within a ~600–1000 nm range and that of Cr^{4+} ions at longer wavelengths, beyond ~1000 nm. Note that the Cr,Er AGCYS fiber presents a fiber heavily doped with both 'active' co-dopants as the maximal extinction in the Er-band (at 1530 nm) is ~30 dB m^{-1} and that in the broad Cr-band in the visible is ~5 dB m^{-1}.

1.3.2 Nonlinear absorption

An important issue for our further analysis is the behavior of nonlinear, i.e. pump power dependent, absorption of the Cr,Er AGCYS fiber. The results revealing this behavior are summarized in figure 1.5 for different excitation wavelengths: (a) falling into the Er-band absorption maximum, at 976 nm (the 'resonant' case) and (b) falling off the Er-band but into the Cr broad absorption band, at 905 and at 1064 nm (the 'non-resonant' case). Note that in figure 1.5(a) we compare the effect of nonlinear (bleaching) absorption in Cr,Er AGCYS fiber and in two commercial purely Er-doped fibers: EDFs L20 and L40 (with nominal extinctions in the Er^{3+} ion maximums at 1530 nm of ~20 and ~40 dB m^{-1}, respectively) [57]. As can be seen, in terms of Er content, the Cr,Er AGCYS fiber takes an intermediate position between the L20 and L40 EDFs.

In the experiments, pump light from a laser diode (LD; with operating wavelengths of 976, 905, and 1064 nm) was launched through a splice to a sample of 'active' fiber with length L_f and its nonlinear transmission was measured, using a power meter (Thorlabs), as $T_{NL} = P_{out}/P_{in}$ (P_{in} is the launched pump power and P_{out} is its unabsorbed part) which was then re-calculated into nonlinear absorption (or

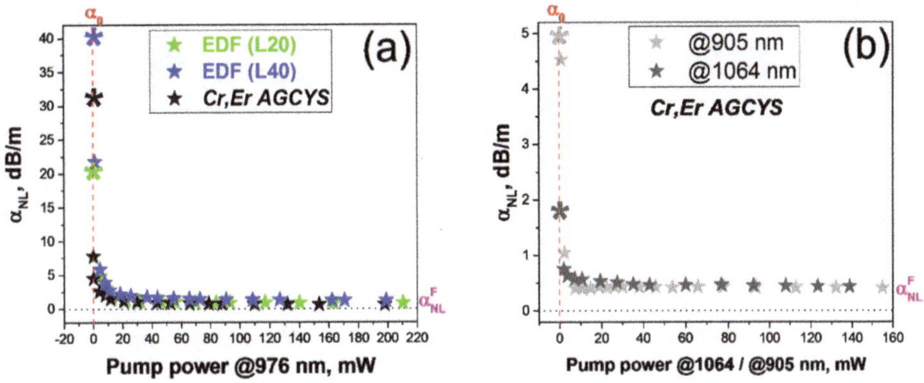

Figure 1.5. (a) Dependences of nonlinear absorption (the bleaching effect) versus pump power at 976 nm (into the Er^{3+} absorption band), for the Cr,Er AGCYS, EDF L20, and EDF L40 fibers. (b) Analogous dependences obtained for the Cr,Er AGCYS fiber at 905 nm/1064 nm (both into the Cr^{3+}/Cr^{4+} absorption band but off the Er^{3+} absorption band). Asterisks designate the fibers' small-signal absorptions, obtained from the loss spectra measured with OSA(α_0); in turn, plateaus ($\alpha_{NL}{}^F$-values) correspond to the unbleached absorptions of the fibers. Copyright (2019) IEEE. Reprinted, with permission, from [56] (CC BY 4.0).

loss) α_{NL} via the relationship $\alpha_{NL} = -\ln(T_{NL})/L_f$. Such dependences were obtained for all the active fibers under study in a broad range of pump powers P_{in}; the results are presented in figure 1.5. Note that the lengths of the fiber samples used were back proportional to small-signal loss α_0, which ensured nearly equal optical densities (OD = $\alpha_0 L_f$) of the samples at either kind of excitation.

As seen from figure 1.5(a), the dependences $\alpha_{NL}(P_{in})$ at 976 nm (at resonant, i.e. in the Er-band, excitation) for all active fibers examined demonstrate strongly bleachable while quite similar behaviors—despite different linear absorption of the fibers (attributed by α_0-values; see the asterisks), all dependences approach nearly the same plateaus $\alpha^F{}_{NL}$ at high pump powers, with $\alpha^F{}_{NL}$ being minimal for the Cr, Er AGCYS fiber.

Furthermore, as seen from figure 1.5(b), this fiber is also easily bleachable at non-resonant (off Er-band) excitation, thus revealing easy bleaching of the broad absorption band, inherent to its co-doping with Cr. Note that the residual (unbleached) loss in the fibers may be of diverse origins. These may arise in both Er and Cr subsystems via excited-state absorption (ESA) and up-conversion (UC) due to the ion clustering effects [57]. In any case, these effects are not significant in the fibers examined, including the Cr,Er AGCYS fiber, at any kind of excitation. Note here that absorption contrast ($\alpha^F{}_{NL}/\alpha_0$) for the Cr,Er AGCYS fiber is measured to be ~0.03 at 976 nm (the Er^{3+} band) and ~0.08 at 905 nm/1064 nm (the Cr^{3+}/Cr^{4+} band).

1.3.3 Emission spectra and lifetimes

In figure 1.6(a) we repeat the absorption spectrum of the Cr,Er AGCYS fiber in the NIR (1200–1650 nm) and compare it to those of the two commercial fibers heavily doped with Er, EDFs L20 and L40. In addition to the features stemming from the presence of Cr ions in the Cr,Er AGCYS fiber, this fiber demonstrates lower loss

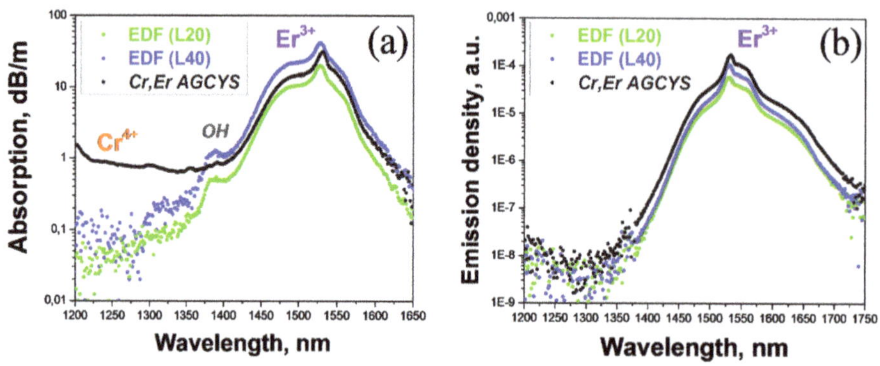

Figure 1.6. Comparison of the NIR (a) absorption and (b) emission spectra of the Cr,Er AGCYS fiber (~30 dB m^{-1}) (black curves) with Cr-free EDFs (L20 and L40) with comparable (~20 dB m^{-1} and ~30 dB m^{-1}, respectively) extinctions in the Er^{3+} peak at 1530 nm (green and blue curves). The emission spectra were measured in frontal counting geometry at 976 nm pumping with the same power (~100 mW), using the fiber lengths providing the same OD (5.3 dB) for all three samples examined. Copyright (2019) IEEE. Reprinted, with permission, from [56] (CC BY 4.0).

at 1380 nm (given by the presence of hydroxyl OH groups) than those in the commercial (L20/L40) EDFs; this points to better overall performance of the former fiber for laser/gain applications.

Accordingly, in figure 1.6(b), we demonstrate the emission spectra (in NIR, 1200–1750 nm) of these three fibers at 976 nm excitation of the same launched power (~150 mW), collected from the tips of the fiber samples with lengths providing nearly the same OD ~5.3 dB. Experimentally, the fiber samples were in-core pumped through splices in the simplest forward geometry by a commercial 976 nm LD (JDSU) with a single-mode output fiber. The same OD of the Cr,Er AGCYS, L20, and L40 fibers, and the entirely same geometry of pump-light launching (100 mW in each case) and emission counting used ensure the applicability of a direct comparison of the plotted emission spectra for the three fibers. Thus, from figure 1.6(b) we conclude that there is a higher emissive potential for the Cr,Er AGCYS fiber in the NIR (1200–1750 nm) compared to commercial L20/L40 (Cr-free) EDFs. Seemingly, this effect stems from co-doping of the former fiber with Cr^{3+}/Cr^{4+} ions and lower NIR fluorescence quenching in it due to the reduced content of OH groups (see figure 1.6(a)).

We also examined the strength of UC emission in these three fibers, using the same pump arrangement at 976 nm excitation. As can be seen from the photographs of the fibers' lateral visible (mostly green) lightening, obtained at equal excitation conditions (figure 1.7), the intensity of UC emission in the Cr,Er fiber (Er^{3+} absorption, ~30 dB m^{-1}) does not exceed that in the EDF L20 (Er^{3+} absorption, ~20 dB m^{-1}) and is much weaker than in the EDF L40 (Er^{3+} absorption, ~40 dB m^{-1}). This may indicate a weaker effect of Er ion clustering in the Cr,Er fiber than in the commercial EDFs, which probably stems from their more homogeneous distribution in the former fiber given co-doping with Y. However, partial reabsorption of UC

Figure 1.7. Snapshots of lateral emission in VIS (UC), for the (a) EDF L40, (b) EDF L20, and (c) Cr,Er AGCYS fibers (excitation at 976 nm, pump power 250 mW), recorded for the fiber segments (10 cm in length) just after splicing with a 976 nm LD output fiber. Copyright (2019) IEEE. Reprinted, with permission, from [56] (CC BY 4.0).

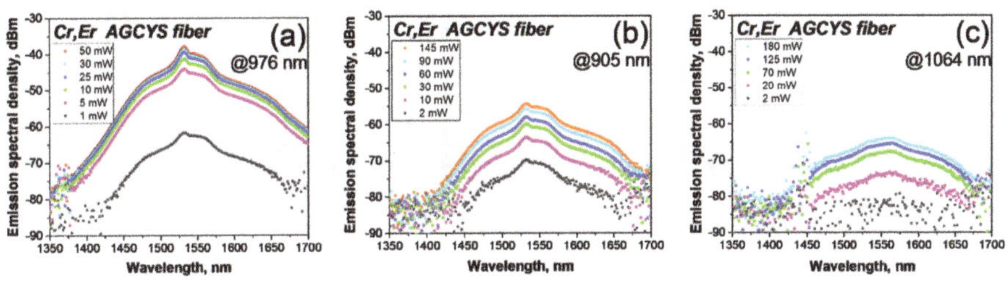

Figure 1.8. NIR frontal emission spectra (Er^{3+}-band) of the Cr,Er AGCYS fiber (L_f = 17 cm) at different excitation wavelengths: (a) 976 nm, (b) 905 nm, and (c) 1064 nm, at different pump powers. Copyright (2019) IEEE. Reprinted, with permission, from [56] (CC BY 4.0).

emission by Cr ions, having notable absorption in the visible (figure 1.4), cannot be dismissed as a factor responsible for the phenomenon.

The emission spectra of the Cr,Er AGCYS fiber within the Er^{3+}-band (1350–1700 nm) are presented in figure 1.8 both for (a) resonant (into the Er^{3+}-band, at 976 nm) and (b) and (c) non-resonant (off the Er^{3+}-band and into the Cr^{3+}/Cr^{4+}-band, at 905 nm/1064 nm) excitations. Experimentally, in each case the fiber was in-core pumped through splices by fiber-coupled commercial LDs and the emission spectra were detected at the Cr,Er AGCYS fiber output. For purposes of comparison, the results are exemplified here for the same length of Cr,Er AGCYS fiber (L_f = 17 cm).

Note that the presence of Er^{3+} related NIR emission at the 905 and 1064 nm excitations ((b) and (c)) points to effective energy transfer from the Cr^{3+}/Cr^{4+} subsystem to the Er^{3+} one, however, it is much less pronounced in the latter case. The revealed effect of indirect excitation of Er^{3+} fluorescence in the mixed Cr,Er system is reported for the first time, to the best of our knowledge.

Regarding the case of resonant (976 nm) excitation, note that Er^{3+} emission adheres to conventional spectral behavior where its maximum shifts to the Stokes side with increasing length of the Cr,Er AGCYS fiber; see figure 1.9(a). Also note that, due to the high amplifying potential of the Cr,Er AGCYS fiber (see section 3.4), amplified spontaneous emission (ASE) easily transforms into spurious lasing within the Er^{3+}

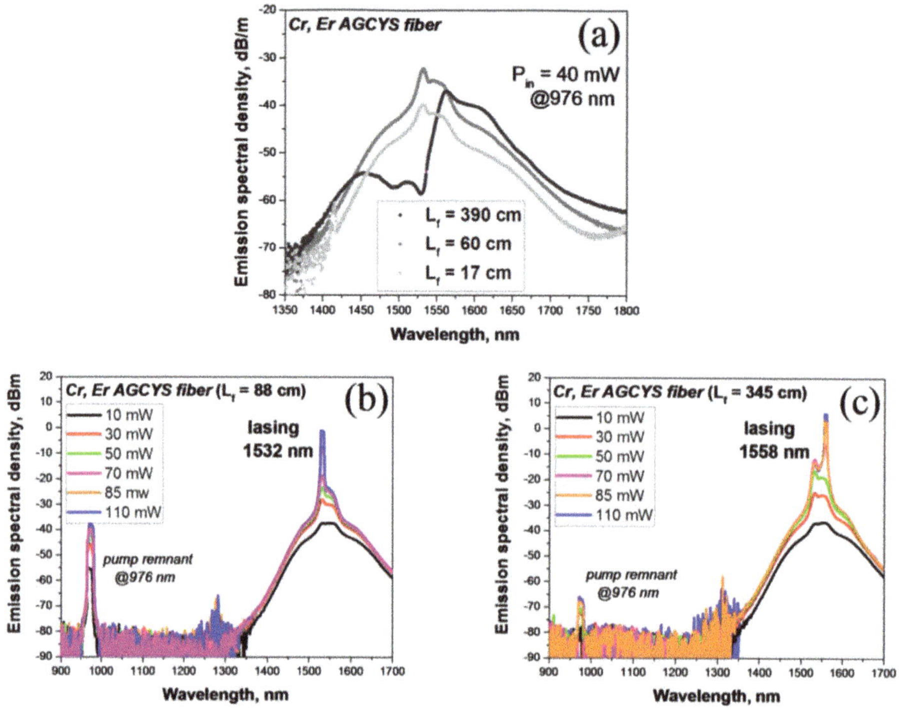

Figure 1.9. (a) Frontal emission spectra of the Cr,Er AGCYS fiber at different lengths of samples and the same pump power at 976 nm (below the threshold of spurious lasing). (b) and (c) Two examples of the spectral response of the fiber (for shorter and longer lengths), which reveal spurious lasing at (b) ~1532 nm (the peak of the emission band of Er^{3+} ions) (L_f = 88 cm) and (c) at ~1558 nm (corresponding to maximal spectral inversion in the Er^{3+} subsystem) (L_f = 345 cm). The data were collected for increased pump powers at 976 nm. Copyright (2019) IEEE. Reprinted, with permission, from [56] (CC BY 4.0).

band (due to Fresnel reflection on the fiber's cleaved ends) at increasing pump power at 976 nm when the length L_f is varied between ~70 and ~380 cm; see the examples of such scenarios presented in figure 1.9(b), (c).

As expected, for shorter pieces of the Cr,Er AGCYS fiber spurious lasing is established near the Er^{3+} fluorescence maximum (~1530 nm, (b)) whereas for longer pieces of the fiber its central wavelength shifts to the Stokes side (up to ~1560 nm, (c)) where optimal conditions for maximal inversion in the Er^{3+} ion system are created.

Meanwhile, the type of lightening of the Cr,Er AGCYS fiber at non-resonant excitation (off the Er^{3+} band but within the Cr^{3+}/Cr^{4+} band) is quite different; see figure 1.10 where we present the results of the experiment on pumping the fiber at 905 nm. As seen from figure 1.10(a), in this case, in addition to the NIR emissions characteristic of Cr^{3+}/Cr^{4+} ions (slightly spectrally structured within the 1.0–1.5 μm domain), there also appears powerful emission inherent to Er^{3+} ions, apparently due to the already mentioned energy transfer from the Cr^{3+}/Cr^{4+} subsystem to the Er^{3+}one (see figure 1.8(b)). Note that, in analogy to the case of resonant excitation at 976 nm (see figure 1.9(a)), the maximum of Er^{3+} emission shifts to longer

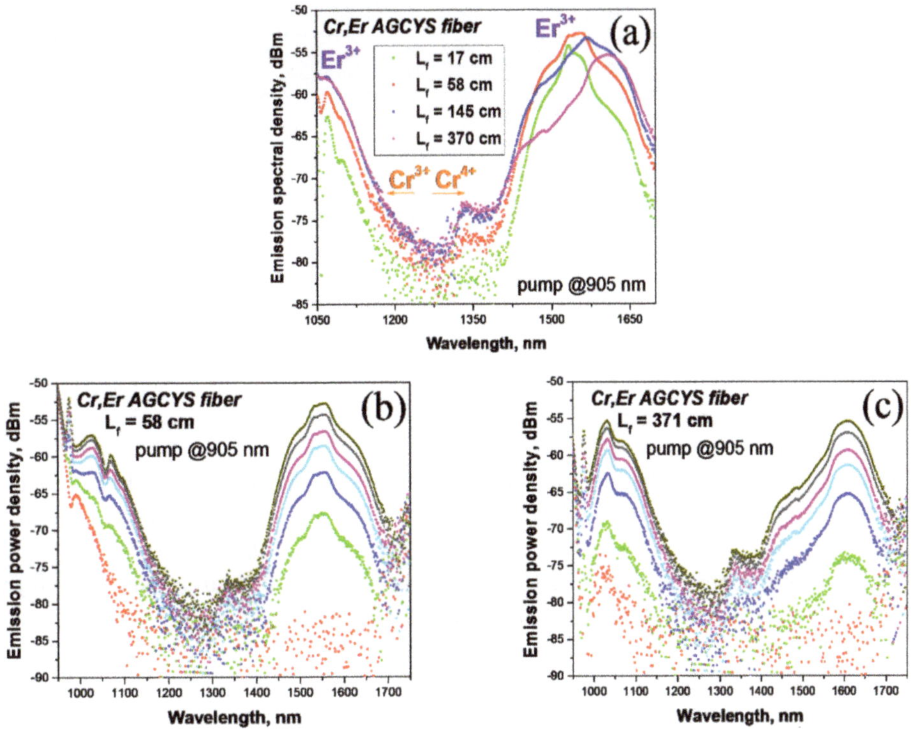

Figure 1.10. (a) Frontal emission spectra of the Cr,Er AGCYS fiber at 905 nm excitation (140 mW) (into the Cr^{3+}/Cr^{4+} absorption band but off the Er^{3+} absorption band) for different fiber lengths L_f; the spectral positions of the NIR emissions contributed by the Cr^{3+}/Cr^{4+} and Er^{3+} subsystems are highlighted. (b) and (c) Frontal emission spectra of the fiber at 905 nm excitation at increasing pump power, exemplified by 1, 4, 16, 35, 57, 98, and 140 mW, obtained at different lengths L_f: (b) 58 cm and (c) 370 cm. Copyright (2019) IEEE. Reprinted, with permission, from [56] (CC BY 4.0).

wavelengths with increasing fiber length L_f as well, as seen from figure 1.10(a), where the data for maximal available pump power at 905 nm are demonstrated. Furthermore, figure 1.10(b) and (c) provide the dependences of the NIR fluorescence of the fiber on pump power at 905 nm for $L_f = 58$ and 370 cm, respectively.

In contrast to resonant excitation (compared to figure 1.9(b) and (c)), no spurious lasing in the Er^{3+} subsystem is achieved in this case, as energy transfer from the Cr^{3+}/Cr^{4+} ions to the Er^{3+} ions is not so effective for establishing a high gain and threshold conditions within the latter subsystem.

In figure 1.11(a) and (b), to complete the pattern of emissive potential of the Cr,Er AGCYS fiber, we present the data of NIR (>1 μm) fluorescence decay measurements, acquired at both kinds of excitation—resonant (at 976 nm) and non-resonant (at 905 nm). The data were collected in backward geometry, using specially designed wavelength-division multiplexors for separating the pump light and the detected NIR emissions. NIR (up to ~1.6 μm) fluorescence signals from the fiber samples were captured at fast switching pump light off. To further reduce the pump background in the measured signal, a long-pass 1000 nm optical filter (Thorlabs

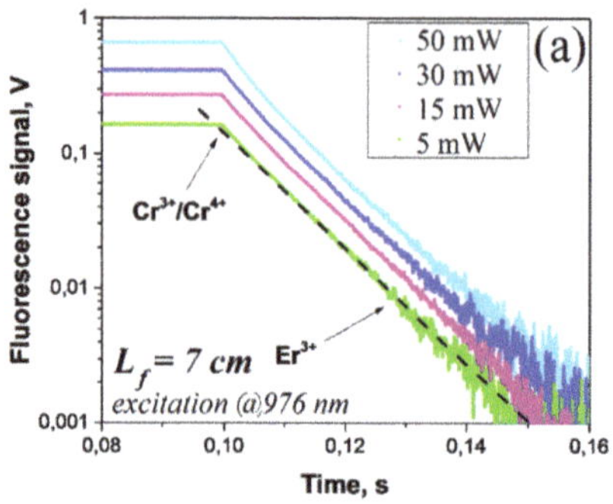

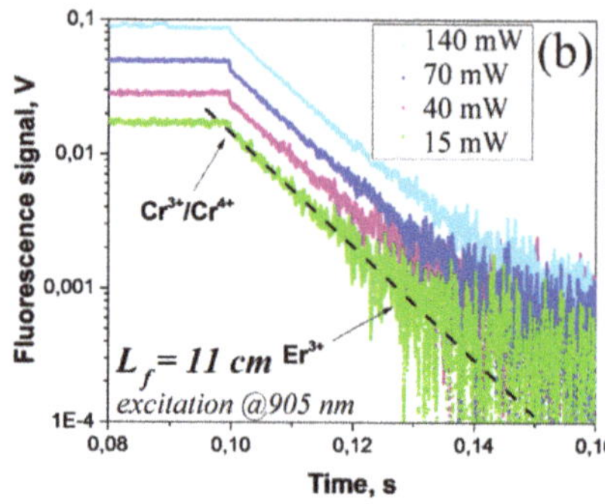

Figure 1.11. NIR fluorescence kinetics in the Cr,Er AGCYS fiber at (a) 976 nm and (b) 905 nm excitations at different pump powers, detected in backward geometry using short fiber samples. Copyright (2019) IEEE. Reprinted, with permission, from [56] (CC BY 4.0).

FEL1000) was placed between a fiber sample and a Ge-based photo-detector (Newport 2033, 200 kHz bandwidth). To avoid any significant role of reabsorption, relatively short fiber samples were used in the experiments: $L_f = 7$ and 11 cm (at 976 and at 905 nm, respectively). The time resolution of the measurements was ~8 μs.

As can be seen from figure 1.11, the decay kinetics is composed of two exponents with decay times of ~9.8 ms and ~40 μs (highlighted in the figure by arrows), which can be ascribed to the Er^{3+} and Cr^{3+}/Cr^{4+} ions, respectively. At both excitation wavelengths the decay times were found to be virtually independent of pump power (at least for their low levels).

1.4 Gain

We also gained insight into the spectral features of amplification (gain) of the Cr,Er AGCYS fiber. First we compared the net gain in this fiber and in the commercial Cr-free EDFs L20 and L40 at resonant (in the Er-band) 976 nm pumping and then examined what happens in this case at non-resonant (off the Er-band but in the Cr-band) 905 nm/1064 nm pumping. All experiments were carried out under the same experimental conditions applying the following procedure. An active fiber sample was pumped from one side by an LD (at 976, 905, or 1064 nm) through a proper WDM while the small signal to be amplified was delivered from the white-light source from its opposite side. By means of measuring the fiber's fluorescence spectra at white light switched on and off, then measuring its attenuation spectrum without pumping (but with white light switched on), and posterior data processing, net-gain spectra (in dB m^{-1}) were found. In the experiments we used the same WDMs as in the lifetime measurements reported above.

The results for 976 nm pumping at increasing pump power are shown in figure 1.12, where the data for commercial L20/L40 EDFs ((a) and (b)) are compared with those for the Cr,Er AGCYS fiber (c). One can reveal two main advantages of the new fiber in the sense of the amplification potential of the fiber. First, as can be seen from figure 1.12(c), maximal gain (at 1530 nm) in the Cr,Er AGCYS fiber at full Er^{3+} absorption saturation (>50 mW) reaches g_0 ~28 dB m^{-1}, i.e. a value only slightly less

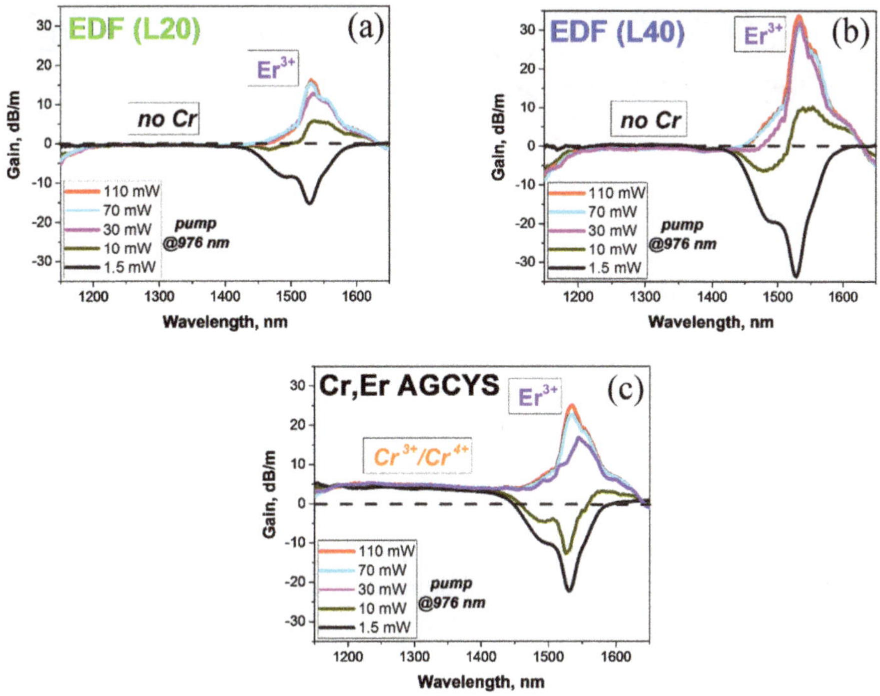

Figure 1.12. Spectral dependences of net gain in (a) EDF L20, (b) EDF L40, and (c) Cr,Er AGCYS fiber in the NIR (1150–1650 nm), measured at excitation by an LD with an operating wavelength of 976 nm; the launched pump powers are specified in the insets. Copyright (2019) IEEE. Reprinted, with permission, from [56] (CC BY 4.0).

than the small-signal absorption of the fiber (α_0 ~32 dB m^{-1}); importantly, the ratio g_0/α_0 approaches ~0.88. This demonstrates quite a large amplification potential of the fiber, even better than those (figure 1.12(a) and (b)) of the commercial EDFs L20/L40 (α_0 ~21/40 dB m^{-1}; g_0~17/32 dB m^{-1}; g_0/α_0~0.82/0.81). Second, in contrast to the latter fibers that do not amplify within 1.15–45 μm (given the absence of Cr ions), the Cr,Er AGCYS fiber has broadband plain positive net gain in this spectral domain (g_0 ~3–4 dB m^{-1}), obviously because of its co-doping with Cr^{3+}/Cr^{4+} ions; compare figure 1.12(c) and figure 1.12(a) and (b). Seemingly, this effect is produced not only by direct excitation of Cr^{3+}/Cr^{4+} ions by pump light at 976 nm, but also via efficient energy transfer from excited states of Er^{3+} ions.

Furthermore, analogous measurements of spectrally dependent gain in the Cr,Er AGCYS fiber were performed using LDs at 905 nm and at 1064 nm as pump sources (see figure 1.13(a) and (b), respectively), namely under excitation off the Er^{3+} absorption band but into the Cr^{3+}/Cr^{4+} band. We limit ourselves here by giving the results for maximal available pump powers: 145 mW and 180 mW, respectively. The plotted spectra were obtained for different lengths of the Cr,Er AGCYS fiber to shed more light on the physics of energy transfer between the Er^{3+} and Cr^{3+}/Cr^{4+} subsystems, involved at these types of non-resonant excitation. As can be seen from figure 1.13, in both cases there is positive, more-or-less flat gain within the range ~1150 to ~1400 nm, given by the presence of Cr ions in the Cr,Er AGCYS fiber,

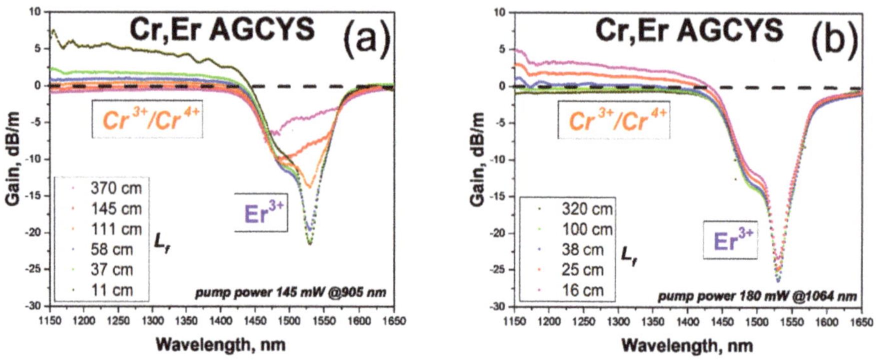

Figure 1.13. Spectral dependences of net gain in the Cr,Er AGCYS fiber in the NIR (1150–1650 nm), measured at excitation by LDs with operating wavelengths of (a) 905 nm and (b) 1064 nm; the launched pump powers at these wavelengths and the lengths L_f of the fiber samples under study are specified in the insets. Copyright (2019) IEEE. Reprinted, with permission, from [56] (CC BY 4.0).

effectively inverted under the action of pump lights. However, with increasing length of the fiber, the gain within this spectral domain becomes notably worse, this effect being explained not only by the common effect of depletion of pumping along the fiber length, but also by leakage of excitation accumulated in the Cr subsystem to the Er subsystem, seen as effective bleaching of absorption within the Er^{3+} band.

1.5 An erbium-doped specialty optical fiber based on phase-separated alumina–silica glass

1.5.1 The basic principle for the phase separation phenomenon of optical fiber preforms

The origin of alumina rich phase-separated glass is in the phenomenon of meta-stable glass-in-glass separation [58] on rapid quenching of Al_2O_3–SiO_2 melts comprised of 4–50 mole% Al_2O_3, indicated by the phase diagram shown in figure 1.14. Such phase separation in the glass matrix through the solution doping process is achieved by incorporating small amounts (<10 mole%) of co-dopant which lies within their miscibility gap. First we try to identify suitable co-dopants which give rise to immiscibility into silica glass with the incorporation of small amounts of such co-dopants. Based on different thermodynamical data, such as the entropy of mixing, enthalpy of mixing, activity coefficient, and Gibbs-free energy change [59–61], the phase diagram of binary SiO_2–Al_2O_3 glass was derived employing FactSage software and a miscibility gap was found to exist within 4–30 mole% of Al_2O_3 into silica glass. Such a miscibility gap was found to match closely with the experimentally determined miscibility gap region of the binary SiO_2–Al_2O_3 system [61]. The shape of the meta-stable binary two-liquid region was found to be asymmetric and to have a tail towards the Al_2O_3 rich composition side of the system in the temperature ranges between 1350 °C–1650 °C (figure 1.14).

In the presence of a small amount of phosphorous, the shape of the meta-stable binary two-liquid region was found to be much more asymmetric in nature,

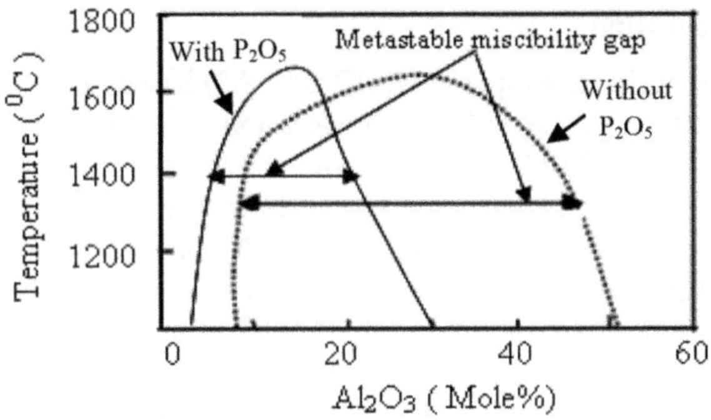

Figure 1.14. The meta-stable immiscibility zone of the SiO_2–Al_2O_3 system with and without P_2O_5. Reprinted from [62] with permission from MDPI.

having a tail towards the lower Al_2O_3 content side than that of the binary SiO_2–Al_2O_3 system in the temperature range 1350 °C–1650 °C (figure 1.14). Accordingly, annealing of the glass in this temperature range induces phase separation without any crystallization of the glass. The pre-requisite to achieve phase separation in multi-component glasses is the presence of an immiscibility region in their phase diagram. It is possible to prepare a clear and transparent glass having a composition within the immiscibility region when their melts are quenched rapidly below the glass transition temperature T_g. It has been found that the glass viscosity is too high and the necessary rearrangements of structural groups for phase separation can be achieved below T_g. However, when these phase-separated glasses are further heated in the range between T_g and the upper critical temperature of the immiscibility region, this leads to formation of sub-liquidus phase separation. The two principal mechanisms of phase separation, namely through nucleation and growth and by spinodal decomposition, are well known. However, the type of mechanism depends on the glass composition as well as the temperature. The phase morphology also depends on the annealing condition of the samples. The opaqueness is a consequence of light scattering on phases with different compositions and refractive indices and depends on the size of the new phases, which depends on the annealing condition. The addition of P^{5+} has an accelerating effect on the phase separation [63]. To induce phase separation, a small amount of P_2O_5, around 0.15–0.20 mole%, is incorporated into such an alumina–silica glass host. The thermal treatment to prevent nucleation and subsequent crystallization of mullite formation within the high-alumina-dispersed glass phase is important either during cooling from the heated melts or on reheating the preform within 1600 °C–1900 °C, effectively to achieve a working temperature within 1350 °C–1650 °C. It is assumed that the temperature difference between the outer surface (10 mm diameter) and central region (0.75 mm diameter) of the preform becomes 250 °C. The meta-stable binary two-liquid region is formed below a temperature of around 1650 °C. The phase-separated alumina–silica glass containing at least 4 mole% of Al_2O_3 appears through heating below 1650 °C.

1.5.2 Sample preparation

Three Er_2O_3 doped preform samples (A1, A2, and A3) based on an alumina–silica host with different contents of Al_2O_3 with a slight variation in Er_2O_3 concentration (table 1.1) were prepared employing MCVD coupled with the SD technique (figure 1.15).

All the process parameters were standardized along with the thermal treatment to obtain an alumina rich phase-separated glass host. A porous SiO_2 soot layer containing small amounts of P_2O_5 (<0.01 mol%) was deposited along the forward direction inside the substrate silica tube to promote the phase separation phenomenon. In the forward deposition method the deposition of the porous silica soot layer was performed at variable temperatures ranging from 1350 °C to 1450 °C along the direction of the reactant flow containing a halide precursor such as $SiCl_4$, $POCl_3$, or O_2 as the carrier gas to adjust the porosity of the deposited SiO_2 soot layer for incorporation of the desired amount of Al_2O_3, varying from 3.5 to 4.5 mole% and Er_2O_3 varying from 0.073 to 0.075 mole%. A list of the fabricated preforms/fibers is given in table 1.1. The Al_2O_3 content of fibers A1, A2, and A3 was 4.5, 4.0, and 3.5 mole% under the deposition of porous silica soot layers at 1350 °C, 1400 °C, and 1450 °C, respectively, containing minor amounts of P_2O_5 using the same strength of $AlCl_3$ $6H_2O$ for a fixed soaking period of one hour; microscopic views are shown in figure 1.16.

For incorporation of the desired quantity of Er^{3+} and Al^{3+} ions, an alcoholic solution of a mixture of suitable strength of the precursor salts was used. The strengths

Table 1.1. The composition of the three Er_2O_3-doped fiber preforms (measured using EPMA) [62].

Fiber number	Er_2O_3 (Mole%)	Al_2O_3 (Mole%)
A1 (Phase separated)	0.075	4.5
A2 (Phase separated)	0.073	4.0
A3 (Non-phase separated)	0.074	3.5

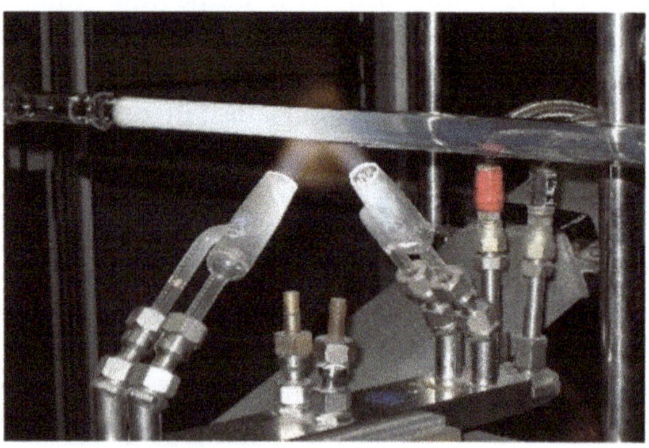

Figure 1.15. Deposition of a porous silica layer within the silica tube along the forward direction using the MCVD process. Reprinted from [62] with permission from MDPI.

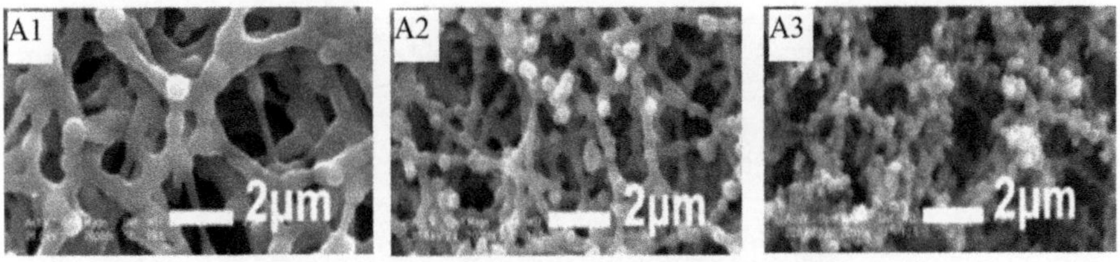

Figure 1.16. FEGSEM images of the porous silica core layer deposited at different temperatures to create the three preforms/fibers : (A1) 1350 °C, (A2) 1400 °C, and (A3) 1450 °C. Reprinted from [62] with permission from MDPI.

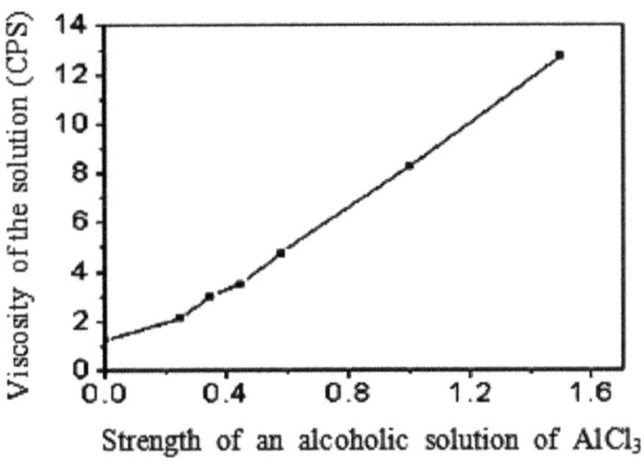

Figure 1.17. The viscosity of an alcoholic solution of a mixture of 0.01(M) $ErCl_3$ $6H_2O$ and variable strength of $AlCl_3$ $6H_2O$. Reprinted from [62] with permission from MDPI.

of the $ErCl_3$ $6H_2O$ and $AlCl_3$ $6H_2O$ were maintained around 0.01 and 1.5 M, respectively. The dipping time was maintained at around 1.0 h. The porosity of the deposited silica soot layer related to the deposition temperature is critical with respect to the viscosity of the solution when the strength of the precursor of the Al^{3+} ions becomes high (>1.25 M). The viscosity of the alcoholic solution, measured using a Brookfield Digital Viscometer (Model LV TDV), was found to increase gradually with increasing strength of $AlCl_3$ $6H_2O$, as shown in figure 1.17. The strength of $AlCl_3$ $6H_2O$ was maintained at 1.5 M to incorporate 3.5–4.5 mole% of Al_2O_3 by adjusting the porosity of the deposited unsintered silica layer. The corresponding viscosity of the solution was around 12.75 centi poise (cP) and lead to disturbances of the coated layer if the porous silica layer was deposited at relatively lower temperatures, say below 12 500 °C, due to a lack of proper thermal adhesion of the porous soot layer with the inner silica surface of the tube. The solution soaking period was maintained at around one hour to allow sufficient soaking time for diffusion through the porous soot layer which takes place due to capillary action. The viscosity as well as the surface tension of the solution increased considerably with increasing strength of the solution of $AlCl_3$ $6H_2O$ from 0.05 to 1.5 M. As a result of this, the deposition temperature of the porous soot layer along the forward direction was adjusted carefully with respect to the strength of the soaking solution to be used during the solution doping process.

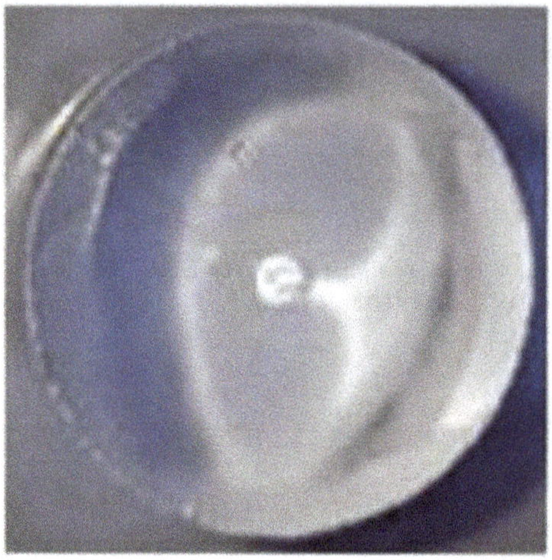

Figure 1.18. Cross-sectional view of an Al_2O_3+Er_2O_3 doped fiber preform section for EPMA where the central bright portion shows the doping region. Reprinted from [62] with permission from MDPI.

1.5.3 Material and optical characterization

Electron-probe microanalysis (EPMA) was used to determine the incorporated dopant amount, using a two-side polished preform sample of thickness 1–1.5 mm taken from the fabricated preform, maintaining spatial resolution of 1 μm after applying a thin graphite coating, as presented in figure 1.18. Field emission gun scanning electron microscope (FEGSEM) images of the three small circular polished preform samples were obtained for identification of the phase-separated region, followed by energy dispersive x-ray (EDX) and x-ray diffraction (XRD) analyses to detect the nature and composition of the phase-separated glass. To evaluate the nature of the phase-separated region, the electron beam was focused on and outside of the phase-separated region when the EDX data were collected. To detect the surrounding coordination numbers of Al^{3+} ions, ^{27}Al nuclear magentic resonance (NMR) spectroscopy of the core glass of the preform samples A1 and A3 was performed after complete etching of the surrounding silica glass, covering the open core end with a Teflon tape using buffer HF solution followed by making a powder of the remaining 0.8 mm thickness of the preform sample.

High resolution optical microscopic views of two of the fabricated preforms are shown in figure 1.19 and the corresponding refractive-index profiles (RIPs) are presented in figure 1.20. No central dip is observed in the RIP presented in figure 1.20, a common features in the case of preforms fabricated using the MCVD process due to slow evaporation of the index-raising elements.

An extended x-ray absorption fine structure (EXAFS) measurement was carried out using Si(111) monochromating crystal and a focusing second mirror in order to obtain the smallest x-ray spot size (150 × 800 μm) available on the sample synchrotron radiation. The experiments were carried out in the fluorescence mode using a multi-element energy-resolving detector. The liquid nitrogen temperature was used in order to enhance the signal from the highest coordination shells. EXAFS spectra of all the samples were taken and compared to those of Er_2O_3 and Al_2O_3.

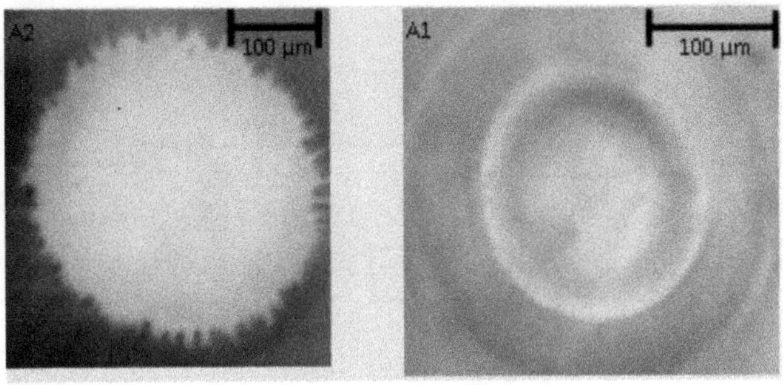

Figure 1.19. Microscopic views of the core–cladding interface of two fiber preform samples: A1 and A2. Reprinted from [62] with permission from MDPI.

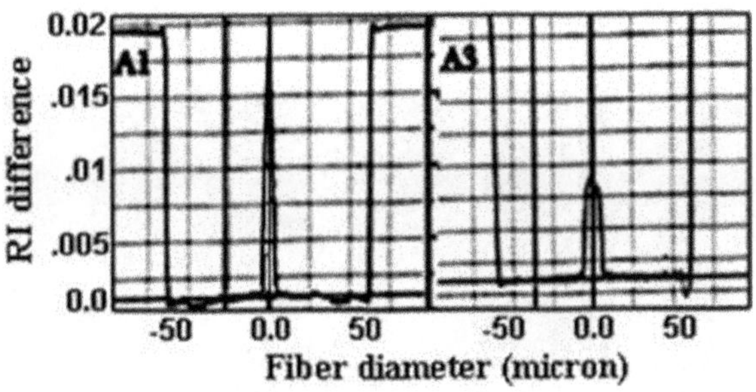

Figure 1.20. RIPs of two EDFs: phase-separated (A1) and non-phase separated (A3). Reprinted from [62] with permission from MDPI.

From the EXAFS analyses a clear structure was observed up to 8 in the *k*-space of the different Er_2O_3 doped fiber preform samples, which provides sufficient information to understand the nearest-neighbor atom structure. Different spectroscopic properties, namely the absorption coefficient, fluorescence spectra, and lifetime of the fabricated fibers drawn from the phase-separated (A1 and A2) and non-phase-separated glass (A3) based preforms were measured. The fluorescence spectra of the fabricated fiber samples were measured with pumping at 980 nm under different pump powers (25, 125, and 245 mW) in a lateral direction.

The optical gain spectra of the fabricated EDFs (A1, A2, and A3) were measured within the C (1530–1565 nm) band employing a tunable distributed feedback (DFB) laser source under optimized conditions; the input signal level was tuned using an attenuator. The EDF was pumped using a Bragg grating stabilized semiconductor laser diode operating at 980 nm, coupled through 980/1550 nm WDM couplers, as shown in figure 1.21. The length of each fiber was optimized to obtain the minimum gain-difference at constant pump power.

1.6 Results and discussion

Phase-separated alumina–silica glass containing at least 4 mole% of Al_2O_3 was achieved by thermal annealing of the fabricated preform below 1900 °C.

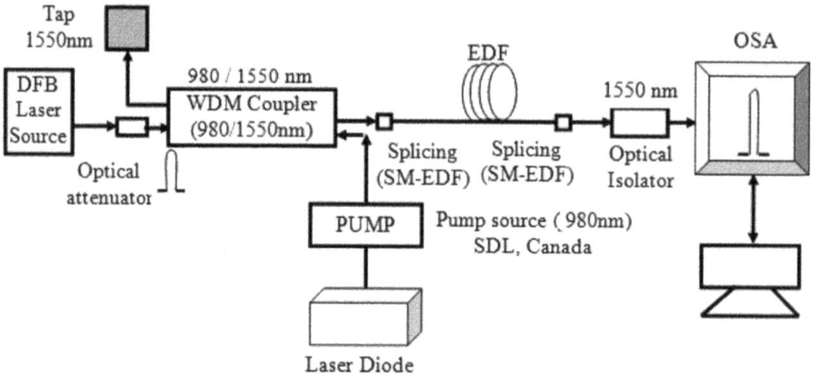

Figure 1.21. Experimental set-up for optical gain measurement of the EDF. Reprinted from [62] with permission from MDPI.

Figure 1.22. FEGSEM images of three alumina–silica preform samples of EDF. Phase-separated: sample A1 and sample A2; non-phase separated: sample A3. Reprinted from [62] with permission from MDPI.

To maintain the required doping level of Al_2O_3, of at least 4 mole%, the porosity of the unsintered silica layer along with the strength of the solution of $AlCl_3$ $6H_2O$ was optimized. During preform production at the collapsing stage no phase-separation occurs as the melt glass was heated to around 2300 °C with rapid cooling at room temperature where no meta-stable immiscibility arises. To achieve the phase-separated alumina–silica glass two conditions need to be satisfied, the doping level of Al_2O_3 and thermal heating treatment. To obtain the phase separation the minimal doping level of Al_2O_3 should be around 4.0 mole% followed by heating of the preform within a 1600 °C–1900 °C temperature range. After fabrication the preform sample was heated to around 1600 °C–1900 °C where the glass forms a meta-stable immiscibility during fire-polishing, which gives rise to phase separation. However, in both cases the heating period and cooling rate become very low, around 5–8 s and 25 °C–30 °C s^{-1} respectively, under normal MCVD conditions. FEGSEM images of two phase-separated alumina–silica based fiber preform samples, A1 and A2, are given in figure 1.22 along with a non-phase-separated preform sample.

The microscopic views clearly show the phase-separated regions. The preform sample A3 containing 3.5 mole% of Al_2O_3 with and without heat treatment within a 1600 °C–1900 °C temperature range does not show any type of phase separation as the glass does not enter into the immiscibility region under such a doping level of Al_2O_3 at any temperature. The microscopic image of such a preform sample is shown in figure 1.22. The fibers A1 and A2 were drawn from the phase-separated preforms melted below 1900 °C so that the doping host experiences a temperature below 1700 °C to retain the phase-separation within the core of the fiber. The FEGSEM micrographs (figure 1.23) of two such large bare-fiber sections clearly show the retention of the phase separation phenomenon in samples A1 and A2. On the other hand, the fiber drawn from the same preform A2 melted at 2100 °C does not produce phase-separated glass (see figure 1.24) as the doped core-glass matrix experiences a temperature above the upper limit temperature of the immiscibility region. The fiber drawn from the non-phase-separated preform A3 containing 3.5 mole% of Al_2O_3 melted below 1900 °C does not show any phase-separation as the doping level is below the minimum required level of the immiscibility region.

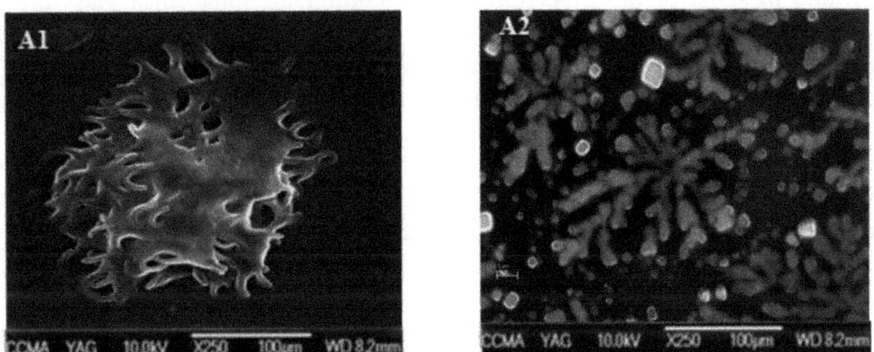

Figure 1.23. FEGSEM images of two alumina–silica bare-fiber samples of EDF: sample A1 and sample A2 drawn at a temperature of around 19 000 °C. Reprinted from [62] with permission from MDPI.

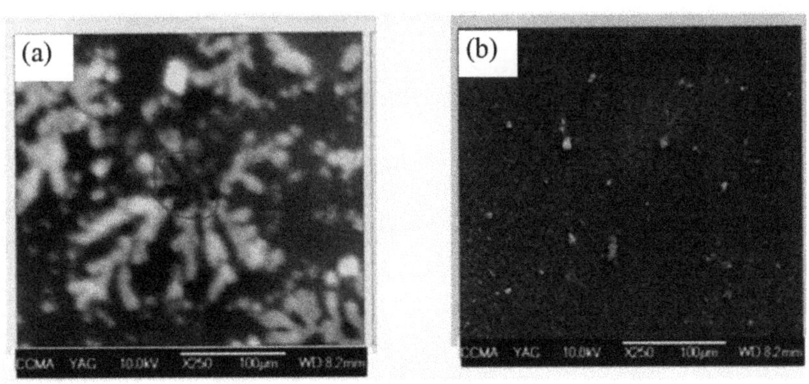

Figure 1.24. FEGSEM images of alumina–silica glass based (a) phase-separated and (b) non-phase-separated Er-doped fiber preforms obtained by heating of the same preform to around 19 000 °C and above 21 000 °C, respectively. Reprinted from [62] with permission from MDPI.

As a result of this, the EDF drawn below 1900 °C from the phase-separated alumina–silica glass based optical fiber preform containing at least a 4.0 mole% doping level of Al_2O_3 retains the phase-separation phenomenon within the core-glass matrix. The core–cladding boundary of the fabricated fiber made using the solution doping process generally exhibits some imperfection if the core–cladding viscosities differ substantially from one another. This results in excess loss attributed to scattering at the core–cladding interface and reduces the pump efficiency of the fiber. Incorporation of a high content of Al_2O_3, to make a greater amount of Er^{3+} ions soluble in germanium–silica glass, generates another problem of imperfection—a star-like pattern at the core–cladding boundary arises due to a significant viscosity difference developed at the collapsing stage [64–66]. It is known that the viscosity of pure SiO_2 based glass becomes higher than that of a GeO_2–Al_2O_3–Er_2O_3 mixture with a SiO_2 content of 85–90 mole%. A large viscosity difference between the core and cladding glass produces high chemical potential differences and results in diffusion of the core glass into the cladding glass. Such a type of imperfection was found in some of the Er_2O_3-doped germanium–alumina–silica glass preform sections, as presented in figure 1.25. On the other hand, such a type of core–cladding imperfection was reduced greatly in the case of the Er_2O_3 doped alumina–silica optical fiber preform section shown in figure 1.19. These imperfections, in the case of the Er_2O_3 doped germanium–alumina–silica glass fiber preform, develop from the local change of viscosity in the vicinity of the core–cladding interface, which arises during deposition of the porous SiO_2–GeO_2 layer. However, in the case of the Er_2O_3 doped alumina–silica glass fiber preform, the local viscosity changes close to the core–cladding interface during the deposition of the pure porous silica layer become much smaller.

The content of Al_2O_3 in the three fiber preform samples (A1, A2, and A3) was found to be distributed uniformly along their whole diameter region (figure 1.26). The nature of their RIPs, as shown in figure 1.20, was found to be very sharp without the formation of any central dip, arising from evaporation of the index-raising elements as gaseous mono-oxide species during the collapsing stage. As the vapor

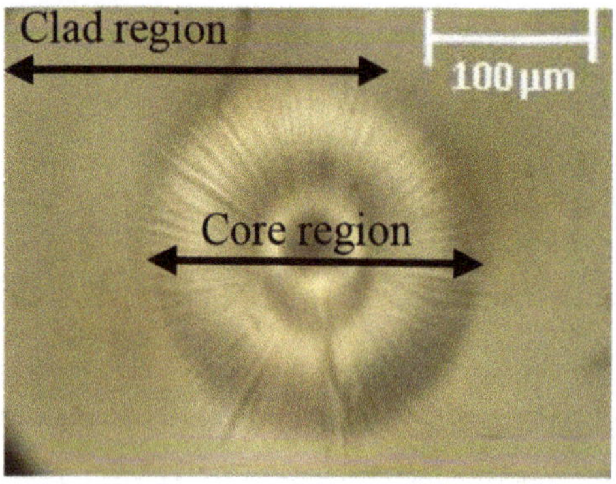

Figure 1.25. A microscopic image of the core–cladding interface of the erbium-doped germanium–alumina–silica fiber preform. Reprinted from [62] with permission from MDPI.

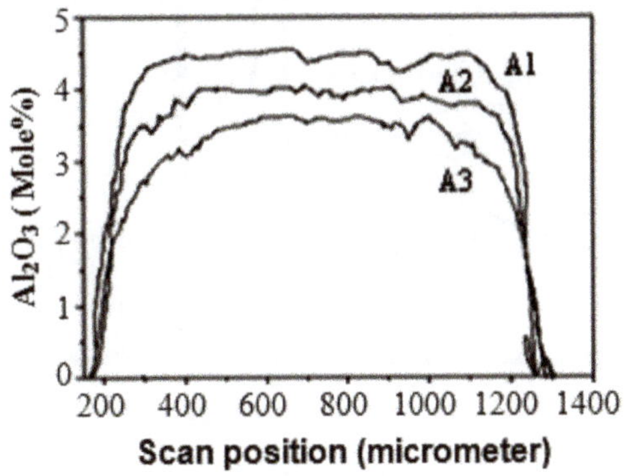

Figure 1.26. EPMA for Al_2O_3 of the three fibers: A1, A2, and A3. Reprinted from [62] with permission from MDPI.

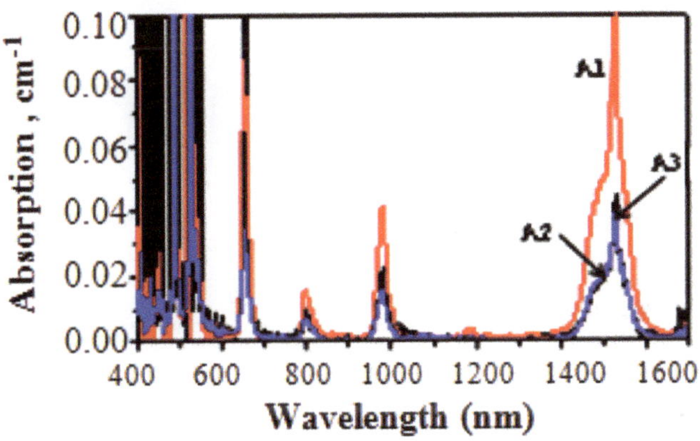

Figure 1.27. Absorption curves of the three EDFs: A1, A2, and A3. Reprinted from [62] with permission from MDPI.

pressure of AlO species is very low even at high temperatures, no central dip will be generated. The optical absorption spectra of three EDFs based on phase-separated and non-phase-separated alumina–silica glass host are shown in figure 1.27. The absorption coefficient of the EDF A1 becomes higher than that of the other two EDFs (A2 and A3), which indicates the higher doping level of Er_2O_3 in EDF A1 compared to the others.

The Al_2O_3 content of the three preform samples A1, A2, and A3 was found to be 4.5, 4.0, and 3.5 mole%, respectively, with 0.075, 0.073, and 0.074 mole% of Er_2O_3. The nature as well as the composition of the phase-separated alumina–silica fiber preform samples were evaluated using EDX followed by XRD analyses

A1, containing 4.5 mole% of Al_2O_3, clearly shows the phase-separated region along with sample A2, whereas sample A3 does not form any phase-separated region within the core-glass matrix. Sample A2 also forms the phase-separated region. The EDX data taken in and out of the phase-separated region of fiber preform A1, as shown in figure 1.28, clearly reveal that the content of Al^{3+} and Er^{3+} in the phase-separated

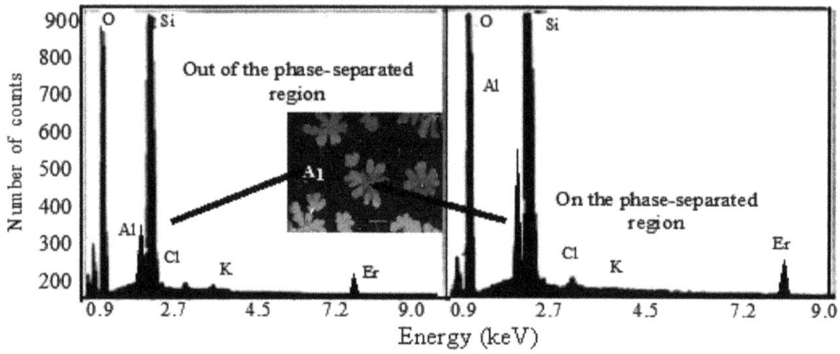

Figure 1.28. EDX curves from in and out of the phase-separated region of the alumino-silicate glass based Er_2O_3 doped preform (A1). Reprinted from [62] with permission from MDPI.

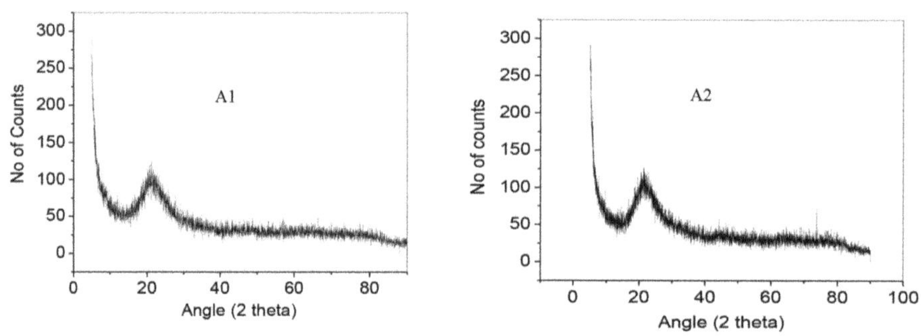

Figure 1.29. The XRD curves of two of the alumino-silicate based preform samples of EDF. Reprinted from [62] with permission from MDPI.

region becomes larger compared to the non-phase-separated region. The corresponding XRD data shown in figure 1.29 indicate that the phase-separated region does not contain any crystalline phases of Al_2O_3 or $ErAl_3O_6$ complexes.

The EXAFSs of EDFs containing different amounts of Al_2O_3 are shown in figure 1.30. From the EXFAS analyses a clear oscillation structure is observed up to 8 in the *k*-space of the different Er_2O_3-doped fiber preform samples and is endowed with sufficient information to achieve understanding regarding the nearest-neighbor atom structure. The radial distribution functions of the three EDFs (A1, A2, and A3) containing different contents of Al_2O_3 and Er_2O_3, along with one EDF sample without Al_2O_3 (A0) and a pure Er_2O_3-doped preform sample are shown in figure 1.31. From these analyses it can be concluded that Er–O bonding exists at approximately 0.2 nm in samples A0, A1, A2, A3, and Er_2O_3. However, the Er–O distance of sample A0 is shorter compared to the other samples. The effect of doping with Al_2O_3 on the Er^{3+} ion coordination number is shown in figure 1.32. The coordination number of the Er^{3+} ions of the EDF (A0) without doping of Al_2O_3 is found to be around 3.75. The effect of Al_2O_3 on the Er–O bond distance is shown in figure 1.33. The Er–O bond distance of EDF (A0) without doping of Al_2O_3 is found to be around 0.205 nm.

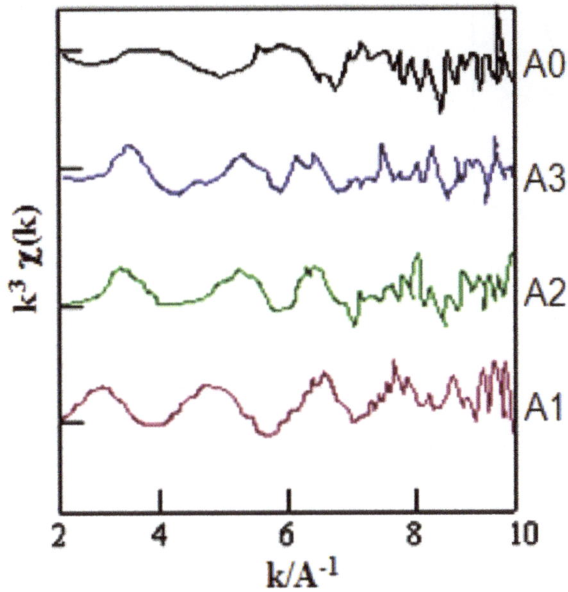

Figure 1.30. EXAFS of the three EDFs (A1, A2, and A3) containing different contents of Al_2O_3 and Er_2O_3 along with one EDF sample without Al_2O_3 (A0). Reprinted from [62] with permission from MDPI.

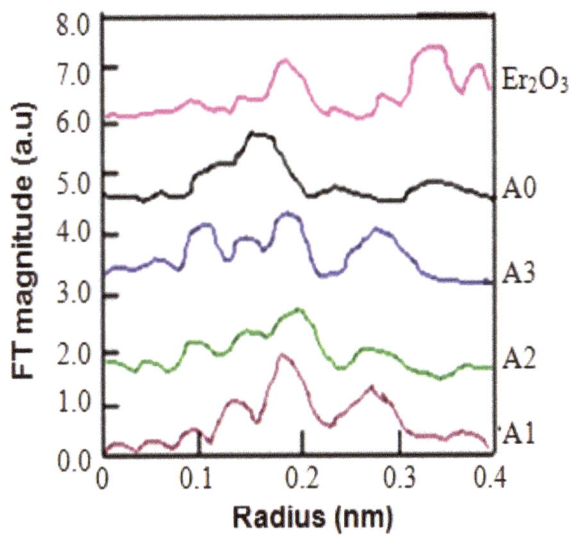

Figure 1.31. The radial distribution function of the three EDFs (A1, A2, and A3) containing different contents of Al_2O_3 and Er_2O_3 along with one EDF sample without Al_2O_3 (A0). Reprinted from [62] with permission from MDPI.

The phase-separated EDFs A1 and A2 show improved spectral broadening of the $^4I_{13/2}$–$^4I_{15/2}$ transition compared to that of the non-phase-separated EDF A3, as shown in figure 1.34. The transition lifetimes of EDFs A1 and A2 (10.87 and 10.71 ms) were found to be larger than that of the EDF A3 (10.3 ms), as shown in figure 1.35.

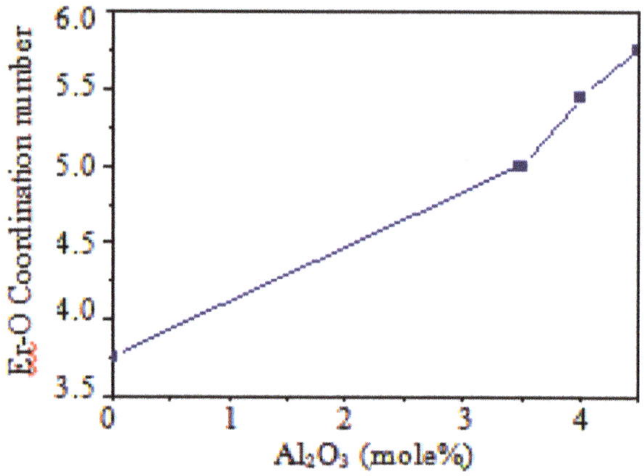

Figure 1.32. The effect of doping with Al_2O_3 on the Er–O coordination number. Reprinted from [62] with permission from MDPI.

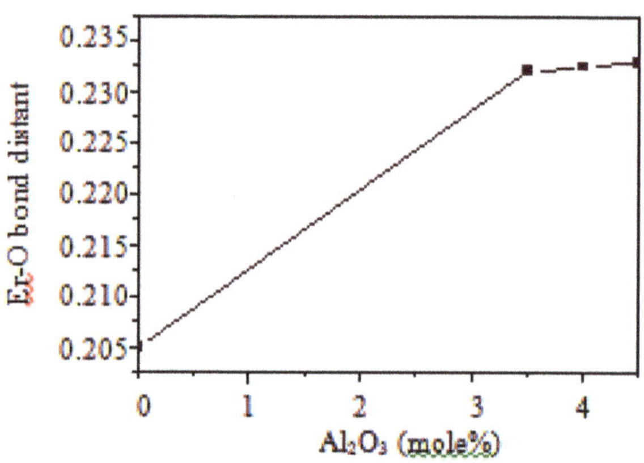

Figure 1.33. The effect of doping with Al_2O_3 on the variation of the Er–O bond distance. Reprinted from [62] with permission from MDPI.

All three fibers show very little change in their lifetime with increasing pump power, as shown in figure 1.36. This indicates that clustering of Er^{3+} ions is almost absent, although all the fiber samples contain more than 0.07 mole% of Er_2O_3, equivalent to 1400 ppm of Er^{3+} ions.

The optical gain spectra of these three EDFs in the C-band region are shown in figure 1.37. The gain differences of the phase-separated EDFs (A1 and A2) are within 1.2 dB whereas that for fiber A3 is 1.70 dB.

In ErO_6 complexes, each Er^{3+} ion is surrounded by six neighboring oxygen atoms forming an octahedron structure possessing C_{3v} point symmetry and producing strong Stark splitting, leading to the characteristic Er^{3+} emission in the 1460–1610 nm range. When incorporated into the silica network the aluminum ions are incorporated either as a network former, having a tetrahedral coordination with a formula of $AlO_{4/2}$, or as a network modifier in an octahedral

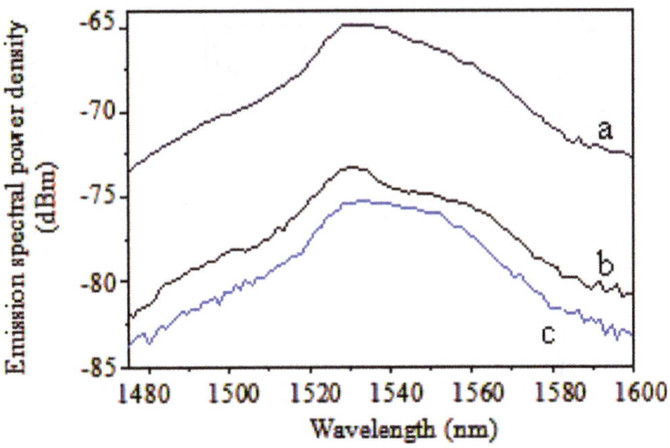

Figure 1.34. The fluorescence spectra of the three EDFs: (a) A1, (b) A3, and (c) A2 under a pump power of 25 mW. Reprinted from [62] with permission from MDPI.

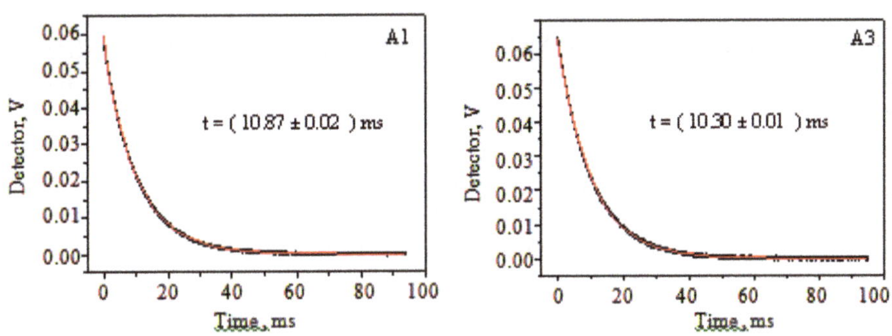

Figure 1.35. The lifetime of two of the EDFs, A1 and A3, at a pump power of 25 mW. Reprinted from [62] with permission from MDPI.

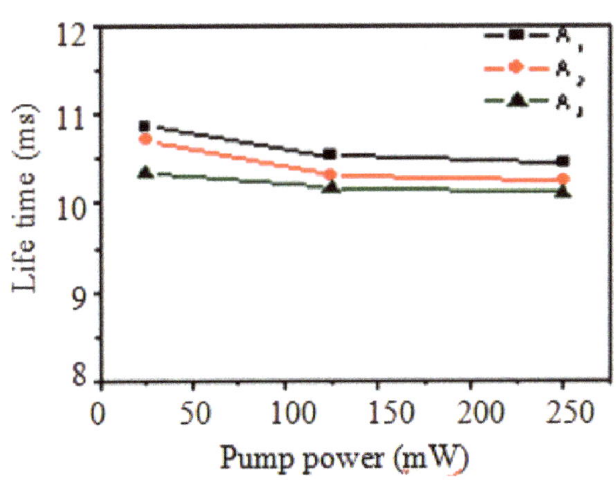

Figure 1.36. The effect of pump power on the lifetime of the three EDFs. Reprinted from [62] with permission from MDPI.

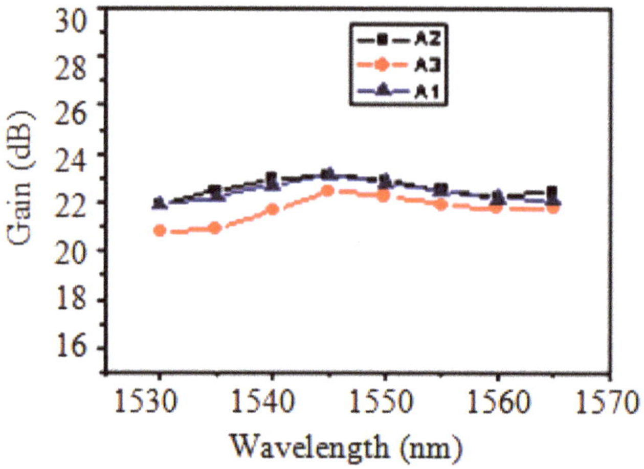

Figure 1.37. The optical gain curves of the three EDFs in the C-band under 0 dBm signal power, under a pump power of 400 mW. Reprinted from [62] with permission from MDPI.

coordination with a formula of $AlO_{6/2}$. These Al-structures create solvation shells in the glass network and provide necessary valiancy to incoming rare-earth ions. When Al functions as a network former due to charge compensation, the Er^{3+} ions are preferentially accommodated near the aluminum sites, while when Al serves as a network modifier, the Al^{3+} ions break the silica structure, producing non-bridging Al–O groups which can coordinate to the Er^{3+} ions. An increase of Al_2O_3 content weakens the covalent character of Er–O and leads to formation of the Si–O–Al bond. It is well known that a glass composition with higher basicity prefers a lower coordination number compared to the formation of six-coordinated Al^{3+} [67]. Moreover, the optical basicity of four-coordinated Al^{3+} in the network of SiO_4 is known to be smaller than that of six-coordinated Al^{3+} [68]. Accordingly, Er^{3+} ions prefer the four-coordinated Al^{3+} ion environment rather than six-coordinated Al^{3+} ions with an increase of Al_2O_3.

However, in our case ^{27}Al NMR–MAS spectroscopic analyses of the Er_2O_3 doped phase-separated alumina–silica fiber preform samples shown in figure 1.38 reveal that most of the Al^{3+} ions are hexa- and penta-coordinated instead of four-coordinated. The result obtained are found to be in good agreement with the available literature [69, 70]. Er^{3+} ions surrounded by more hexa- and penta-coordinated Al^{3+} ions will experience a larger site-to-site variation of the crystal field strength which exhibits more and more homogeneous and inhomogeneous broadening in order to achieve a large gain bandwidth with better flatness of the gain spectrum, specifically in the C-band region. The coordination number of Al^{3+} increases from four to six under incorporation of a large amount of Al_2O_3 (more than 3.0 mole%). The substitution of six-coordinated Al^{3+} instead of four-coordinated Al^{3+} can increase the Al_2O_3 content and create more ionic environments surrounding the Er^{3+} ions. Therefore, high Al_2O_3 doping is likely to increase the distortion of the ligand field of the Er^{3+} site caused by the existence of six-coordinated Al^{3+} which gives rise to strong interaction effects on the Stark splitting of the $^4I_{15/2}$ and $^4I_{13/2}$ manifolds, which further results in

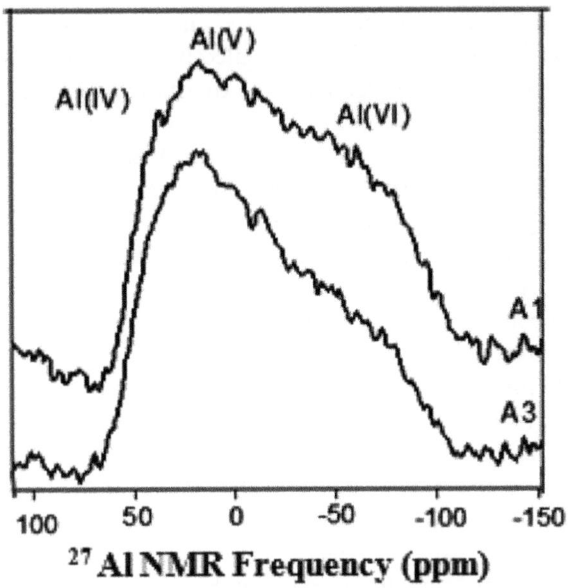

Figure 1.38. ^{27}Al NMR study of the doping region of Al content in two preform samples of EDF, A1 and A3. Reprinted from [62] with permission from MDPI.

broader gain spectra with much better flatness in the C-band region as well as in their fluorescence curves. The phase-separated EDFs (A1 and A2) possess a slightly longer lifetime than the non-phase-separated EDF (A3) due to the presence of most of the Er^{3+} ions in the alumina rich phase-separated region which possesses lower phonon energy than that of the silica matrix.

1.7 An erbium-doped specialty optical fiber based on zirconia–yttria–alumina–barium silica glass

1.7.1 Fiber fabrication and characterization

Erbium-doped nano-engineered zirconia–yttria–alumina–barium (Zr–Y–Al–Ba) silica glass based fiber (ZYAB-EDF) was drawn from a post-thermal optical preform made using an MCVD process in combination with the SD technique followed by suitable thermal annealing. Initially a porous silica layer was deposited on the inner surface of a silica tube at 1550 °C using the MCVD process, which was followed by presintering at 1500 °C. The porous silica layer was soaked with an alcoholic solution of suitable strength containing $ErCl_3.6H_2O$, $AlCl_3.6H_2O$, $YCl_3.6H_2O$, $ZrOCl_2.8H_2O$, and $BaCl_2.2H_2O$ for one hour. Then the porous layer was dried using air through draining out of the solution. The dried layer was sintered gradually, increasing the heating temperature from 1500 °C to 1850 °C in order to achieve a transparent layer. Finally, such a tube coated with a transparent layer was collapsed above 20 000 °C to convert it into a solid rod known as a preform. A small piece of the preform (~ 10 cm) was annealed in a closed furnace at 1250 °C for 3 h under a heating and cooling rate of 10 °C min^{-1} maintaining an optimized heating cycle. The fiber was drawn from such an annealed preform with on-line resin coating after heating the preform at 2000 °C. Here the thermal annealing is the main step to develop a nano-engineered zirconia–yttrium–alumina–silica glass based optical fiber. The nano-crystalline host of ZrO_2 is retained in the fiber's core during the fiber drawing step [72] as confirmed by TEM analyses as well as EDX spectra and electron diffraction patterns, which are described below.

The sizes of the phase-separated particles in the core region change through heating of the preform at 1250 °C for 3 h to drawn the fiber. Figure 1.39 shows the TEM analyses of the annealed preform sample showing particle sizes of 7–8 nm. The inset in the high resolution TEM image (figure 1.39(b)) shows the electron diffraction pattern (figure 1.39(c)) indicating the crystalline nature of the particles. The sizes of the particles decreased to 3–4 nm when the fiber was drawn from the annealed preform, as shown in figure 1.40(a). The larger sized particles (7–8 nm) present in the core of the annealed preform transformed into very fine particles which diffused into the parent glassy phase with an average size of 3–4 nm and minimize the scattering loss. The decrease in particle size during fiber drawing [27, 73] may occur due to re-melting and stretching of the phase-separated region.

When the fiber is drawn from the annealed preform, the shapes of the particles become very irregular compared to a circular structure. The crystalline nature of the particles is destroyed partially during the fiber drawing at high temperature, as confirmed by the electron diffraction pattern shown in figure 1.40(c). Such a surrounding crystalline environment of the Er ions will influence the erbium spectroscopy as well as the EDFA performance through the broadening of emission spectra with better gain flatness.

The fiber's core appears homogeneous at optical wavelengths, but not at a nanoscale level. In order to determine the nanoscale heterogeneities, we performed TEM analyses of the optical fiber samples using HRTEM (model Tecnai G2 30ST, FEI Company, USA). The particle sizes become 3–4 nm, confirmed from the TEM

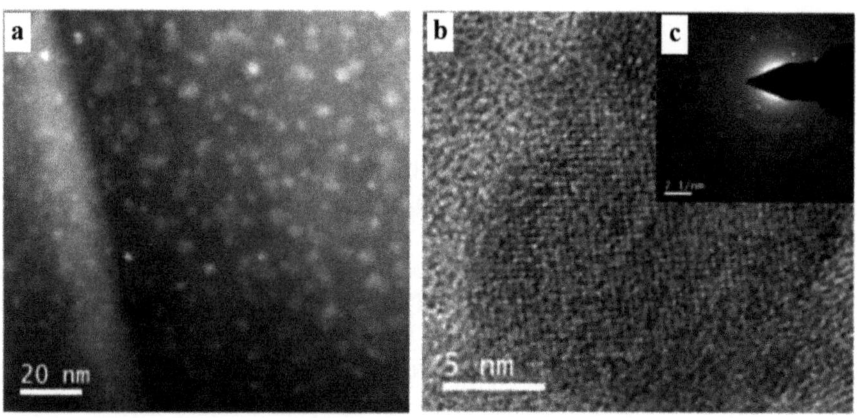

Figure 1.39. (a) TEM image, (b) HTEM image, and (c) electron diffraction pattern of the optical fiber preform sample. (Reproduced with permission from [71]. Copyright 2019 the Optical Society of America.)

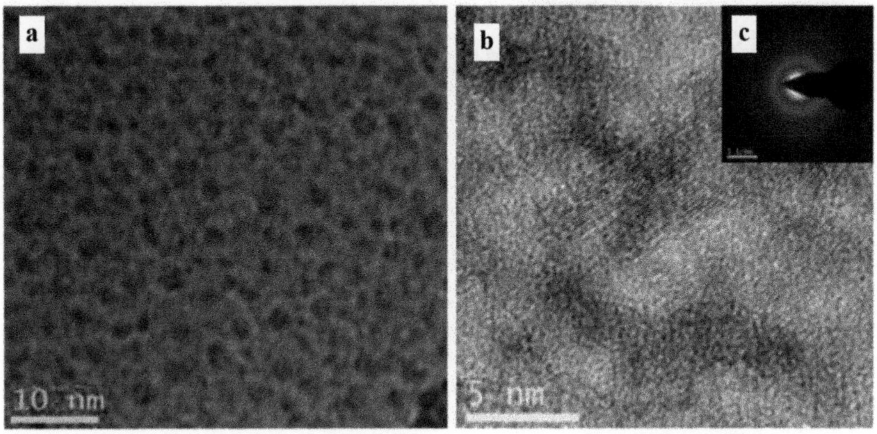

Figure 1.40. (a) TEM image, (b) HTEM image, and (c) electron diffraction pattern of the optical fiber sample. (Reproduced with permission from [71]. Copyright 2019 the Optical Society of America.)

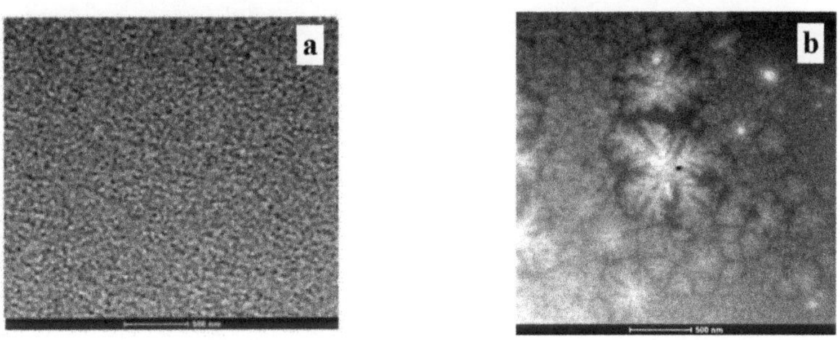

Figure 1.41. FESEM image of optical fiber preform samples (a) without and (b) with doping with BaO. (Reproduced with permission from [71]. Copyright 2019 the Optical Society of America.)

image of optical fiber samples shown in figure 1.40(a). The nature of the particles becomes partial crystalline, as observed from the HRTEM (figure 1.40(b)) along with the electron diffraction pattern in the inset (figure 1.40(c))

We produced optical preforms with and without incorporation of BaO, keeping all of the other fabrication parameters the same, including the solution strength. The presence of BaO improves the optical transparency of the core glass through minimization of the formation of the phase-separated region, as shown in the field emission SEM (FESEM) image (figure 1.41(a)) of the optical fiber preform sample, which reduces the background loss of the fiber. On the other hand, without incorporation of BaO into the erbium-doped zirconia–yttria–alumina–silica glass based fiber preform phase-separated zones are formed, as shown in figure 1.41(b).

An optical microscope (Olympus BX51) was used to inspect the core and cladding geometry of the fiber. No defects were observed at the interface between the core and the silica cladding. The fiber cross-section is shown in figure 1.42(a). The core and cladding diameters of the fiber were found to be 10.84 and 125.75 μm, respectively.

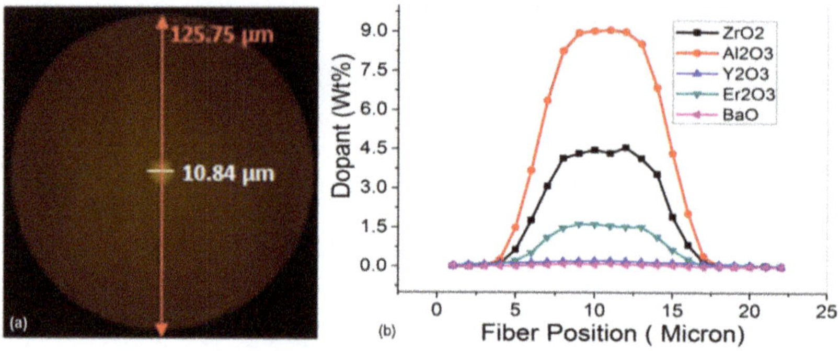

Figure 1.42. Cross-sectional view of (a) the fiber and (b) the elemental distribution profile along the core diameter of the fiber. (Reproduced with permission from [71]. Copyright 2019 the Optical Society of America.)

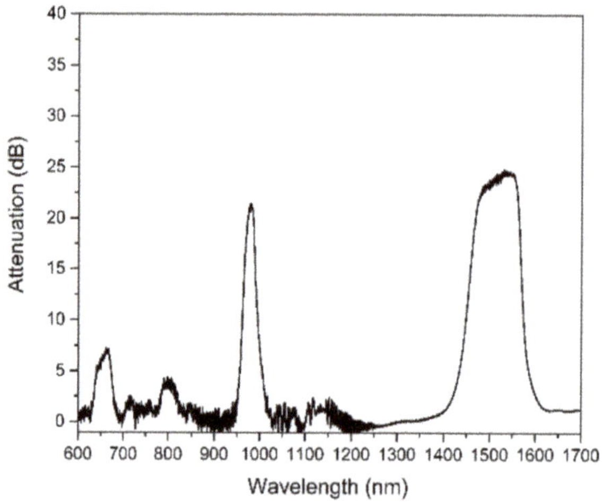

Figure 1.43. The attenuation spectrum of ZYAB-EDF. (Reproduced with permission from [71]. Copyright 2019 the Optical Society of America.)

The distributions of the doping levels of Al, Zr, Y, Ba, and Er were measured from electron-probe microanalyses (EPMA) using a JEOL EPMA instrument. The fiber core glass contained 9.0 wt.% Al_2O_3, 4.5 wt.% ZrO_2, 0.225 wt.% Y_2O_3, 0.10 wt.% BaO, and 1.5 wt.% Er_2O_3, measured from the EPMA data. The elemental distribution along the diameter of the core of the fiber is presented in figure 1.42(b). Co-doping with Zr and Al ions helps to increase the incorporation of erbium ions (1.5 wt.%) without any clustering effects. Incorporation of 0.10 wt.% BaO increases the glass forming zone of the doping host of Er_2O_3 through weakening or breaking the Si–O and Al–O bonds, with formation of non-bridging oxygen (NBO) [74]. Generally, BaO tends to reinforce the three-dimensional nature of the glass network with higher density due to the large size of the Ba ions, leading to a higher T_g [75].

The attenuation spectrum of the fiber, shown in figure 1.43, was measured using a broadband source with a spectral range from 410 to 2400 nm (model Super K EXTREME from Fianium) and an optical spectrum analyzer (OSA) (model Q8384

Table 1.2. Fiber data and composition [71].

Core composition	Core diameter (μm)	NA	Absorption loss at 980 nm	Absorption loss at 1531 nm
SiO_2–Al_2O_3–ZrO_2–Y_2O_3–BaO–Er_2O_3	10.84	0.195	~ 20 dB m^{-1}	~25 dB m^{-1}

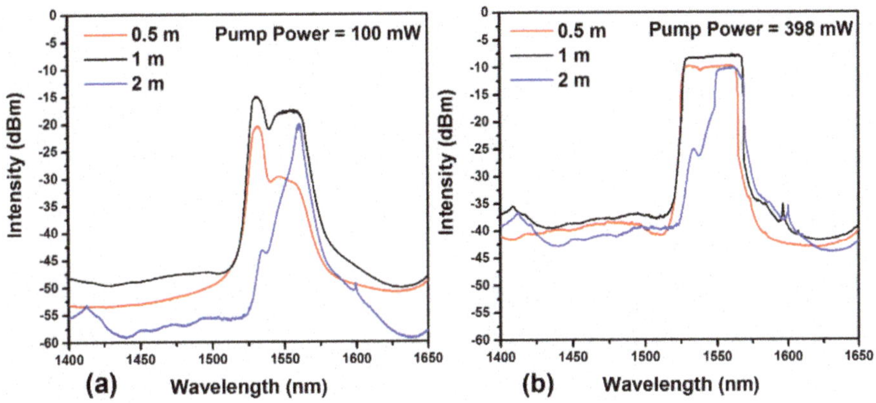

Figure 1.44. The emission spectra of various lengths of ZYAB-EDF under 980 nm laser excitation with powers of (a) 100 mW and (b) 398 mW. (Reproduced with permission from [71]. Copyright 2019 Optical Society of America.)

from Advantest). The spectrum shows two main absorptions bands at 980 nm with a maximum value of 21.5 dB and another within 1450-1570 nm with a maximum value of 24.7 dB.

The fiber parameters are summarized in table 1.2.

The emission spectrum of the ZYAB-EDF fiber was measured under two different excitation powers of 100 and 398 mW using a laser diode with 980 nm wavelength as the excitation source. The light emitted from the source was injected into the fiber which connected with an OSA (model Q8384 from Advantest). The results for different fiber lengths are shown in figure 1.44, which shows a broad emission spectrum with a strong emission around 1550 nm having the maximum emission intensity for a fiber length of one meter. The emission spectrum became flatter with an increase of power at the C-band. We observed a deviation of the maximum intensity peak towards the L-band region with 2 and 4 m fiber lengths. This indicates that it may be possible to produce gain in the L-band with longer fibers and higher pump powers.

1.7.2 Optical amplification

The gain and noise figures of the proposed fibers were measured for amplifiers using short pieces of the ZYAB-EDF as the gain medium. Since the emission was

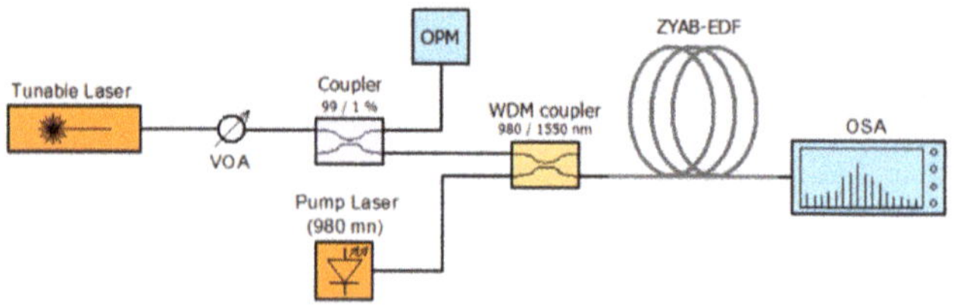

Figure 1.45. Scheme of the experimental set-up of the optical amplifier. (Reproduced with permission from [71]. Copyright 2019 the Optical Society of America.)

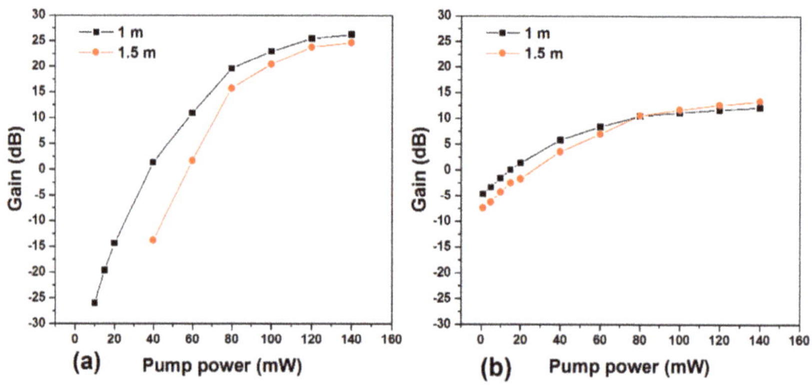

Figure 1.46. Gain as a function of the pump power for two ZYAB-EDF lengths using a transmission signal of (a) 1550 nm and (b) 1590 nm wavelength and −25 dBm of power. (Reproduced with permission from [71]. Copyright 2019 the Optical Society of America.)

maximum for a length of 1 m, we tested the fiber using two different fiber lengths of 1 and 1.5 m. The experimental set-up used is displayed in figure 1.45, in which the amplifier was forward pumped by a 980 nm laser diode via a 980/1550 nm WDM coupler. A tunable laser (model OSICS 1560 from Anritsu) from 1500 to 1620 nm with a maximum output power of 13 dBm was used as the input signal. The input signal power was controlled using a variable optical attenuator (VOA) and a 99%–1% coupler which was placed between the laser output and the WDM coupler. An optical power meter (OPM) (model AXS-100 from EXFO) was used to measure the laser output power. The gain and noise figure were characterized using an OSA (FTB-500 from EXFO).

We analyzed the gain variation using an input signal power of −25 dBm (small signal) with 1550 and 1590 nm wavelengths. The results are displayed in figure 1.46. It was expected that initially the gain would increase exponentially with the pump power but this was not achieved. This indicates that the near-complete inversion regime is not achieved with the used pump power values. Therefore, it is possible to improve the performance of this amplifier with higher pumping power values. A gain higher than 27 dB was achieved for the 1550 nm signal wavelength using 1 m fiber length. On the other hand, a gain of 13 dB was achieved for the 1590 nm signal wavelength using 1.5 m fiber length, showing better performance.

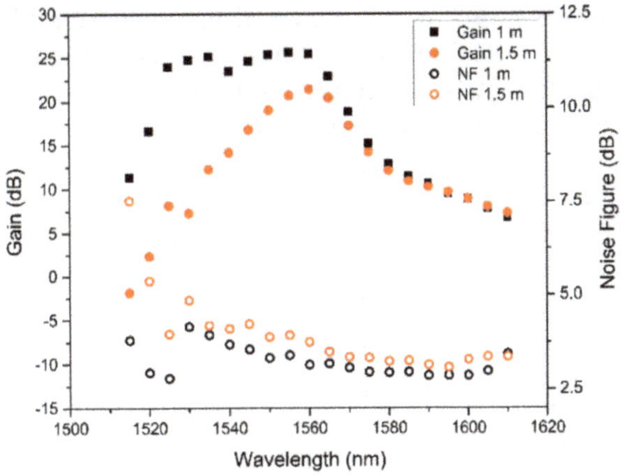

Figure 1.47. Gain and noise figure spectra for an input signal power of −25 dBm and a pump power of 100 mW. (Reproduced with permission from [71]. Copyright 2019 the Optical Society of America.)

We evaluated the gain and noise performance in the C-band and L-band for a small signal. We used different signals of −25 dBm power in the range from 1515 to 1610 nm under a fixed pump power of 100 mW. We achieved a maximum gain of 25.6 dB using 1 m fiber length, as shown in figure 1.47. The results show that the gain is higher in the C-band with a flat band from 1525 to 1565 nm, with flatness less than 3 dB. The gain is slightly higher in the L-band for the 1.5 m fiber. This suggests that the gain performance in the L-band can be improved with a longer fiber and higher pump power. We can also observe that the noise shows good performance in the two lengths of fiber, as shown in figure 1.47. However, the 1 m fiber shows the best results for the noise figure with values lower than 4.2 dB.

1.8 Conclusions

We report the fabrication and complex analysis of the basic materials and optical properties of chromium, erbium co-doped alumina–germania–calcia–yttria–silica (AGCYS) fiber. The fabricated fiber is compared to commercial purely Er-doped fibers (EDFs) of L20 and L40 type (Thorlabs). Such a novel Cr,Er AGCYS fiber demonstrates amplification properties in the NIR range inherent to doping with Er^{3+} ions, as is well known for EDFs, as well as a set of advantages stemming from its co-doping with Cr^{3+}/Cr^{4+} ions. This concerns a set of characteristics relevant to NIR (>1 μm) applications, that is, absorption and fluorescence spectra, fluorescence decay, resonant-absorption bleaching, and gain spectra. Also, we measured the amplification characteristics at different wavelengths (i.e. at 976, 905, and 1064 nm), falling into the resonant-absorption bands of Er^{3+} and Cr^{3+}/Cr^{4+} ions, which allows the details of the direct excitation of these two subsystems and the energy transfer between the two to be uncovered. Among the observed prospective features of the Cr,Er AGCYS fiber, note its high quality in terms of the ratio of bleached to residual resonant absorption (for both kinds of excitation) and the net-gain to small-signal absorption (at 976 nm pumping), both relevant for NIR applications with in-core laser-diode pumping. We also found a reduced up-conversion phenomenon in the fiber and relatively long NIR fluorescence lifetimes, inherent to both Er^{3+} and Cr^{3+}/Cr^{4+} ions. Seemingly, the latter advantage can be ascribed to the almost ideal core-glass structure of the fiber, which is capable of facilitating (because of the presence of Al and Y precursors) homogeneous dispersion of both 'active' co-dopants (Cr/Er) in the core glass and, eventually, diminished clustering effects.

Another EDF based on a phase-separated alumina–silica doping host for Er^{3+} ions made using the solution doping technique followed by the MCVD technique exhibits better optical performance in terms of broadening and flattening of the gain spectra in the C-band region and a longer lifetime compared to non-phase-separated silica based glass hosts. In the phase-separated region most of the Al^{3+} ions are hexa- and penta-coordinated instead of four-coordinated compared to non-phase-separated glass. In the phase-separated region, the coordination number of most of the Er^{3+} ions is greater than four, compared to the non-phase-separated glass matrix, and the bond length of Er–O becomes larger. This in turn leads, for the phase-separated region, to a larger site-to-site variation of the crystal field strength which exhibits more and more homogeneous and inhomogeneous broadening in order to achieve a large gain bandwidth for better flatness of the gain spectrum. The lifetime of the $^4I_{13/2}$ level of the Er^{3+} ion also increases within the phase-separated alumina–silica based glass matrix of the optical fiber. The optical gain of the phase-separated EDFs (A1 and A2) shows good flatness with variation of ±0.60 dB gain within the C-band region compared to the gain flatness of the non-phase-separated EDF A3 with variation of ±0.85 dB. An EDFA module based on this type of EDF will be useful in current metro networks.

The amplification performance of a new zirconia–yttria–alumina–barium (Zr–Y–Al–Ba) silica glass based fiber was also described here. A flat gain of 25 dB was achieved with a flatness less than 3 dB and a width of 40 nm within the C-band using 1 m YZAB-EDF fiber length as a gain medium for the small-signal gain region and 100 mW pump power. The fiber shows an impressive noise figure of less than 4.2 dB under these conditions. Such results suggest that the performance of the amplifier can be improved with higher pump powers. The gain in the L-band region can be enhanced by using longer lengths of ZYAB EDFs. These results demonstrate that a short length of ZYAB-EDF is highly suitable in C + L-band optical amplification and is cost effective. Such specialty erbium-doped optical fibers will be highly effective for use as optical fiber amplifiers in broadband communication systems compared to the standard erbium-doped fibers.

Acknowledgements

The Indian team acknowledges with thanks the financial assistance from DST-Nano Mission, Government of India. We are grateful to DST, Government of India, for providing the fellowship of one of the co-authors, Mr Debjit Dutta. The Mexican authors are grateful to the Consejo Nacional de Ciencia y Tecnologia (CONACyT, Mexico: Project 242 221) for partial financial support of this research. The authors would like to thank the SEM-ESCA Lab for performing the morphological analysis of the porous layers, and the University of Minnesota, USA, for performing the EPMA. The funding from the Indo-Portuguese Programme for Cooperation in Science & Technology under DST, New Delhi, is also acknowledged by all the Indian and Portugese co-authors. The authors are grateful to the Director of CGCRI for his continuous encouragement, guidance, and support of this work.

References

[1] Foschini G J and Habbab I M I 1995 *J. Lightw. Technol.* **13** 507–16
[2] Otani T, Goto K, Kawazawa T, Abe H and Tanaka M 1997 *J. Lightw. Technol.* **15** 737–42
[3] Giles C R, Desurvire E, Talman J R, Simpson J R and Becker P C 1989 *J. Lightw. Technol.* **7** 651–6
[4] Desurvire E, Giles C R and Simpson J R 1989 *J. Lightw. Technol.* **7** 2095–104
[5] Lumholt O, Povlsen J H, Schusler K, Bjarklev A, Dahl-Petersen S, Rasmussen T and Rottwitt K 1993 *J. Lightw. Technol.* **11** 1344–52
[6] Pal M, Bandyopadhyay S, Biswas P, Debroy R, Paul M C, Sen R, Dasgupta K and Bhadra S K 2007 *Opt. Quant. Electron.* **39** 1231–43
[7] Vengsarkar A M, Bergano N S, Davidson C R, Pedrazzani J R, Judkins J B and Lemaire P J 1996 *Opt. Lett.* **21** 336–38
[8] Kim H S, Yun S H, Kim H K, Park N and Kim B Y 1998 *IEEE Photon. Technol. Lett.* **10** 790–2
[9] Wang B, Pub G, Osnato R and Palsdottir B 2004 *Proc. SPIE* **5280** 162–6
[10] Yamada M, Kanamori T, Terunuma Y, Oikawa K, Shimizu M, Sudo S and Sagawa K 1996 *IEEE Photon. Technol. Lett.* **8** 882–4
[11] Jha A, Shen S and Naftaly M 2000 *Phys. Rev.* B **62** 6215–27
[12] Snoeks E, Kik P G and Polman A 1996 *Opt. Mater. (Amst.)* **5** 159–67
[13] Gill D M, McCaughan L and Wright J C 1996 *Phys. Rev.* B **53** 2334–44
[14] Dvoyrin V V, Mashinsky V M, Dianov E M, Umnikov.A A, Yashkov M V and Guranov A N 2005 *ECOC Proc.* **4** 949–50
[15] Dianov E M 2012 *Quant. Electron.* **42** 754–61
[16] Dianov E M 2013 *J. Lightw.Technol.* **31** 681–8
[17] Felice V, Dussardier B, Jones J K, Monnom G and Ostrowsky D B 2000 *Eur. Phys. J. Appl. Phys.* **11** 107–10
[18] Yeh S M *et al* 2011 *J. Lightw.Technol.* **30** 921–7
[19] Unger S, Schwuchow A, Dellith J and Kirchhof J 2007 *Proc. SPIE* **6469** 646913
[20] Luo Y, Wen J, Zhang J, Canning J and Peng G D 2012 *Opt. Lett.* **37** 34473449
[21] Zhang J, Sathi Z M, Luo Y, Canning J and Peng G D 2013 *Opt. Express* **21** 7786–92
[22] Sathi Z M, Zhang J, Luo Y, Canning J and Peng G D 2015 *Opt. Mater. Express* **5** 2096–105
[23] Yan B, Luo Y, Zareanborji A, Xiao G, Peng G D and Wen J 2016 *J. Opt.* **18** 105705
[24] Luo Y, Yan B, Ding M, Zhao Q, Xiao G, Wen J and Peng G D 2017 *Optik* **133** 132–9
[25] Luo Y, Yan B, Zhang J, Wen J, He J and Peng G D 2018 *Front. Optoelectron.* **11** 37–52
[26] Hatipoğlu M and Barutcigil C 2015 *Niger. J. Clin. Pract.* **18** 213–20
[27] Dutta D, Dhar A, Kir'yanov A V, Das S, Bysakh S and Paul M C 2015 *Phys. Status Solidi* A **212** 1836–44
[28] Kir'yanov A V, Dutta D, Barmenkov Y O, Dhar A, Das S, Koltashev V V, Plotnichenko V G and Paul M C 2016 *IEEE J. Quantum Electron.* **52** 6800112
[29] Bayart D, Clesca B, Hamon L and Beylat J L 1994 *IEEE Photon. Technol. Lett.* **6** 613–15

[30] Guan B O, Tam H Y, Liu S Y, Wai P K A and Sugimoto N 2003 Optoelectronics *Proc. Sixth Chinese Symp. (Hong Kong)* pp 147–49

[31] Fibercore 2007 www.fibercore.com/isogaindownload.pdf?PHPSESSID=e34627a5fe664e1d84697f48d9b3022e (Accessed: 27 May 2007)

[32] Wang S and Andrejco M J 2005 *Proceedings of the Asia-Pacific Optical Communications Conference* (Shanghai, China)

[33] Miniscalco W J 1991 *IEEE J. Lightw. Technol.* **9** 234–50

[34] Miniscalo W J and Quimby R S 1991 *Opt. Lett.* **16** 258–60

[35] Barnes W L, Laming R I, Tarbox E J and Morkel P R 1991 *IEEE J. Quantum Electron.* **27** 1004–10

[36] Wang S, Pub G, Osnato R and Palsdottir B 2003 *SPIE Proc.* **5280** 161–63

[37] Yoshida S, Kuwano S and Iwashita K 1995 *Electron. Lett.* **31** 1765–67

[38] Yamada M, Mori A, Kobayashi K, Ono H, Kanamori T, Oikawa K, Nishida Y and Ohishi Y 1998 *IEEE Photon. Technol. Lett.* **10** 1244–46

[39] Millar M C, Brierley M C and France P W 1988 *Proc. 14th Eur. Conf. Optical Communication (Brighton, UK)* pp 66–9

[40] Yamada M, Ono H, Kanamori T, Sakamoto T, Ohishi Y and Sudo S 1996 *IEEE Photon. Technol. Lett.* **8** 620–22

[41] Dhar A, Kasik I, Dussardier B, Podrazky O and Matejec V 2012 *J. Appl. Ceram. Technol* **9** 341–48

[42] Kiryanov A V, Paul M C, Barmenkov Y O, Das S, Pal M and Escalante L 2013 *IEEE J. Quant. Electron.* **49** 528–44

[43] Halder A, Paul M C, Bhadra S K, Bysakh S, Das S and Pal M 2015 *Sci. Adv. Mat* **7** 631–42

[44] Paul M C, Bysakh S, Das S, Dhar A, Pal M, Bhadra S K, Sahu J K and Kiryanov A V 2016 *Trans. Ind. Ceram. Soc.* **75** 195–208

[45] Reddy H P, Das S, Dutta D, Dhar A, Kiryanov A V, Pal M, Bhadra S K and Paul M C 2018 *Phys. Status Solidi* A **215** 1700615

[46] Iskhakova L D, Milovich F O, Yu E D, Vel'miskin V V and Semenov S L 2018 *Glass Phys. Chem.* **44** 137–44

[47] Costa V C, Lochhead M J and Bray K L 1996 *Chem. Mater.* **8** 783–90

[48] Paul M C *et al* 2010 *Opt. Lett.* **35** 2882–84

[49] Courrol L C, Kassab L R P, Fukumoto M E, Wetter N U, Tatumi S H and Morimoto N I 2003 *J. Lumin.* **102–103** 91–5

[50] Dhar A, Paul M C, Chowdhury S D, Pal M, Pal A and Sen R 2016 *Phys. Status Solidi* A **213** 3039–45

[51] Yuliantini L, Djama M, Hidayat R, Boonin K and Kaewkhao J 2018 *Mater. Today Proc.* **5** 15076–80

[52] Pokhrel M, Kumar G A, Balaji S, Debnath R and Sardar D K 2012 *J. Lumin.* **132** 1910–6

[53] Pal M, Paul M C, Bhadra S K, Das S, Yoo S, Kalita M P, Boyland A J and Sahu J K 2011 *J. Lightw. Technol.* **29** 2109–15

[54] Nagel S R, MacChesney J B and Walker K L 1982 *IEEE Trans. Microw. Theory Tech.* **30** 305–22

[55] Townsend J E, Poole S B and Payne D N 1987 *Electron. Lett.* **23** 329–31

[56] Kir'yanov A V, Dutta D, Das S, Dhar A and Paul M C 2019 *IEEE Photon. J.* **11** 7105013

[57] Kir'yanov A V, Barmenkov Y O, Sandoval-Romero G E and Escalante-Zarate L 2013 *IEEE J. Quantum Electron.* **49** 511–21

[58] Aksaf A and Pask J A 1975 *J. Am. Ceram. Soc.* **58** 507–12
[59] Roth R S 1995 *OxidesPhase Equilibria Diagrams* vol 11 (Westerville OH: American Ceramic Society)
[60] McHale A E and Roth R S (ed) 1996 *OxidesPhase Equilibria Diagrams* vol 12 (Westerville OH: American Ceramic Society)
[61] Besmann T M and Spear K E 2002 *J. Am. Ceram. Soc.* **85** 2887–94
[62] Paul M, Kir'yanov A, Barmenkov Y, Pal M, Youngman R, Dhar A and Das S 2018 *Fibers* **6** 67–82
[63] Tomazawa M and Doremus R H (ed) 1979 *Phase Separation in Glass* (Cambridge, MA: Academic)
[64] Bubnov M M *et al* 2004 *IEEE Photon. Technol. Lett.* **16** 1870–76
[65] McNamara P, Lyytikainen K J, Ryan T, Kaplin I J and Ringer S P 2004 *J. Opt. Commun.* **230** 45–53
[66] Honglin A, Tang Y, McNamara P and Felming S 2004 *Opt. Express* **12** 6153–58
[67] Kawazoe H 1980 *J. Non-Cryst. Solids* **42** 281–85
[68] Duffy J A and Ingram M D 1976 *J. Non-Cryst. Solids* **21** 373–410
[69] Tsomala N, Brantley S L, Hamilton J P, Pantano C G and Mueller K T 2003 *Am. Mineral.* **88** 54–67
[70] Sen S and Youngmen R E 2004 *J. Phys. Chem.* B **108** 7557–64
[71] Duarte J, Paul M C, Das S, Dhar A, Leitao J P, Ferreira M F and Rocha A M 2019 *Opt. Mater. Express* **9** 2652–61
[72] Paul M C, Sobon G, Sotor J, Abramski K M, Jagiello J, Kozinski R, Lipinska L and Pal M 2013 *Laser Phys.* **23** 035110
[73] Paul M C, Bysakh S, Das S, Pal M, Bhadra S K, Yoo S, Boyland A J and Sahu J K 2012 *Sci. Adv. Mater.* **4** 292–321
[74] Shen S X and Jha A 2008 *Adv. Mater. Res.* **39–40** 159–64
[75] Ray N H 1974 *J. Non. Cryst. Solids* **15** 423–34

Chapter 2

A wideband optical amplifier with a hafnia–bismuth–erbium co-doped fiber as the active medium

Alabbas A Al-Azzawi, Mukul Chandra Paul and Sulaiman Wadi Harun

Compact and wideband optical amplifiers with flat-gain characteristics using a hafnia–bismuth–erbium co-doped fiber (HB-EDF) as the gain medium are demonstrated. The HB-EDF was fabricated using a modified chemical vapor deposition (MCVD) process in conjunction with a solution doping (SD) technique. The fiber has an erbium ion concentration of 12 500 wt. ppm, which was realized due to the co-doping with hafnium and aluminum ions. The proposed wideband HB-EDF amplifier (HB-EDFA) was realized by utilizing two short lengths of HB-EDF to fulfill amplification in both the C- and L-bands. The wideband HB-EDFA was achieved using two stages in both series and parallel structures. It is found that both the parallel and backward pumping based series HB-EDFAs achieved flat-gain characteristics over a wideband operation wavelength. For example, in the parallel HB-EDFA using a 1.72 m long HB-EDF, a flat gain of 12.1 dB was realized with a gain ripple of less than 2 dB along the wavelength region of 80 nm from 1525 to 1605 nm. Within the flat-gain region the noise figure (NF) values varied from 6 to 11.8 dB. In addition, in the series backward pumping HB-EDFA using a 2 m long HB-EDF, a flat gain of 14.6 dB was realized with a gain ripple of less than 2 dB along the wavelength region of 65 nm from 1530 to 1595 nm. Within the flat-gain region the NF values varied from 6.8 to 10.2 dB.

2.1 Introduction

The massive expansion of optical communication systems in recent years has resulted in a huge demand for efficient optical fiber amplifiers with a wideband operating region and flat-gain characteristics, to cover the whole range of the dense wavelength division multiplexing (DWDM) spectrum [1, 2]. In addition to the conventional band (C-band), from 1530 to 1565 nm, the optical amplification in the

doi:10.1088/978-0-7503-3243-9ch2

long wavelength band (L-band) permits the capacity of DWDMs to be upgraded with an additional 40 nm bandwidth (1565–1605 nm) [3, 4]. Wideband erbium doped fiber amplifiers (EDFAs) are commonly constructed utilizing dual stages in series or parallel structures, with different lengths of erbium doped fibers (EDFs) [5, 6]. This is due to the amplification in the L-band region which requires longer EDFs to obtain the same attainable gain in the C-band region.

Previously, a wideband bismuth–erbium doped fiber amplifier (Bi-EDFA) utilizing dual stages in series structure was introduced [7]. Unfortunately, high noise figures (NFs) were observed compared to the achieved gain. The average gain was about 7 dB while the NF fluctuated from 6 to 10 dB. In addition, Bi-EDFs have a problem in splicing the standard single mode fibers utilizing a standard splicing machine. In another study a wideband optical amplifier using a silica–erbium doped fiber (Si-EDF) as a gain medium was proposed, in parallel and series configurations [8]. However, the total length of the Si-EDFs used was 10.5 m, which is very long. In addition, the gain ripple was high in the parallel amplifier and thus this amplifier could not achieve flat gain over the wideband operation region.

In this chapter, a new wideband and flat-gain hafnia–bismuth–erbium co-doped fiber amplifier (HB-EDFA) is demonstrated and developed by utilizing two short lengths of HB-EDF sections to fulfill amplification in the C- and L-band telecommunication regions. The proposed two-stage amplifier is investigated for both series and parallel structures in conjunction with a double-pass scheme. The performance of the amplifier is explored for two different lengths of HB-EDFs, as well as various pump powers, to determine the optimum design. The proposed series amplifier is examined in both backward and forward pumping schemes.

2.2 The fabrication and characteristics of HB-EDF

Two major steps were adopted to fabricate the HB-EDF. The first step was forming a fiber preform of hafnium–bismuth–erbium co-doped yttria–aluminum–silica glass by precipitation on a porous silica layer at a temperature of 1500 °C. This was achieved by using two techniques, modified chemical vapor deposition (MCVD) and solution doping (SD). In the SD technique $Al(NO_3)_3{\cdot}9H_2O$, $Y(NO_3)_3{\cdot}6H_2O$, $HfCl_3$, and $Bi(NO_3)_3{\cdot}xH_2O$ were used for around one hour to soak the porous layer to integrate all the co-doped materials, such as HfO_2, Al_2O_3, Y_2O_3, Bi_2O_3, and Er_2O_3. All the salts of nitrate and halide were held on the porous layer after exhaustion of the solution. To oxidize these salts, the core layer was first dried through the flow of nitrogen gas (N_2) at around 20 °C to 25 °C. Then it was dried by heating at around 800 °C to 900 °C through the flow of a combination of oxygen (O_2) and helium (He) gases. Sintering the porous layer in order to achieve transparent glass was performed by raising the temperature gradually from 1300 °C to 1900 °C. Collapsing stages were used to convert the glass into a solid preform by covering it with a thick silica tube. The second step to fabricate the HB-EDF was to draw the final fiber preform using the conventional method of a fiber drawing tower at a temperature of about 2000 °C.

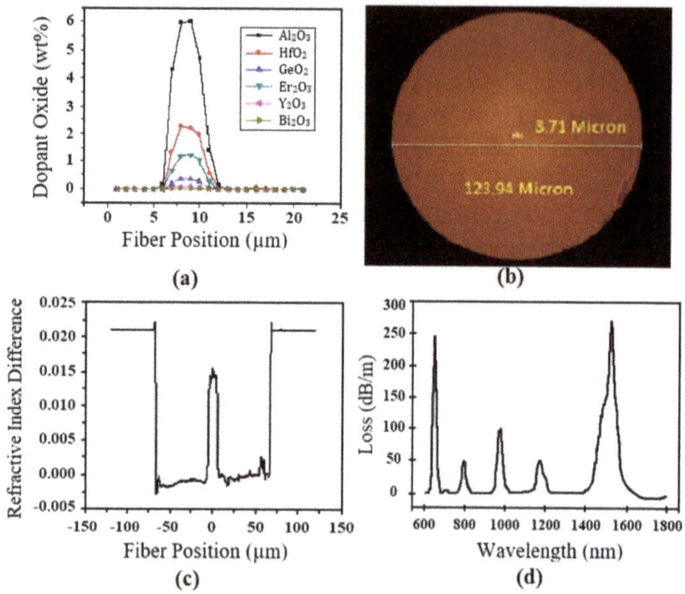

Figure 2.1. HB-EDF characteristics: (a) distributions of the elements constituting the core-glass of the HB-EDF, (b) microscopic view, (c) refractive index profile, and (d) absorption loss curve. Reprinted from [9], copyright (2019), with permission from Elsevier.

Figure 2.1(a) illustrates the curves of the doping level along the core diameter for the fabricated HB-EDF, which were measured using electron probe microanalysis (EPMA). This process was carried out by grinding and polishing a 1–2 mm fiber sample on both sides, after which the polished sample was coated by a thin graphite layer and then EPMA was performed with a maximum spatial resolution of 1 μm. It was found that the element distribution comprises 2.2 wt.% HfO_2, 0.035 wt.% Bi_2O_3, 6.0 wt.% Al_2O_3, and 1.23 wt.% Er_2O_3. Figure 2.1(b) depicts a microscopic view of the fabricated HB-EDF which has a core diameter of 3.71 μm and a cladding diameter of 123.94 μm. The numerical aperture (NA) of the fiber is calculated to be 0.2. Figures 2.1(c) and (d) illustrate the refractive index profile and the absorption loss curves, respectively. It can be seen that the absorption loss at 980 nm was 100 dB m^{-1}, while at 1530 nm it was 270 dB m^{-1}.

2.3 The wideband HB-EDFA using a parallel structure

The two-stage double-pass HB-EDFA which is connected in combination with a parallel configuration can be seen in figure 2.2. The amplifier utilizes two sections of HB-EDFs that are engaged in parallel to obtain an amplification for the wavelength span of both the C- and L-bands. A 980 nm laser diode (LD1) is used to pump the C-band HB-EDF, while the L-band HB-EDF is forward pumped using a 1480 nm laser diode (LD2). This is due to the fact that the 1480 nm pumping provides a higher optical power conversion efficiency (PCE) than that of 980 nm pumping for the L-band region. Wavelength division multiplexing (WDM) is used in each stage to

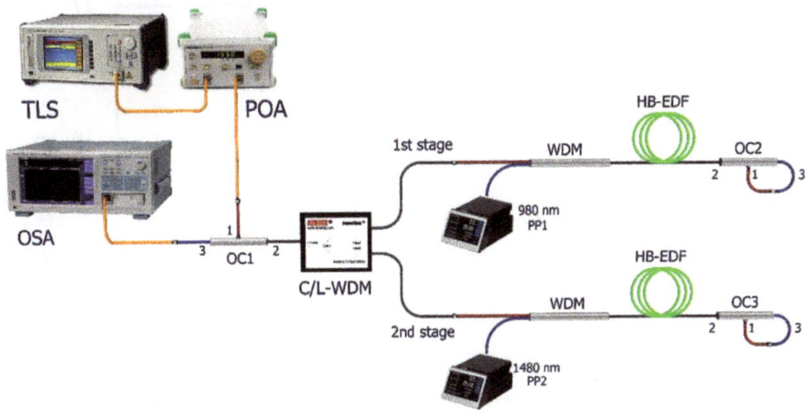

Figure 2.2. Two-stage double-pass HB-EDFA in conjunction with the parallel configuration. Reprinted from [9], copyright (2019), with permission from Elsevier.

combine the pump laser and the input signal. To isolate and integrate the C- and L-band wavelength signals into the amplifier stages, a coarse wavelength division multiplexing (C/L-WDM) filter is used. At the amplifier input and output ports, three optical circulators are used to allow double signal propagation in the HB-EDFs.

The first optical circulator is used at the input port to supply the input signal into the C/L-WDM and to route the amplified signal into the optical spectrum analyzer (OSA). The other optical fiber circulators at the output ports are used as a loop mirror by splicing port 1 with port 3, thereby allowing double signal amplification routed back in the same port. The input signals are provided by a tunable laser source (TLS), while the amplified signal is tested by OSA. The optical power of the input signal entering the amplifier was adjusted by using a programmable optical attenuator (POA). All the experiments were carried out at room temperature, ~25 °C. The optimum length of the HB-EDF for the C-band wavelength region was a 0.5 m long HB-EDF. However, the short HB-EDF length of about 0.2 m also achieved an efficient amplification performance. Thus two wideband HB-EDFAs are investigated using various lengths of HB-EDFs in a parallel configuration to achieve the best performance.

2.3.1 A parallel HB-EDFA with a total erbium fiber length of 2 m

In this amplifier the HB-EDFs are adjusted to 50 and 150 cm long for amplification in the C- and L-bands, respectively, such that the total length is 2 m. Figures 2.3 and 2.4 illustrate the performance of the parallel HB-EDFA at various L-band pump powers for input signal powers of −30 and −10 dBm, respectively. In the experiments the LD1 power is fixed at the maximum power of 170 mW. However, the LD2 power which controls the L-band gain is varied from 110 to 170 mW. It is obvious that the variation of the LD2 power gives the same effect for both input signal powers.

It can be observed that the L-band gain spectrum increases as the LD2 power increases. However, the increment in the gain spectrum decreases when the LD2

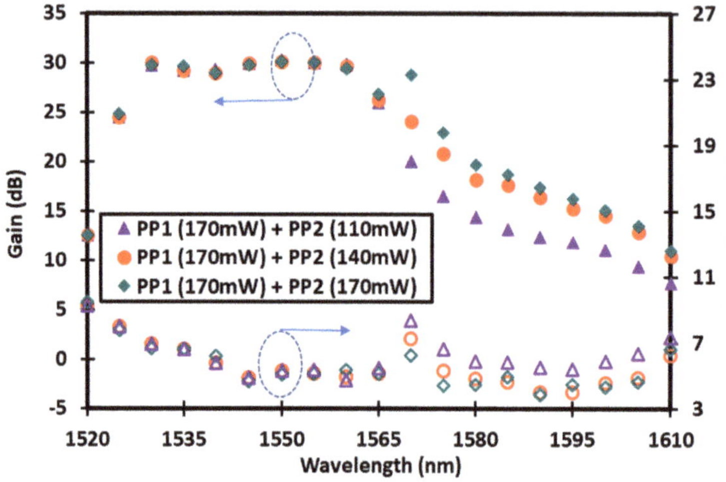

Figure 2.3. Measured gain (solid symbol) and NF (hollow symbol) spectra of the parallel HB-EDFA using a total of 2 m of active fiber at various laser diode powers for an input signal power of −30 dBm.

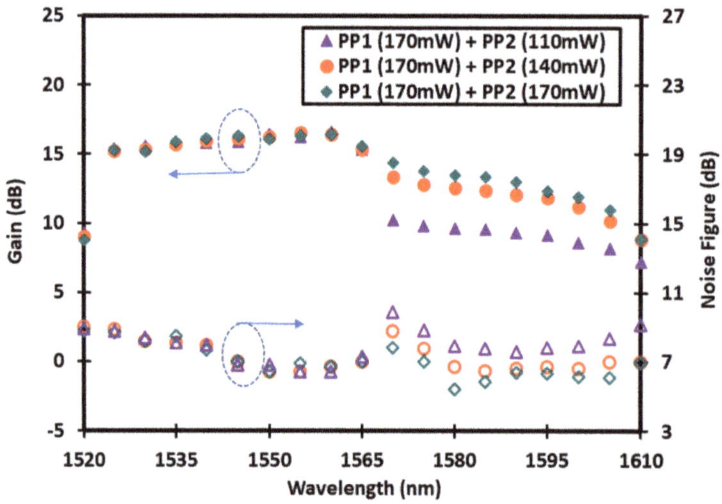

Figure 2.4. Measured gain (solid symbol) and NF (hollow symbol) spectra of the parallel HB-EDFA using a total of 2 m of active fiber at various laser diode powers for an input signal power of −10 dBm.

power increases from 140 to 170 mW due to the saturation effect. For example, at a wavelength of 1580 nm and −30 dBm input signal power, the gain increases from 14.4 to 18.2 and 19.7 dB as the LD2 power changes from 110 to 140 and 170 mW, respectively. The L-band NF is relatively higher when the LD2 power is fixed at 110 mW. This is due to the lower gain and higher loss related to the lower LD2 power. On the other hand, no significant changes in the gain and NF spectra are observed in the C-band region. This is due to the LD1 power, which is fixed, and the LD2 power variation that affects only the L-band region.

It is found that the optimum laser diode power to achieve higher gain and lower NF characteristics over a wideband operation region is 170 mW for both LD1 and LD2. At an input signal power of −30 dBm with optimum LD power the gain fluctuates from 11 to 30.1 dB along the wavelength span of 85 nm from 1525 to 1610 nm. In addition, the NF is obtained at less than 7.8 dB within this 85 nm wavelength region. However, at an input signal power of −10 dBm, the gain fluctuates from 11 to 16.4 dB along the wavelength span of 80 nm from 1525 to 1605 nm. Within this 80 nm wavelength span the NF values vary from 5.4 to 8.7 dB.

2.3.2 A parallel HB-EDFA with a total erbium fiber length of 1.72 m

The above experiment was repeated with a slightly shorter length of total active fiber. In this amplifier the HB-EDFs are adjusted to 22 and 150 cm long for amplification in the C- and L-bands, respectively, such that the total length is 1.72 m. Figures 2.5 and 2.6 illustrate the performance of the parallel HB-EDFA at various L-band pump powers for input signal powers of −30 and −10 dBm, respectively. In the experiment the LD1 power is fixed at a maximum power of 170 mW. However, the LD2 power which controls the L-band gain is varied from 110 to 170 mW.

It is clearly observed that the L-band gain increases as the LD2 power increases before it is saturated as the pump power is increased above 140 mW. Therefore it is concluded that the optimum laser diode powers to achieve a higher and flatter gain spectrum over a wideband operation region are 170 and 140 mW for LD1 and LD2, respectively. At an input signal power of −30 dBm with the optimum LDs powers, the gain fluctuates from 10.5 to 24 dB along a wavelength span of 90 nm from 1520 to 1610 nm. On the other hand, an NF of less than 10.1 dB is obtained within the wavelength region from 1520 to 1610 nm.

At an input signal power of −10 dBm a flat gain of 12.1 dB is realized with a gain ripple of less than 2 dB along the wideband wavelength region of 80 nm from 1525 to

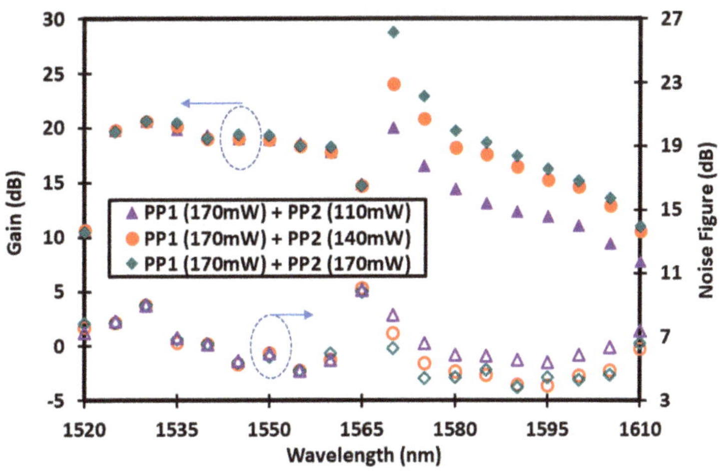

Figure 2.5. Measured gain (solid symbol) and NF (hollow symbol) spectra of the parallel HB-EDFA using a total of 1.72 m for the active fiber, at various LD2 powers when the input signal power is set at −30 dBm. Reprinted from [9], copyright (2019), with permission from Elsevier.

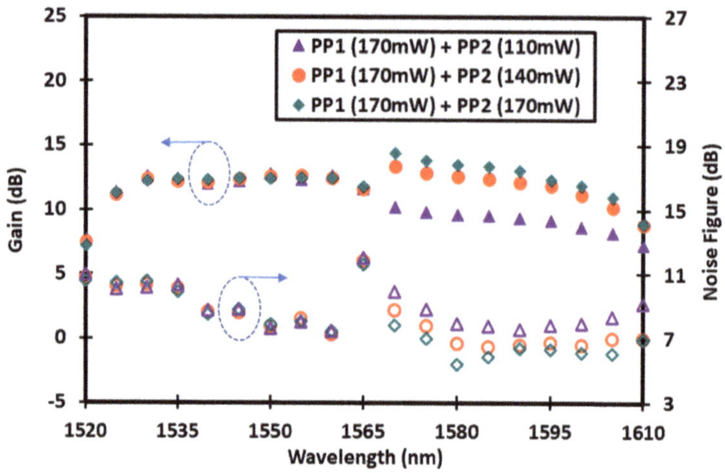

Figure 2.6. Measured gain (solid symbol) and NF (hollow symbol) spectra of the parallel HB-EDFA using a total of 1.72 m the active fiber at various LD2 powers for an input signal power of −10 dBm. Reprinted from [9], copyright (2019), with permission from Elsevier.

1605 nm. Within the flat-gain region the NF values vary from 6 to 11.8 dB. It can be inferred that the NF is higher at shorter wavelength spans and it decreases gradually when it goes to a longer wavelength span. This is attributed to the decrement in the erbium absorption to emission cross section as it goes from a shorter to longer wavelength span. Furthermore, a higher NF is obtained at the 1565 nm wavelength. This is attributed to two reasons: the lower gain and higher loss characteristics at the tail of the C-band region owing to the short 22 cm long erbium fiber HB-EDF, as well as the insertion loss of the C/L-WDM which is high when the amplification medium shifts from 22 to 150 cm.

2.4 Performance comparison of the proposed parallel HB-EDFAs

The performance of the proposed parallel HB-EDFA using a 2 m long HB-EDF is compared to that using a 1.72 m long HB-EDF. It is found that both amplifiers achieve efficient amplification with a wideband operation region from 1520 to 1610 nm. The use of a 2 m long HB-EDF achieves a higher gain and lower NF compared to the 1.72 m long fiber. However, the 1.72 m long HB-EDF provides a better flat-gain characteristic along the wideband wavelength span of 80 nm. This result indicates that the use of a 1.72 m long HB-EDF is preferable in the parallel configuration. Unfortunately, the NFs at 1525, 1530, and 1565 nm wavelengths are higher with a shorter gain medium. Further investigations aimed at improving the NFs are required. Table 2.1 summarizes the performance comparison of the proposed parallel HB-EDFA for two different total lengths of the gain medium, the 2 and 1.72 m long HB-EDFs.

2.5 A wideband HB-EDFA using a series configuration structure

In a series configuration, the amplification characteristic at each stage is affected by the other stage. Due to this the performance of the series HB-EDFA is investigated

Table 2.1. Performance comparison of the proposed parallel HB-EDFAs.

Feature	Parallel HB-EDFA using a 1.72 m long HB-EDF	Parallel HB-EDFA using a 2.0 m long HB-EDF
Erbium fiber length	Total length: 1.72 m. 22 cm for the C-band region. 150 cm for the L-band region.	Total length: 2 m. 50 cm for the C-band region. 150 cm for the L-band region.
Optimum LD power	LD1 power: 170 mW. LD2 power: 140 mW.	LD1 power: 170 mW. LD2 power: 170 mW.
Wideband amplification	At an input signal power of −30 dBm: • The gain fluctuates from 10.5 to 24 dB along the wavelength span of 90 nm. • The NF is less than 10.1 dB. At an input signal power of −10 dBm: • A flat gain of 12.1 dB with a gain ripple of ± 2 dB along a wavelength span of 80 nm is realized. • The NF values vary from 6 to 11.8 dB.	At an input signal power of −30 dBm: • The gain fluctuates from 11 to 30.1 dB along the wavelength span of 85 nm. • The NF is less than 7.8 dB. At an input signal power of −10 dBm: • The gain fluctuates from 11 to 16.4 dB along a wavelength span of 80 nm. • The NF values vary from 5.4 to 8.7 dB.

for different pumping schemes, forward and backward pumping. Schematic diagrams of the two series stage HB-EDFA with forward and backward pumping are depicted in figures 2.7(a) and (b), respectively. Both structures consist of a C/L-WDM filter, three optical circulators, and two pieces of HB-EDF sections. The optical fiber circulator (C1) is used to transfer the input signal inside the first stage and to pull out the returned output signal to the OSA. In the experiment a novel structure utilizing a C/L-WDM filter, instead of a C-band fiber Bragg grating (FBG), is proposed. The C/L-WDM is placed between the two stages to separate and combine the C- and L-band signals. An optical circulator (C2) is used to reflect the C-band signal, which then passes over the C/L-WDM to the first stage for double amplification. The L-band signal first undergoes single amplification by the first and second stages. Another optical circulator (C3) is used to reflect the L-band signal to undergo double amplification in both stages. A TLS is used as a signal source while an OSA is utilized to test the gain and NF characteristics. A programmable optical attenuator (POA) is utilized to provide an accurate input power to the amplifier. All the experiments were carried out at room temperature, ~25 °C. The performance of

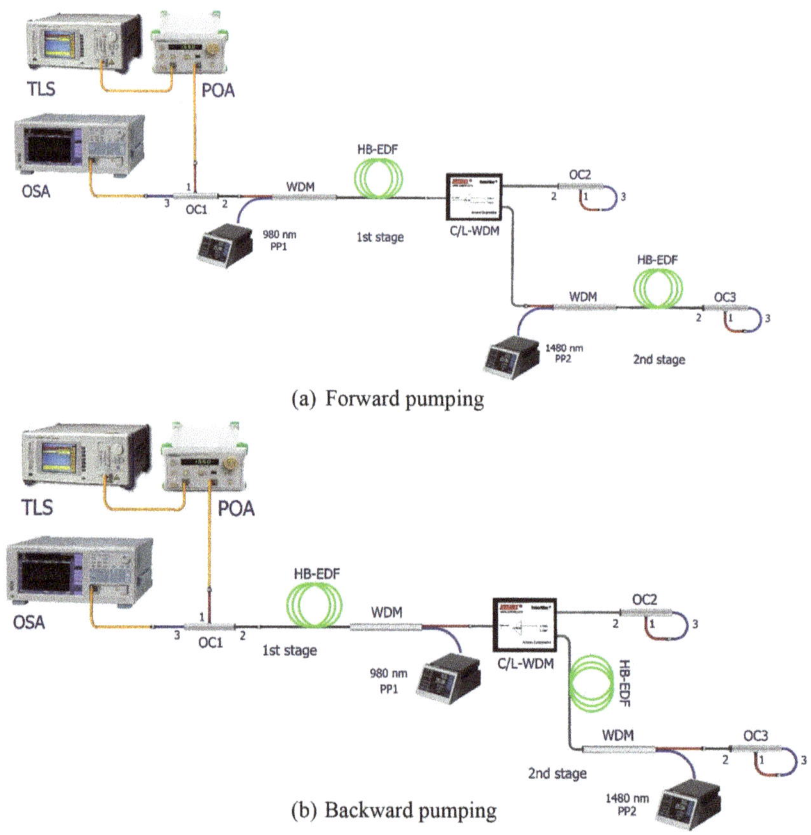

Figure 2.7. A two-stage double-pass HB-EDFA based on a series configuration using (a) forward and (b) backward pumping.

the proposed series HB-EDFA is investigated for two different lengths of the HB-EDFs, where the total lengths of HB-EDF used are 2 and 1.72 m.

2.5.1 Forward pumping

In this amplifier the HB-EDFs are adjusted to 22 and 150 cm long for amplification in the C- and L-bands, respectively, such that the total length is 1.72 m. Like for the parallel HB-EDFA the C-band HB-EDF is forward pumped using a 980 nm LD1. However, the L-band HB-EDF is forward pumped using a 1480 nm LD2. Figures 2.8 and 2.9 illustrate the performance of the series forward pumping HB-EDFA at various L-band pump powers for input signal powers of −30 and −10 dBm, respectively. In the experiment the LD1 power is fixed at a maximum power of 170 mW. However, the LD2 power which controls the L-band gain is varied from 110 to 140 and 170 mW. Like for the parallel type amplifier, the variation of the LD2 power produces the same effect for both input signal powers.

Overall, it can be inferred that the optimum laser diode powers to achieve even better performance in term of gain, NFs, and amplification bandwidth are 170 and 110 mW for LD1 and LD2, respectively. At an input signal power of −30 dBm with the optimum LDs powers, the gain fluctuates from 10 to 20.4 dB along a wavelength span of 55 nm from 1525 to 1580 nm. In addition, the NF is less than 12.1 dB within this 55 nm wavelength region. However, at an input signal power of −10 dBm, the gain fluctuates from 9.5 to 11.4 dB along a wavelength span of 50 nm from 1525 to 1575 nm. Within this 50 nm wavelength span the NF is less than 12.8 dB.

It can be seen in figures 2.8 and 2.9 that the L-band gain spectrum increases as the LD2 power increases. However, the increment in the gain spectrum decreases when the LD2 power increases from 140 to 170 mW. This is most probably due to the gain saturation effect. In addition, the C-band gain spectrum decreases when the LD2 power increases above 110 mW, although the LD1 power is fixed. This is due to the

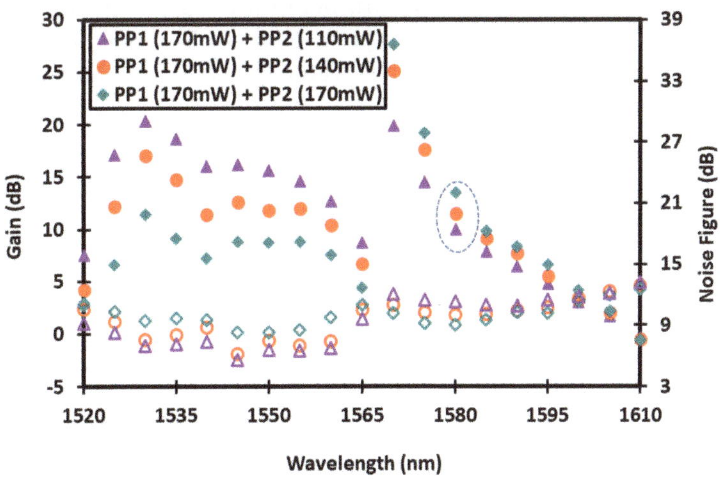

Figure 2.8. Measured gain (solid symbol) and NF (hollow symbol) spectra of the series forward pumping HB-EDFA using 1.72 m of total active fiber at various LD2 powers for an input signal power of −30 dBm.

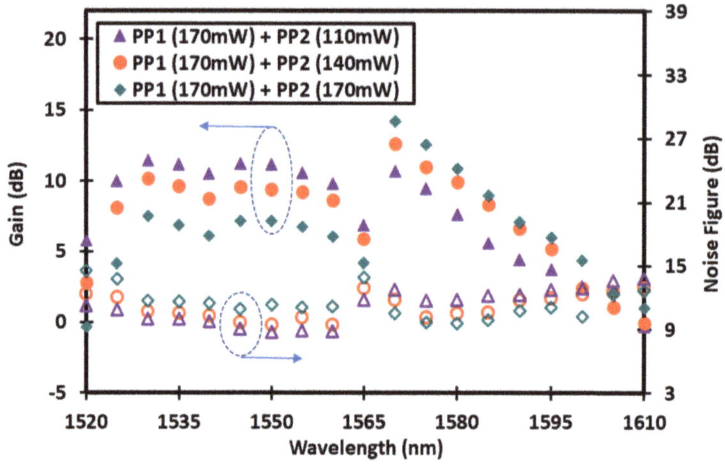

Figure 2.9. Measured gain (solid symbol) and NF (hollow symbol) spectra of the series forward pumping HB-EDFA using 1.72 m of total active fiber at various LD powers for an input signal power of −10 dBm.

high forward amplified spontaneous emission (ASE) effect from the second stage which decreases the population inversion and hence decreases the C-band gain spectrum. On the other hand, the L-band NF spectrum decreases as LD2 power increases. This is due to the higher gain and lower loss resulting from higher LD2 power. In addition, the C-band NF spectrum is increased as LD2 power is increased, due to the increment of the ASE noise effect.

2.5.2 Backward pumping

In this amplifier configuration the HB-EDFs are also adjusted to 22 and 150 cm long for amplification in the C- and L-bands, respectively, such that the total length is 1.72 m. The shorter HB-EDF is backward pumped using a 980 nm LD1 while the longer HB-EDF is backward pumped using a 1480 nm LD2. Figures 2.10 and 2.11 show the gain and NF spectra of the series type HB-EDFA with backward pumping at various L-band pump powers for input signal powers of −30 and −10 dBm, respectively. In the experiment the LD1 power is fixed at 170 mW while the LD2 power which controls the L-band gain is varied from 110 to 140 and 170 mW. It is clearly shown that the variation of the LD2 power changes the gain and NF performance of the amplifier for both input signal powers. The gain at the L-band region increases as the LD2 power increases before it saturates as the LD2 power exceeds 140 mW. In addition, the C-band gain spectrum decreases when the LD2 power increases above 110 mW, although the LD1 power is fixed. This is because of the high backward ASE from the second stage which decreases the population inversion and hence decreases the C-band gain spectrum. On the other hand, the L-band NF spectrum decreases as LD2 power increases. This is due to the higher gain and lower loss resulting from higher LD2 power. In addition, the C-band NF spectrum is increased as the LD2 power is increased due to the increment of the ASE noise effect. Overall it can be inferred that the optimum laser diode powers to

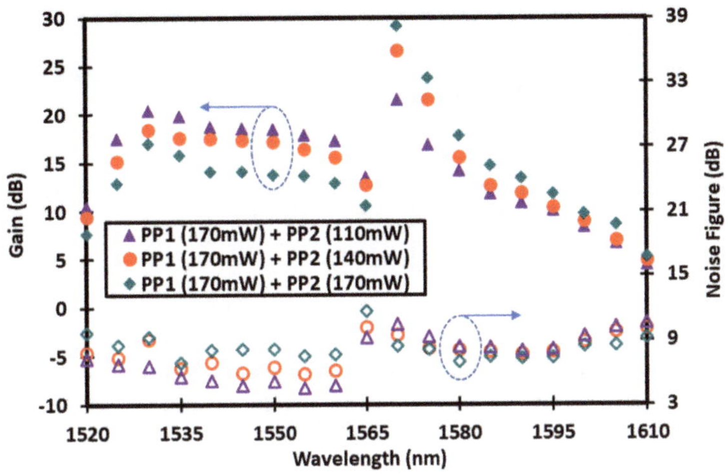

Figure 2.10. Measured gain (solid symbol) and NF (hollow symbol) spectra of the series backward pumping HB-EDFA using 1.72 m of total active fiber at various laser diode powers for an input signal power of −30 dBm.

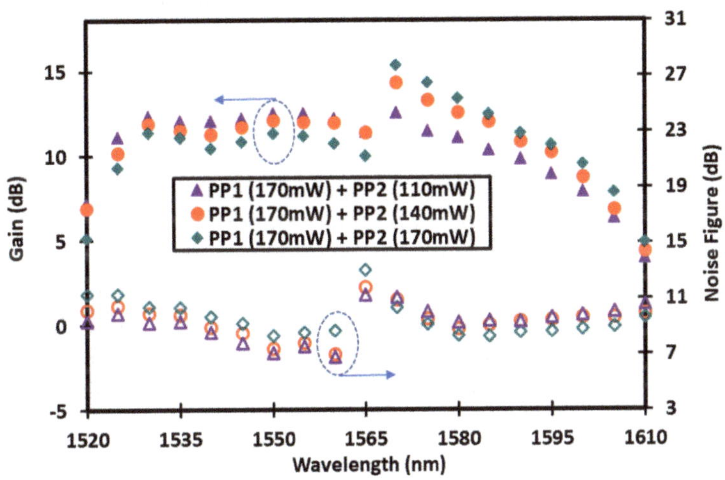

Figure 2.11. Measured gain (solid symbol) and NF (hollow symbol) spectra of the series backward pumping HB-EDFA using 1.72 m of total active fiber at various laser diode powers for an input signal power of −10 dBm.

achieve even better performance in terms of gain, NFs, and bandwidth are 170 and 110 mW for LD1 and LD2, respectively.

At an input signal power of −30 dBm with the optimum LD powers, the gain fluctuates from 10.1 to 21.6 dB along a wavelength span of 75 nm from 1520 to 1595 nm. It can be observed that the gain suddenly increases beyond 1565 nm. This is attributed to the change in the input signal from the C-band to L-band wavelength and thus the signal is amplified in both the first and second stages. On the other hand, the NF is less than 10.5 dB within the wavelength region from 1520 to 1600 nm. At a high input signal power

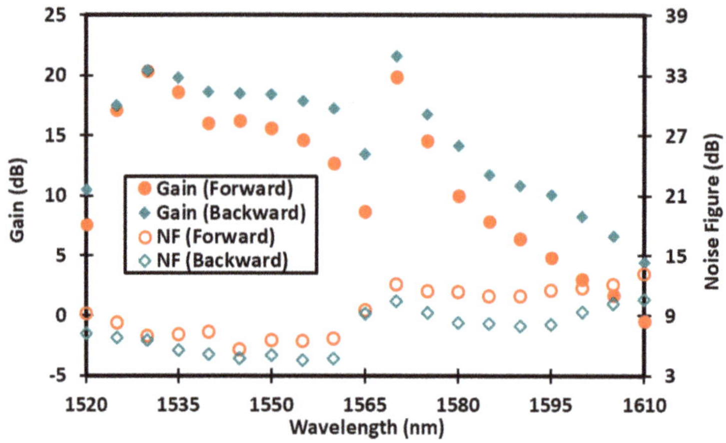

Figure 2.12. Comparison of the gain and NF characteristics between the forward and backward pumping series HB-EDFAs at an input power of −30 dBm.

of −10 dBm, a flat gain of 11.7 dB is realized with a gain ripple of less than 2 dB along the wavelength region of 65 nm from 1525 to 1590 nm. Within the flat-gain region the NF values vary from 6.7 to 11.2 dB.

2.6 Comparison of the performance of forward and backward pumping

The performance comparison between forward and backward pumping for the series HB-EDFA is investigated at the optimum LD powers when the length of the HB-EDFs are fixed at 22 and 150 cm in the first and second stages, respectively. Figures 2.12 and 2.13 show the amplification performance comparisons at input signal powers of −30 and −10 dBm, respectively. The comparisons show that the backward pumping amplifier obtains a higher gain and lower NF than that of the forward pumping amplifier. This is attributed to the saturation effect of ASE, which is weaker with backward pumping

With the backward pumping amplifier at a −30 dBm input signal power, average gain enhancements of around 2.5 and 4.1 dB are observed along the C-band and L-band wavelength regions, respectively. However, at an input signal power of −10 dBm average gain enhancements of around 1.7 and 4.1 dB are observed along the C-band and L-band wavelength regions, respectively. Furthermore, the backward pumping amplifier shows a more efficient wideband and flatter gain spectrum compared to that of the forward pumping amplifier. On the other hand, a lower NF is obtained with the backward pumping amplifier.

2.7 The effect of total HB-EDF length on the backward pumped amplifier

Since the backward pumping of the series HB-EDFA displays a preferable performance, this technique is adopted to accomplish a wideband HB-EDFA using another

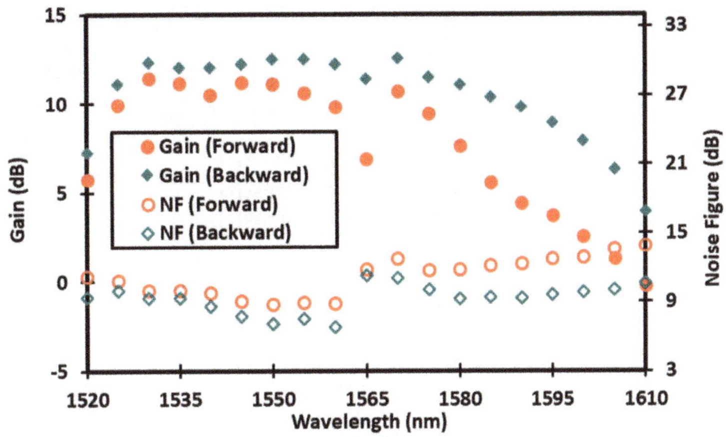

Figure 2.13. Comparison of the gain and NF characteristics between the forward and backward pumping series HB-EDFAs at an input power of −10 dBm.

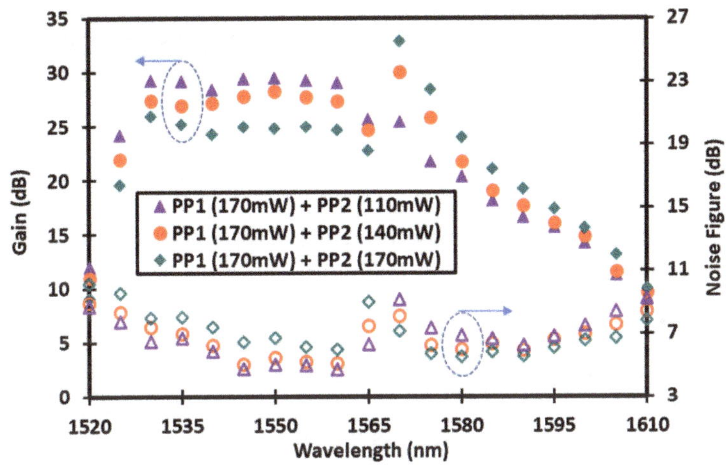

Figure 2.14. Measured gain (solid symbol) and NF (hollow symbol) spectra of the series backward pumping HB-EDFA using 2 m of total active fiber at various laser diode powers for an input signal power of −30 dBm.

length of HB-EDF. In the experiment, the HB-EDFs are adjusted to 50 and 150 cm long for amplification in the C- and L-bands, respectively, so that the total length is 2 m. The C-band HB-EDF is backward pumped using a 980 nm LD1, while the L-band HB-EDF is backward pumped using a 1480 nm LD2. Figures 2.14 and 2.15 present the performance of the series backward pumping HB-EDFA at various L-band pump powers for input signal powers of −30 and −10 dBm, respectively. In the experiment the LD1 power is fixed at a maximum power of 170 mW. The LD2 power which controls the L-band gain is varied from 110 mW to 140 and 170 mW. It is found that the optimum laser diode powers to achieve even better performance in term of gain, NFs, and bandwidth are 170 and 140 mW for LD1 and LD2, respectively. At an input signal power of −30 dBm with the optimum LD powers,

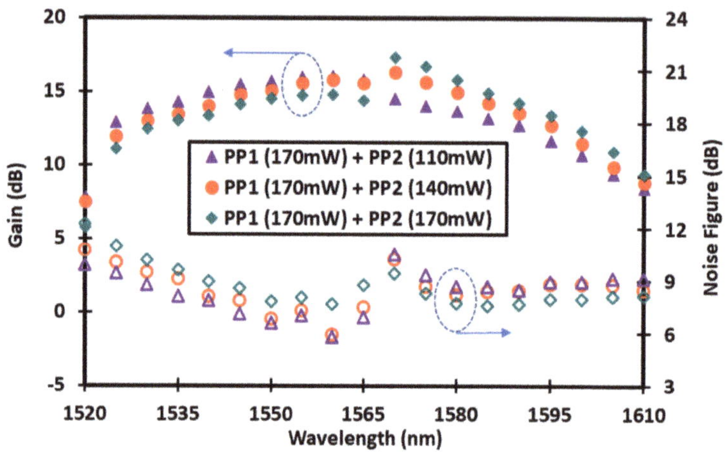

Figure 2.15. Measured gain (solid symbol) and NF (hollow symbol) spectra of the series backward pumping HB-EDFA using 2 m of total active fiber at various laser diode powers for an input signal power of −10 dBm.

the gain fluctuates from 10.9 to 30 dB along the wavelength span of 85 nm from 1520 to 1605 nm. In addition, the NF is less than 9 dB within this 85 nm wavelength region. However, at a high input signal power of −10 dBm a flat gain of 14.6 dB is realized with a gain ripple of less than 2 dB along the wavelength region of 65 nm from 1530 to 1595 nm. Within the flat-gain region the NF values vary from 6.8 to 10.2 dB.

Like the backward pumping based HB-EDFA configured with a 1.72 long HB-EDF, the L-band gain spectrum of the amplifier increases as the LD2 power increases, as shown in figures 4.14 and 4.15. However, this increment rate declines drastically when the LD2 power increases from 140 to 170 mW. The C-band gain spectrum is also observed to decrease when LD2 power increases above 110 mW. On the other hand, the L-band NF spectrum decreases as LD2 power increases. However, the C-band NF spectrum is increased as LD2 power is increased. Finally, the performance comparison of the proposed series backward pumping HB-EDFA with two different total HB-EDF lengths is summarized in table 2.2. The comparison shows that both amplifiers obtain an efficient amplification with a wideband operation region from 1520 to 1610 nm. The series HB-EDFA using a 2 m long HB-EDF achieves a higher gain, lower NF, and broader wavelength amplification as well. However, the series HB-EDFA with a 1.72 m long HB-EDF requires lower LD power to operate due to its lower erbium concentration.

2.8 Conclusions

In this chapter, wideband and compact HB-EDFAs have been successfully demonstrated utilizing a newly fabricated HB-EDF as the gain medium. The proposed amplifier is investigated by utilizing two short lengths of HB-EDF sections to achieve amplification in the C and L telecommunication bands. The wideband HB-EDFA is achieved using two stages in both series and parallel structures in conjunction with a double-pass scheme. The performance of the amplifier is

Table 2.2. Performance comparison of the proposed series HB-EDFAs using the backward pumping technique.

Feature	Series HB-EDFA using a 1.72 m long HB-EDF	Series HB-EDFA using a 2.0 m long HB-EDF
Erbium fiber length	Total length: 1.72 m. 22 cm for the C-band region. 150 cm for the L-band region.	Total length: 2 m. 50 cm for the C-band region. 150 cm for the L-band region.
Optimum LD power	LD1 power: 170 mW. LD2 power: 110 mW.	LD1 power: 170 mW. LD2 power: 140 mW.
Wideband amplification	At an input signal power of −30 dBm: • The gain fluctuates from 10.1 to 21.6 dB along the wavelength span of 75 nm. • The NF is less than 10.5 dB. At an input signal power of −10 dBm: • A flat gain of 11.7 dB with a gain ripple of ± 2 dB along the wavelength span of 65 nm is realized. • The NF values vary from 6.7 to 11.2 dB.	At an input signal power of −30 dBm: • The gain fluctuates from 10.9 to 30 dB along the wavelength span of 85 nm. • The NF is less than 9.0 dB. At an input signal power of −10 dBm: • A flat gain of 14.6 dB with a gain ripple of ± 2 dB along the wavelength span of 65 nm is realized. • The NF values vary from 6.8 to 10.2 dB.

investigated for two different lengths of HB-EDFs, as well as various pump powers to determine the optimum design. During the experiments, first the HB-EDFs are drastically adjusted to 50 and 150 cm long for amplification in the C- and L-bands, respectively, so that the total length is 2 m. Second, the HB-EDFs are adjusted to 22 cm and 150 cm long for amplification in the C- and L-bands, respectively, so that the total length is 1.72 m. On the other hand, the LD1 power is fixed at a maximum power of 170 mW. However, the LD2 power which controls the L-band gain is varied from 110 mW to 140 and 170 mW. For comparison purposes the series HB-EDF is examined in both the backward and forward pumping schemes. It can be concluded that both the parallel and backward pumping series HB-EDFAs achieve wideband operation within 1520 to 1610 nm. In addition, a flat-gain characteristic is achieved throughout the C- and L-band wavelength regions. For example, in parallel HB-EDFA using a 1.72 m long HB-EDF, a flat gain of 12.1 dB is realized with a gain ripple of less than 2 dB along the wideband wavelength region of 80 nm from 1525 to 1605 nm. Within the flat-gain region the NF values vary from 6 to 11.8 dB. However, in the series backward pumping HB-EDFA using a 2 m long HB-EDF, a flat gain of 14.6 dB is realized with a gain ripple of less than 2 dB along the wavelength region of 65 nm from 1530 to 1595 nm. Within the flat-gain region the NF values vary from 6.8 to 10.2 dB. It can be inferred that the series backward pumping HB-EDFA achieves higher flat gain with a lower NF. However, the flat-gain region is larger for the parallel HB-EDFA than that for the series backward pumping HB-EDFA.

References

[1] Firstov S, Riumkin K, Khegai A, Alyshev S, Melkumov M, Khopin V, Afanasiev F, Guryanov A and Dianov E 2017 *Laser Phys. Lett.* **14** 110001

[2] Hsu H-Y, Yu Y-L, Liaw S-K, Liu R-Y and Shin C-S 2014 *Opt. Laser Technol.* **56** 307–12

[3] Durak F E and Altuncu A 2017 *Opt. Commun.* **386** 31–6

[4] Yusoff N M, Abas A, Hitam S and Mahdi M 2012 *Laser Phys.* **22** 1252–6

[5] Almukhtar A A, Al-Azzawi A A, Reddy P, Das S, Dhar A, Paul M C, Ahmad H and Harun S 2019 *Optik* **182** 194–200

[6] Ali M H, Abdullah F, Jamaludin M Z, Al-Mansoori M H, Al-Mashhadani T F and Abass A K 2014 *J. Opt. Soc. Korea* **18** 657–62

[7] Cheng X, Hamida B, Naji A W, Ahmad H and HarunS W 2011 *Laser Phys. Lett.* **8** 814

[8] Hamida B, Cheng X, Harun S W, Naji A, Arof H, Khan S, Alkhateeb W and Ahmad H 2012 *Microwave Opt. Technol. Lett.* **54** 629–31

[9] Al-Azzawi A A, Almukhtar A A, Reddy P H, Das S, Dhar A, Paul M C, Arof H, Ahmad H and Harun S W 2019 *Opt. Fiber Technol.* **48** 186–93

IOP Publishing

Optical Fiber Technology and Applications
Recent advances
Mário F S Ferreira and Mukul Chandra Paul

Chapter 3

Photon noise in a continuous-wave ytterbium-doped fiber laser

Yuri Barmenkov, Alexander V Kir'yanov, Pablo Muniz-Cánovas, José L Cruz and Miguel V Andrés

In this chapter the noise properties of a continuous-wave ytterbium-doped fiber laser with moderate optical power are discussed. The laser is implemented in the conventional Fabry–Perot cavity configuration with two fiber Bragg gratings as reflectors. The laser operates in the regime of random noise pulses with the statistics intrinsic to narrowband amplified spontaneous emission; no continuous-wave or quasi-continuous-wave operation is observed. The properties of laser noise are addressed in terms of optical and RF spectra, the probability density function for photons, the autocorrelation function for oscilloscope traces of the laser signal, and the distributions of the noise pulses' power. It is shown that the coherence of the laser depends on its power, namely the laser coherence time drops from a few nanoseconds to tens of picoseconds with increasing laser power. In addition, the laser coherence is repeated over each even number of the cavity length; this number also depends on the laser power. Furthermore, it is demonstrated that the laser noise causes broadening of the laser line through the effect of self-phase modulation.

3.1 Introduction

Ytterbium-doped fiber lasers (YDFLs) are widely used laser devices owing to their excellent energy budget resulting from their high optical and electrical to optical efficiency. Due to the absence of excited state absorption (ESA) in the system of Yb^{3+} ions, a YDFL reaches an optical efficiency of up to ~80%–83% [1, 2] when pumped at ~975 nm. The efficiency is limited basically by the Stokes shift and by the concentration [3–5] and photo darkening [6–8] phenomena, inherent to fibers heavily doped with ytterbium. The output power of the YDFL based on a double-clad active fiber and Fabry–Perot cavity is scalable up to several kilowatts [1, 2, 9] and a

doi:10.1088/978-0-7503-3243-9ch3

few hundred kilowatts [10–12] in the single- and multi-mode operation regimes, respectively, which is important for many technological applications.

The optical spectrum of the YDFL significantly broadens with increasing laser power [13–15]. This phenomenon is usually attributed to spatial hole burning (SHB) and self-phase modulation (SPM) of fluctuating intensity [13–15]. The spectral broadening in the YDFL with a cavity made of two fiber Bragg gratings (FBGs) results in partial leakage of laser power on a high-reflection (HR) FBG coupler outside its reflection spectrum and in spectral rounding on a low-reflection (LR) FBG coupler [14]. As discussed in [16], excessive YDFL intensity noise can lie behind quasi-continuous-wave (QCW) laser operation, established due to de-phasing of the longitudinal modes.

In this chapter we discuss the spectral and noise features of a double-clad YDFL with a Fabry–Perot cavity based on two FBGs as selective mirrors. It is demonstrated that the YDFL optical spectrum gradually broadens with increasing laser power; at low laser power the slope of broadening increase is an order of magnitude less than that at higher powers.

Apart from broadening of the YDFL optical spectrum, the RF spectrum varies with laser power: at low power it contains powerful spectral peaks at the frequencies attributable to mode-beating, whilst at increasing power these peaks vanish and the noise floor moves upward, which makes the resultant RF spectrum flattened. Thus at high power the laser operates in completely random noise pulsing (NP) with a photon distribution inherent to narrowband amplified spontaneous emission (ASE).

We demonstrate that in a broad range of powers the laser oscillates in the NP regime, described by the probability distribution function (PDF), which is usually applicable for characterizing narrowband ASE [17]; neither the true CW nor quasi-CW regime is observed [18]. Such NP was recently reported for an actively Q-switched erbium-doped fiber laser and was also shown to be responsible for supercontinuum (SC) generation [19, 20]. Note that CW and gain-switched CW Fabry–Perot YDFLs are promising pump sources for SC generation [21, 22], thus the importance of knowledge about the noise features in such lasers is difficult to overestimate.

A statistical analysis of lasing is performed here for the cases of polarized and unpolarized laser signals, which is important for better understanding the physics behind the laser intensity noise. NP is analyzed in terms of an autocorrelation function that is a useful tool for studying the laser time coherence. The zero-order peak of the autocorrelation function, and thus also the laser coherence time, dramatically decays from a few nanoseconds to a few tens of picoseconds with increasing laser power.

We also discuss the effect of repeating noise structure (i.e. the temporal pattern of NP), established during one round-trip (RT) time interval of the optical wave traveling along the laser cavity, over a large number of subsequent RTs with a high degree of correlation. This effect depends on the laser power—the number of intervals over which the laser repeats the noise structure drops from tens of RTs at low powers to approximately one RT at high powers.

Finally, we focus on the effect of self-phase modulation (SPM), resulting from the nonlinear interaction of noise pulses with optical fiber, on the spectral shape of the laser-line broadening. We reveal that this shape is defined by the shape of the histogram of the laser noise signal derivative, which is triangular when presented in a lin–log plot.

3.2 The ytterbium-doped fiber laser: structure and basic characteristics

3.2.1 Ytterbium-doped fiber laser arrangement

The experimental set-up of a typical YDFL is illustrated in figure 3.1. The laser is assembled in the commonly used Fabry–Perot cavity with two FBGs as in-fiber mirrors. An FBG with high reflectivity (100%, HR-FBG) is used as the rear mirror and an FBG with low reflectivity (10%, LR-FBG) as the output coupler, both centered at the same wavelength ($\approx$ 1.061 μm). The reflection bands measured as the full width at half maximum (FWHM) of the HR-FBG and LR-FBG are $\approx$ 400 and 30 pm, respectively.

The YDF used as an active laser medium is a standard double-clad (DC) fiber (Nufern, SM-YDF-5/130-VIII) with cladding absorption of ~1.65 dB m^{-1} at 975 nm and mode-field diameter (MFD) of 6.5 μm at the laser wavelength. The YDF length is 15 m, which corresponds to ~25 dB pump absorption; the laser cavity length is 19 m for which the RT interval is $\approx$ 188 ns. The YDF is pumped by two commercial 975 nm laser diodes (LDs) with multi-mode fiber output through a $(2 + 1) \times 1$ pump combiner. The residual pump is filtered by a cladding mode stripper (CMS). A fiber polarizer (POL) is added into the scheme to study statistical properties of polarized light; when assessing laser signal with random polarization (unpolarized light) it is removed. The fibers from both laser outputs (the rear on the left side and the output on the right side of the set-up) are 7° angle-cleaved to protect the laser cavity against parasitic optical feedback.

The laser signal is detected by a 25 GHz InGaAs photodetector (Newport, model 1414), connected to a 16 GHz real-time oscilloscope (Tektronix, DPO71604C). The cumulative 3 dB RF band (B_{el}) of the set photodetector/oscilloscope is from DC to 15.5 GHz, permitting one to detect noise pulses as short as ~60 ps. Before detecting the laser signal is attenuated to a level that ensures the photodetector's operation below saturation.

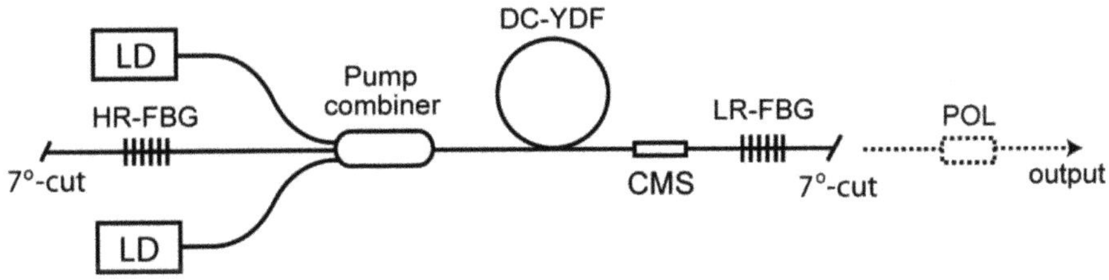

Figure 3.1. The YDFL set-up.

3.2.2 Basic characteristics of the ytterbium-doped fiber laser

The operation wavelength of the laser is 1061 nm. The threshold pump power is 2 W; the slope efficiency at laser powers below 15 W is ~65% and above 16 W it increases up to ~78% (figure 3.2(a)), probably due to decreasing of the effective reflectivity of the output FBG, resulting from the laser-line broadening [23]. The broadband optical spectra were captured using an optical spectrum analyzer (OSA, Yokogawa AQ6370D) with 1 nm resolution (figure 3.2(b)). The spectra reveal the presence of a narrow laser line at ~60 dB above ASE; the small peak at 975 nm corresponds to the residual pump.

The narrowband laser spectra, measured by the OSA with 34 pm resolution at the laser wavelength, are shown in figure 3.2(c). These spectra are normalized to the maximum of that measured at a laser power of 22 W and presented as a function of detuning from the laser spectral peak found at low laser powers (1061.235 nm). From figure 3.2(c) one can see that the spectra broaden with laser power growth; the origin of this effect will be discussed below. All spectra are fit by the Gaussian function with the residual sum of squares $R^2 > 0.995$, an important topic for further discussion. It is also seen that the laser spectrum shifts to the longer wavelengths with increasing laser power, with an averaged slope of ~1.4 pm W^{-1}. This shift results from heating of a section of the photosensitive fiber with the FBG due to additional loss induced in it on writing the FBG by the UV laser at 244 nm (see [24, 25]).

Figure 3.2(d) shows the width of the laser spectra, measured as FWHM, as a function of pump power. Curve 1 corresponds to the original experimental data. Since the spectral response of the OSA to the optical signal with a very narrow optical spectrum (spectrally the δ-function) is Gaussian with FWHM = 34 pm at

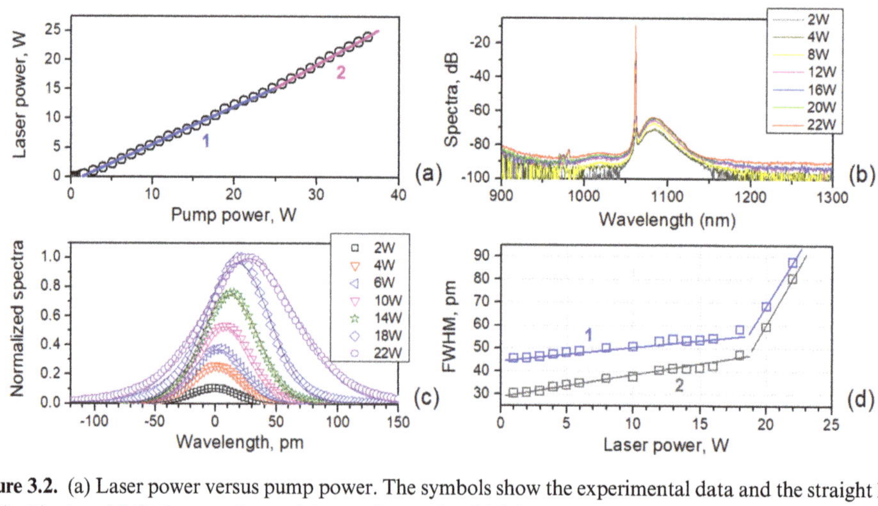

Figure 3.2. (a) Laser power versus pump power. The symbols show the experimental data and the straight lines marked by 1 and 2 fit the experimental data at low and at high laser power, respectively. (b) Broadband laser spectra measured with 1 nm resolution. The laser powers at which the spectra were measured are specified in the legend. (c) Narrowband laser optical spectra measured with 34 pm resolution. (d) Width of optical spectra as FWHM versus laser power.

1061 nm and the YDFL spectra are also fit by a Gaussian function, the deconvoluted spectral width can be found as $(\text{FWHM}^2 - 34^2)^{1/2}$, both in pm.

The deconvolution result is plotted in figure 3.2(d) by curve 2. As can be seen, the laser optical spectrum gradually broadens with laser power. At laser powers below 18 W the slope of the dependence is ~1 pm W^{-1}, whereas that at laser powers above 18 W increases to ~15 pm W^{-1}, i.e. by more than an order of magnitude higher. Note that at $P = 18$ W, above which the slope changes strongly, the nonlinear phase φ_{NL}, found as in [26],

$$\varphi_{NL} = \frac{2\pi}{\lambda} n_2 \frac{P}{A_{eff}} L_{eff} \tag{3.1}$$

is equal to π. In this formula λ is the laser wavelength, $n_2 = 4.3 \times 10^{-20}$ m^2 W^{-1} is the nonlinear refractive index of fused silica at 1.06 μm [27], $A_{eff} = \pi(\text{MFD}/2)^2 = 34$ μm^2 is the effective area of the propagating mode, $R = 0.1$ is the reflection coefficient of the output (LR-FBG) coupler (see the next equation), and L_{eff} is the effective fiber length, found as

$$L_{eff} = 2L_{YDF} \frac{1 - R}{\ln(1/R)} \tag{3.2}$$

under consideration that the YDF gain, g, found for threshold of a CW laser (the active fiber gain compensates the cavity loss [28]) is distributed homogenously along the fiber:

$$g = \frac{1}{2L_{YDF}} \ln\left(\frac{1}{R}\right), \tag{3.3}$$

where L_{YDF} is the YDF length and YDF passive loss is negligible. Above $\varphi_{NL} = \pi$ the effect of SPM is much more prominent; hence the laser optical spectrum becomes strongly broadened with further increase of laser power.

Figure 3.3 shows the typical oscilloscope trains of NP measured on the laser output for polarized and unpolarized light at different laser powers. It is seen that in all cases the noise pulses are of random magnitude and developing time; some of the pulses reach powers exceeding the mean laser power (P_{mean}), as shown by the dashed orange lines. With increasing laser power the noise pulses arise more frequently and their average width decreases. The polarized NP is more dispersive in pulse magnitudes than unpolarized NP; CW lasing is never observed. Similar behavior of the laser signal was recently reported for an erbium-doped fiber amplifier with narrowband ASE [17]. Note that the phenomenon of excessive laser noise is important for better understanding the physics behind CW operation of YDFLs and also for applications based on nonlinear optical effects in fibers.

3.3 Coherence properties

3.3.1 Autocorrelation function of the laser signal

Optical intensity autocorrelator is an important tool for laser signal characterization. It is used usually for pulse-width measuring of narrow (picosecond to

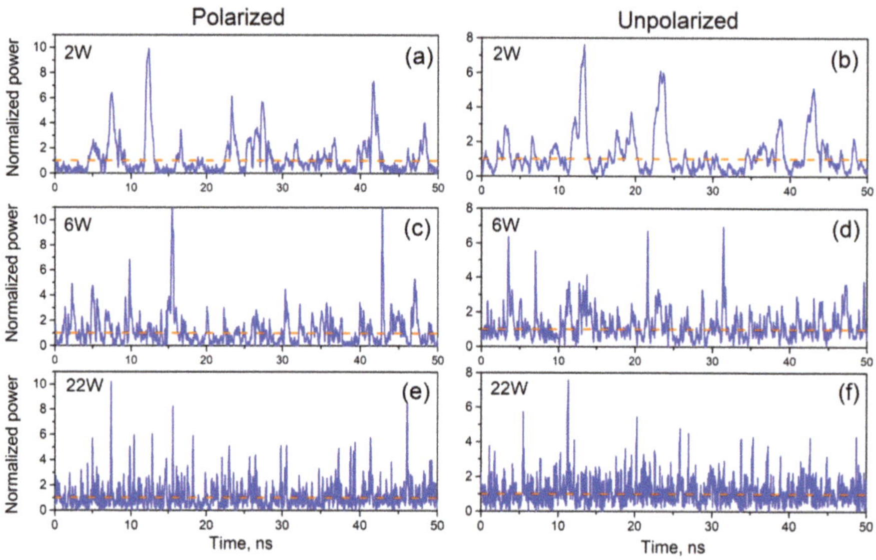

Figure 3.3. 50 ns traces of laser power normalized to the mean power. (a), (c), (e) Polarized and (b), (d), (f) unpolarized laser signals. The mean laser powers at which traces were captured are indicated in each plot. The orange horizontal lines indicate the mean laser power.

femtosecond) pulses since the response time of photodiodes and oscilloscopes is too slow for this purpose. Optical autocorrelators are usually based on a tunable/scanning Michelson interferometer with a nonlinear crystal as a second harmonic generator, the output signal of which is detected by a slow photodetector [29]. In the case of relatively large laser pulses and fast photo-receiving equipment, autocorrelation can be realized numerically using large trains of laser signal. In our current study large oscilloscope trains of 31.25 mega samples (MS) with a sample rate of 100 GS s^{-1} (the timing resolution is 10 ps) were used for simulating autocorrelation functions.

The autocorrelation function, inherent to the second-order coherence of the laser signal, permits one to estimate the laser coherence time. The normalized second-order coherence, $g^{(2)}(\tau)$, is found as [30]

$$g^{(2)}(\tau) = \frac{\langle I(t)I(t+\tau)\rangle}{\langle I(t)\rangle^2} = \frac{\langle P(t)P(t+\tau)\rangle}{(P_{\text{mean}})^2} = \frac{\langle U(t)U(t+\tau)\rangle}{(U_{\text{mean}})^2}, \quad (3.4)$$

where $I(t)$ is the light intensity, $P(t)$ is the instantaneous laser power, $U(t)$ is the photodetector voltage that is proportional to the light power and intensity, P_{mean} and U_{mean} are the mean laser power and the mean detector voltage, respectively, and τ is the shift with respect to the current time t. In our case, the averaging time was 312.5 μs, which corresponds to the limit for the length of the oscilloscope traces (31.25 MS). The first-order coherence, $g^{(1)}(\tau)$, is found from $g^{(2)}(\tau)$ as [31]

$$|g^{(1)}(\tau)|^2 = g^{(2)}(\tau) - 1. \quad (3.5)$$

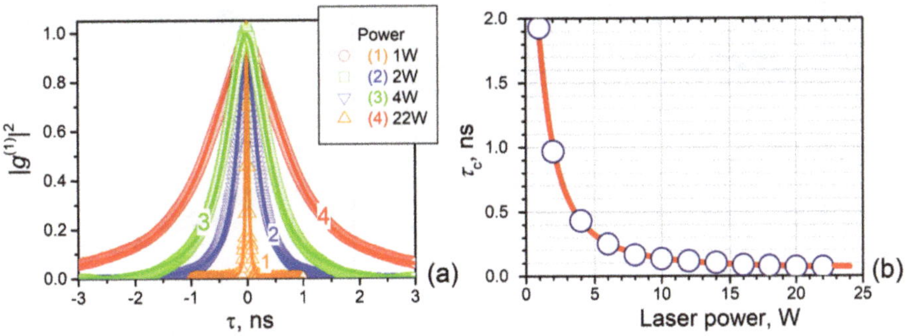

Figure 3.4. (a) Zero-order peaks of the squared first-order coherence. Laser powers are specified in the legend. (b) τ_c as a function of P_{mean}.

Figure 3.4(a) exemplifies the zero-order peaks of $|g^{(1)}(\tau)|^2$ simulated for four laser powers: 1, 2, 4, and 22 W (zeroth order means that the peak is centered at zeroed detuning). The experimental data shown by symbols are precisely fit by the Lorentzian function (solid lines) related to collisionally broadened thermal light [31]. The dependences of the coherence time, τ_c, found from the complete set of experimental data versus laser power are plotted in figure 3.4(b), where the symbols show τ_c values found from $|g^{(1)}(\tau)|^2$ and the curve is the best fit obtained using the generalized logistic function. At high laser powers the coherence time tends to the level of ~70–80 ps. Note that this value corresponds to the coherence time found from the laser Gaussian linewidth measured at low powers (30 pm; see figure 3.2(c), (d)) employing the formula [32]

$$\tau_c = \sqrt{\frac{2\ln 2}{\pi}}\frac{1}{\Delta\nu} = \sqrt{\frac{2\ln 2}{\pi}}\frac{\lambda^2}{\pi\Delta\lambda}. \tag{3.6}$$

One more important feature of the autocorrelation function is uncovered when studying a large interval of delays τ. Three examples of $|g^{(1)}(\tau)|^2$ within the time interval, measured from −20 to +20 ns, are presented in figure 3.5(a) for different laser powers (indicated in each plot). The red lines envelop the trains of peaks of the autocorrelation function, which are separated by the RT interval (188 ns, see figure 3.5(b)). The peaks correspond to partially overlapped noise pulses at the delays $\tau_k = k\text{RT}$, where k is an integer; in the areas between the peaks, the first-order coherence is close to zero. The zero-order peak ($k = 0$) shown in detail in figure 3.5(a) was discussed above.

The magnitudes of the higher-order (lateral) peaks drop with increasing laser power; figure 3.5(c) highlights this trend. The envelopes of the data obtained by calculating the FWHM of the lateral peaks are described by the sigmoidal function (see figure 3.5(c)). As can be seen, the FWHM found at the lowest power is attained up to thirty RT intervals whereas at the highest power it is reduced to approximately one RT.

In other words, at low laser powers the pattern of NP, established during one RT interval, correlates with those established for up to 30 nearby intervals, whereas at

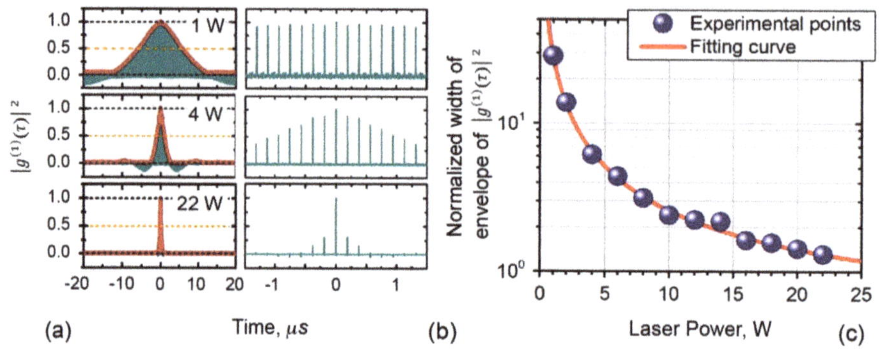

Figure 3.5. Examples of the autocorrelation function $|g^{(1)}|^2$ obtained for different laser powers (1 W, 4 W, and 22 W) in (a) wide and (b) narrow time ranges. (c) Dependence of the $|g^{(1)}|^2$ envelope's width normalized to the RT interval versus laser power. Circles show the experimental data and the solid line is the sigmoidal fit. Adapted with permission from [18]. Copyright 2019 Springer Nature.

high laser powers this number drops to one RT interval. Note that a similar coherence feature (visibility of the interference pattern) was reported for a He–Ne laser at a path difference in a tunable Michelson interferometer, equal to an integer number of RT intervals [33].

3.3.2 Repeatability of the noise pattern

Figure 3.6 shows the typical oscilloscope trains of NP, collected within ~3 RT intervals, for low (2 W, upper traces), medium (6 W, middle traces), and high (22 W, lower traces) laser powers; the left and right plots are for polarized and unpolarized light, respectively.

As can be seen, at low laser powers the trains of noise pulses established during one RT interval repeat the noise pattern in consecutive RT intervals (compare the areas delimited by the vertical dash lines); some discrepancy between the patterns in the adjacent areas appears because of digitalization by the oscilloscope and due to internal noise of the photodetector. In other words, the laser keeps ('remembers') the pattern of NP, established in the cavity during one RT interval, retaining it in the subsequent RT intervals. In contrast, at high laser powers noise pulses do not demonstrate any periodicity in consecutive RT intervals; thus the laser noise in this case is fully random. The traces obtained at middle laser powers demonstrate partial repeatability. Such behavior of laser NP confirms the phenomenon of laser coherence discussed above, namely that the laser time coherence appears at each integer number of RT time of the laser wave inside the cavity and that the degree of coherence depends on the laser power: the higher the laser power, the lower the degree of coherence.

3.4 Laser photon statistics

In this section the photon statistics of YDFL noise is discussed. It is shown that laser noise appearing in the form of random NP is described by the statistics, intrinsic to

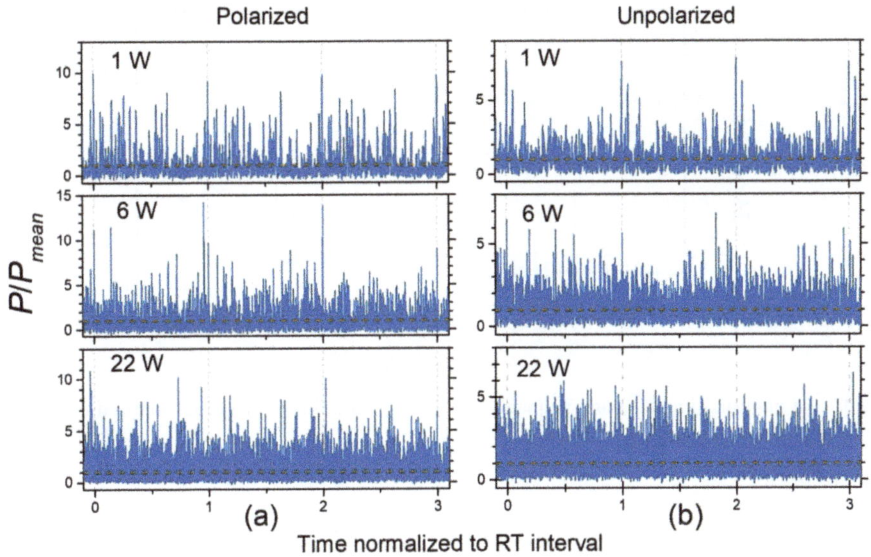

Figure 3.6. Oscilloscope traces measured for (a) polarized and (b) unpolarized YDFL signal. The horizontal dashed lines mark the mean levels of laser power; the vertical dash lines delimit the RT intervals. Each window is labeled by the laser power at which the oscilloscope traces were collected. Adapted with permission from [18]. Copyright 2019 Springer Nature.

narrowband ASE [17]. This type of statistics is characterized by a probability density function (PDF) that is a function of the number M of independent states of narrowband ASE inside the laser cavity.

3.4.1 Theoretical background

M number of ASE with a Gaussian narrowband optical spectrum is found as follows [32, 34]:

$$M = s\frac{\pi\left(B_{\mathrm{opt}}/B_{\mathrm{el}}\right)^2}{\pi\left(B_{\mathrm{opt}}/B_{\mathrm{el}}\right)\mathrm{erf}\left[\sqrt{\pi}\left(B_{\mathrm{opt}}/B_{\mathrm{el}}\right)\right] - \left[1 - \exp\left(-\pi\left(B_{\mathrm{opt}}/B_{\mathrm{el}}\right)^2\right)\right]}, \tag{3.7}$$

where B_{opt} is the deconvoluted optical bandwidth taken from line 2 in figure 3.2(d) and s is the number of polarization states (1 or 2, for polarized or unpolarized light, respectively). The extreme cases, $M = 1$ for polarized light and $M = 2$ for unpolarized light, are only considered when the optical spectrum is very narrow ($B_{\mathrm{opt}}/B_{\mathrm{el}} \ll 1$). Note that values of M less than unity are not possible.

Figure 3.7 shows the M number as a function of the mean laser power. It was simulated using equation (3.7) with B_{opt} presented in figure 3.2(d) by curve 2. At low laser powers the M number is ≈ 1.1 and ≈ 2.2 for polarized and unpolarized light, respectively, which corresponds to nearly one or two independent ASE states. At increasing laser power the laser spectrum broadens steadily and the M number is increased.

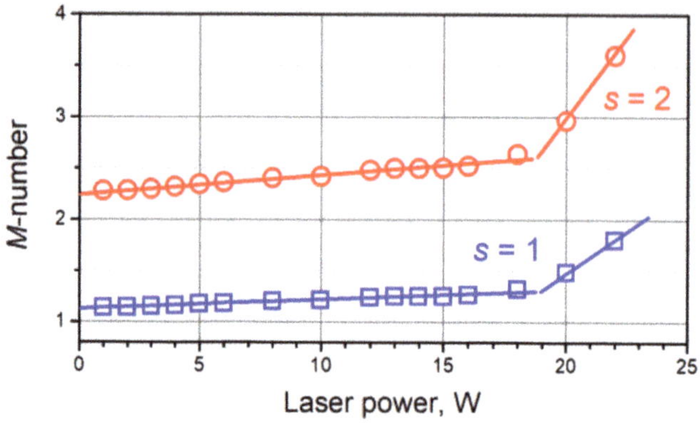

Figure 3.7. *M* numbers simulated using equation (3.7) for polarized ($s = 1$) and unpolarized ($s = 2$) YDFL signals versus mean laser power.

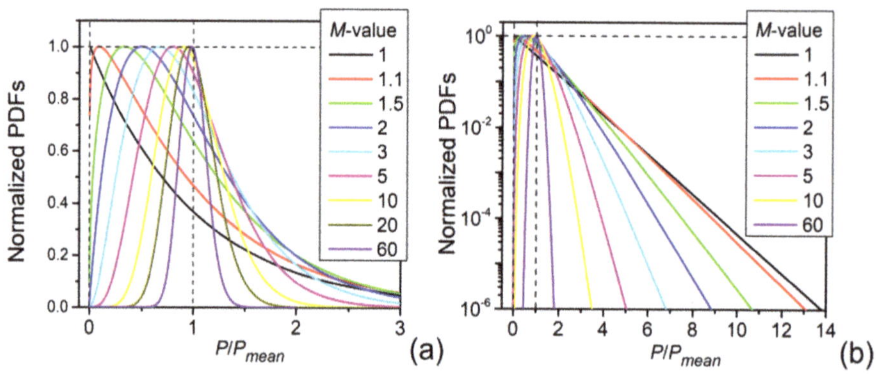

Figure 3.8. Simulated PDFs presented in (a) linear and (b) semi-logarithmic plots. *M* values are shown in the legends.

The photon statistics of narrowband ASE is characterized by an *M*-fold degenerate Bose–Einstein distribution, described as follows [32, 34]:

$$P(n, \bar{n}, M) = \frac{(n + M - 1)!}{n!(M - 1)!} \frac{(\bar{n})^n}{(1 + \bar{n})^{n+M}}, \tag{3.8}$$

where $P(n, \bar{n}, M)$ is the probability of counting n photons during counting (average) time $T = 1/B_{\text{el}}$ in each short interval of laser powers to which the whole range of instantaneous laser power fluctuations is divided and $\bar{n}$ is the mean photon number. Experimentally, n and $\bar{n}$ are found from the instantaneous laser power, $P(t)$, and as the mean laser power, P_{mean}, respectively, attenuated to the level at which the photodetector is not saturated.

Figure 3.8 shows examples of PDFs, simulated in the broad range of *M* values, from 1 to 60, in (a) linear and (b) semi-logarithmic plots. The PDFs (vertical axes) are normalized to the maximal values of the histograms and the powers (horizontal axes) to the mean values; the M numbers used for simulating PDFs are shown in the insets. The power P_{peak} at which the PDF peak is centered depends on M as

$$P_{peak}(M) = \frac{M-1}{M} P_{mean}. \tag{3.9}$$

With increasing M value the PDF peak moves from zero to P_{mean}. At fixed M the normalized PDF does not vary with increasing $\bar{n}$ from ~10^3 to the upper values (see figure 3.3(b) in [19]); thus, despite the mean photon number detected during the time interval $T = 1/B_{el} = 65$ ps (our case) is 3.5×10^5 at an optical power of 1 mW, counting of only 10^3 photons results in the same shape of the normalized PDF.

As can be seen from this figure, at low M numbers ($\leqslant 10$), the PDFs are very asymmetric. Moreover, at the extreme case of the polarized light ($s = 1$) with a very narrow optical spectrum and a very broad RF band ($M = 1$), the normalized PDF is described by the exponential decay function found from equation (3.8) as follows:

$$P(n, \bar{n}, \ M = 1) = \left(\frac{\bar{n}}{1+\bar{n}}\right)^n \cong e^{-\frac{n}{\bar{n}}}, \tag{3.10}$$

where $\bar{n}$ is large. The curves simulated using equation (3.10) are shown by the black curves in figures 3.8(a),(b).

At M numbers above 10, the PDFs are more symmetrical. Interestingly, at very large M numbers the PDFs can be adjusted with a high accuracy by the symmetrical Gaussian function; for example, at $M = 60$ it is fit by the Gaussian curve with residual sum $R^2 = 0.997$.

3.5 Experimental results

The experimental PDFs for laser powers ranged from 1 to 22 W are demonstrated in figure 3.9 by the symbols and the simulated results are presented by solid and dashed curves. The sets of data marked with 1 and 2 correspond to polarized and unpolarized light, respectively; M values were taken from figure 3.7. The fits for the experimental PDFs were built using equation (3.8) whilst n and $\bar{n}$ were found from the instantaneous laser power, $P(t)$, and the mean laser power, P_{mean}, respectively.

It is seen that when the laser signal is polarized the experimental PDFs of photon count (red circles) decay with increasing instantaneous laser power—at low and medium laser powers (~1 W and $\geqslant$ 8 W), the PDFs are fit by the exponential decay function ($M = 1$) better than for the ones found from the optical spectrum width. At laser powers between 1 and 8 W the absolute slope of the histograms is notably less than that found for $M = 1$. The histograms obtained for unpolarized light ($s = 2$) demonstrate similar features, namely they are fit by the PDFs simulated for the extreme case ($M = 2$) better than the ones found from the laser optical spectrum.

To assess the discrepancies between the experimental PDFs and their fits, it deserves attention first that at lower laser powers (< 8 W) the effective length of nonlinear interaction of the laser field with the optical fiber is long as the noise pattern arising during one RT is preserved for the large number of RTs; see

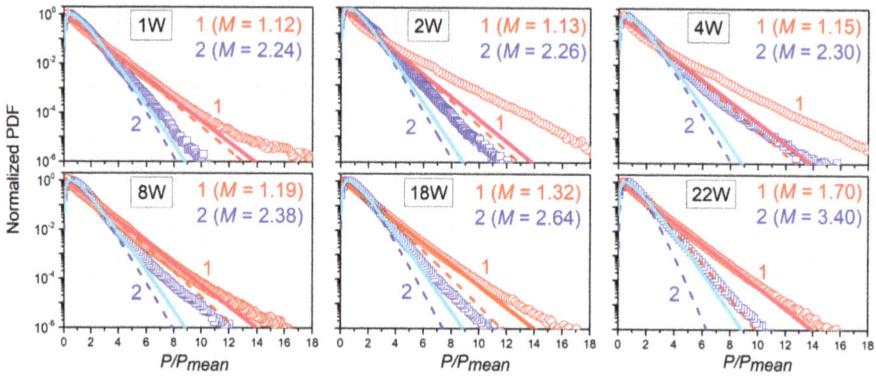

Figure 3.9. Normalized PDFs of laser signals. The red symbols and lines, marked with 1, correspond to polarized light and the blue ones, marked with 2, to unpolarized light. Symbols are experimental points. Dashed lines are theoretical PDFs calculated using equation (3.8) with M values taken from figure 3.7; the M values are marked in each plot. Solid lines show the results of simulations for $M = 1$ (polarized light) and $M = 2$ (unpolarized light). Adapted with permission from [18]. Copyright 2019 Springer Nature.

figure 3.5. Even for low mean laser powers but long effective lengths of nonlinear interaction (measured by tens of cavity lengths), the nonlinear phase φ_{NL} is multiplied by a large number (see equation (3.1)). This effect results in strong optical nonlinearity and, accordingly, the photon statistics with excessive probability of high-power noise pulses.

At higher laser powers, from 8 to 16 W, the effective nonlinear length is shortened, whereas the laser mean power is still not so high. In this case the histograms of laser noise are fit well by the Bose–Einstein distribution. At the highest laser powers (>18 W), regardless of the effective nonlinear length becoming shorter (~ 1 RT), the laser field of powerful noise pulses strongly interacts with the active fiber, which causes the noticeable discrepancies between the experimental PDFs and those modeled using M numbers found from the widths of the optical spectra. In the latter case the precise fitting is reached for the extreme cases of $M = 1$ for polarized light and $M = 2$ for unpolarized light. This effect can be explained by two reasons: (i) the photon statistics is really determined by the narrowband reflection from the output (LR-FBG) coupler and (ii) the broadening of the optical laser spectrum does not affect the PDFs (and the equivalent M numbers), given that the dispersion length of the intra-cavity fiber ($L_D \approx 250$ km) is much larger than the cavity length (19 m).

An interesting property of laser noise relates to the distribution of NP repetition intervals/frequencies with the magnitude of noise pulses. Figure 3.10 shows the PDF of the number of the polarized noise pulses with magnitudes falling into the intervals between kP_{mean} and $(k+1)P_{mean}$ (k is an integer) for a laser power of 22 W. As can be seen, the right slope of the histogram is precisely fit by the exponential function at NP power exceeding $5P_{mean}$ (see the linear function in the semi-logarithmic plot shown by the blue line). Above this power the pulse number decreases by approximately 5 dB per each step of increasing the pulses' power.

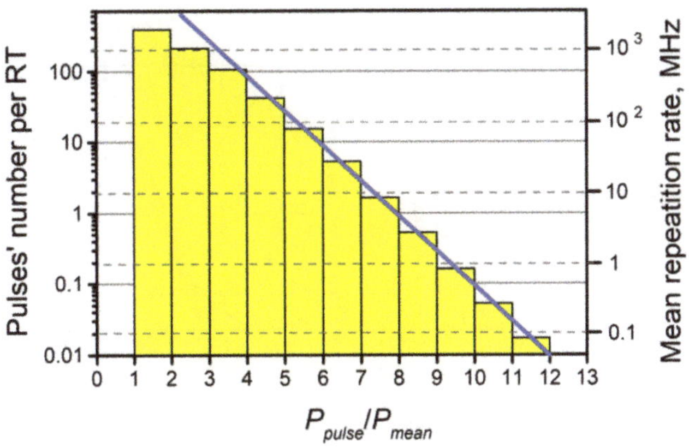

Figure 3.10. Pulse count per one RT (the left scale) and the pulses' mean repetition rate (the right scale) versus pulse power.

The mean repetition frequency of NP with low powers, from P_{mean} to $2P_{\text{mean}}$, is ~2 GHz that corresponds to approximately 400 noise pulses per each RT interval (188 ns), whereas the repetition frequency of NP at laser powers varied from $10P_{\text{mean}}$ to $11P_{\text{mean}}$ is ~90 kHz, that is, one pulse is generated per every 60 RT intervals. Bearing in mind the exponential law of the changes (see the blue line in figure 3.10), we infer that, for example, noise pulses with magnitude $20P_{\text{mean}}$ (440 W) arise for every 1.4×10^6 RT intervals, corresponding to approximately four pulses per second. Note that the strong noise pulses arising with a very low probability can be assigned to optical rogue waves [35–38] and extreme events [38, 39].

3.6 The effect of photon statistics on optical spectrum broadening

Here we discuss the effect of photon statistics on sharpening of the broadened laser spectrum through SPM, which is an important detail of the physics behind NP, as SPM originates from the nonlinear interaction of powerful noisy ASE with both active YDF and sections of passive fiber that are always present in the laser cavity.

It is known that the magnitude of spectral broadening is interrelated with the nonlinear optical chirp, induced by laser pulses via SPM, and is defined by the laser signal derivative [26]. As NP is inherent to narrowband thermal light, it is proportional to the histogram of the derivative count of an ASE pulse train. The instant magnitude of spectral broadening is found as

$$\Delta\lambda(t) = \frac{n_2 \lambda L_{\text{eff}} P_{\text{mean}}}{c A_{\text{eff}}} \frac{\Delta P(t)}{P_{\text{mean}} \Delta t} = \xi \frac{\Delta P(t)}{P_{\text{mean}} \Delta t}, \tag{3.11}$$

where $\Delta P / P_{\text{mean}}$ is the small difference in normalized laser power, Δt is the corresponding, also small, time interval, equal to the oscilloscope time resolution (in our case 10 ps), and $\xi = n_2 \lambda L_{\text{eff}} P_{\text{mean}} / (c A_{\text{eff}})$ is the proportionality coefficient.

In figure 3.11 are shown (a) the derivative of the polarized laser signal measured at 22 W and (b) the normalized histogram of the derivative count, gathered with the optical laser spectrum obtained at this power. The slope of the optical spectrum is

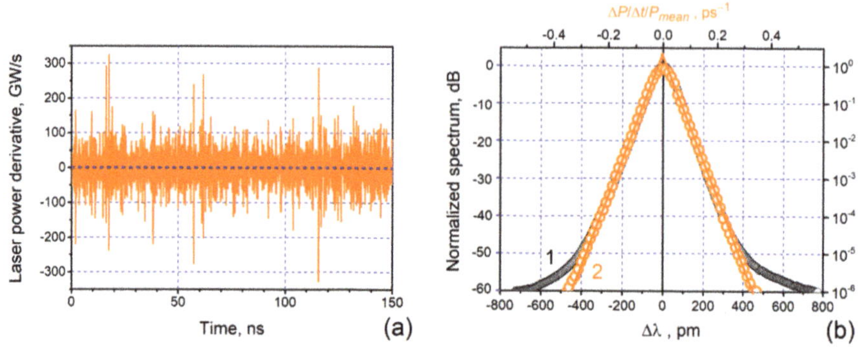

Figure 3.11. (a) Derivative of YDFL instantaneous power obtained from the oscilloscope trace measured for $P_{\rm mean}$ = 22 W and s = 1. (b) YDFL optical spectrum measured with 34 pm resolution (black circles 1, the bottom horizontal and the left vertical axes) and the histogram of the laser signal derivative (orange circles 2, the top horizontal and the right vertical axes).

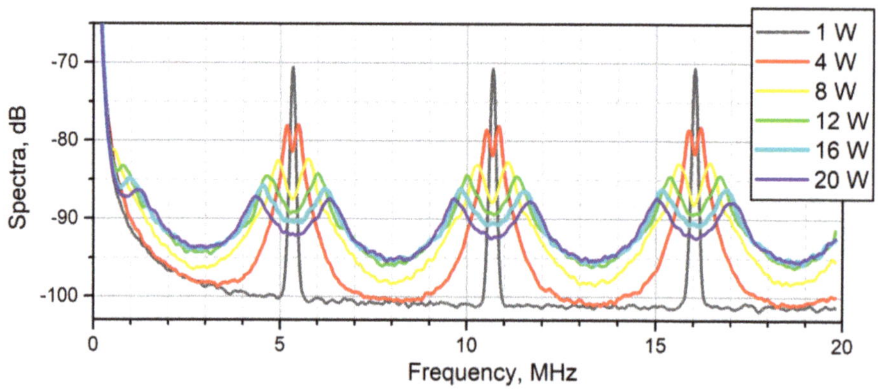

Figure 3.12. RF spectra of the polarized YDFL signal measured at different laser powers (specified in the legend).

~0.14 dB pm^{-1} and that of the histogram is ~0.19 dB ps^{-1}. These two dependences are matched well when the coefficient of proportionality ξ in equation (3.11) is ~10^3 pm ps^{-1} (in fact, it is the experimental value found as the ratio between the lower and upper horizontal axes in figure 3.11(b)). Note that the theoretical ξ value coincides with the experimental one at $L_{\rm eff}$ = 12.3 m ≈ $L_{\rm YDF}$.

3.7 Laser RF spectra

Finally, we examine the YDFL's RF spectra. The spectra are demonstrated in figure 3.12. Experimentally, we measured the RF spectra after optical attenuating of the laser power to a ~1 mW level, at which the photodetector operates in the linear regime. As can be seen from the figure the shape of the RF spectra changes dramatically with increasing laser power. At low laser power ($P_{\rm mean}$ = 1 W), one observes the narrow spectral peaks, separated by the cavity's proper frequency

$f_c = 5.32$ MHz (reciprocal RT time) and, hence, could be explained by the mode-beating (at QCW operation) or mode-locking effect. However, in our case these peaks are better explained by the presence in the YDFL of an NP pattern, repeatable every RT time (refer to figure 3.6). At laser powers equal to or above 2 W, each peak is split into a couple of sub-peaks, located symmetrically around f_c and its harmonics. With further increasing laser power these sub-peaks become progressively separated while the intensity level at f_c is simultaneously lowered, thereby demonstrating the effect of 'hole burning' in the RF spectra or, in other words, loss by the YDFL of a property such as the longitudinal modes.

Figure 3.13 demonstrates this feature in more detail. One can see that the spectral levels found at the proper frequency of the cavity (green filled circles) and, also, at the sub-peaks' maxima (blue filled squares) decay dramatically with increasing laser power, by approximately 1 and 0.8 dB/W, respectively. At the same time, the level of the valley between the sub-pulses' couples (filled orange symbols) grows by approximately 6 dB within the whole range of laser powers (see the open orange circles, showing the valley levels at the noise floor of the electronic equipment).

The peak-to-valley ratio measured as the difference between the peak magnitude at f_c and the valley level drops from ~30 to ~2 dB with increasing laser power from 1 W to 22 W; accordingly, the ratio measured as the difference between the sub-peaks' maxima and the valley level drops from ~30 down to ~8 dB. Thus one can expect that for very high laser powers (inaccessible in the current YDFL version), the RF spectrum will become almost flattened and lacking any periodicity, which is inherent to narrowband thermal light that in our case is equivalent to completely random NP.

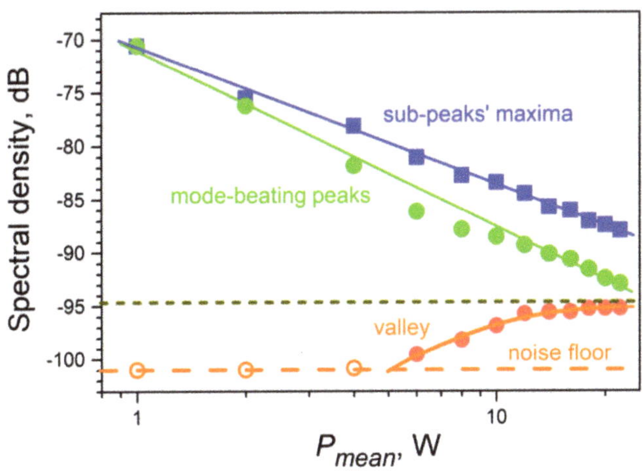

Figure 3.13. Dependence of the spectral density measured at the cavity proper frequency f_c (green symbols), the magnitude of sub-peaks (blue symbols), and the valley level (orange symbols) on laser power. The dashed line is the value of spectral density to which all shown dependencies reach at high laser power. Symbols are experimental points and solid lines are fits.

3.8 Conclusions

In this chapter the noise properties of a continuous-wave double-clad ytterbium-doped fiber laser (YDFL) with a Fabry–Perot cavity are comprehensively examined. It is shown that the laser operates in the regime of random noise pulses (i.e. in neither true CW nor quasi-CW mode) as one might expect; moreover, at the highest available laser powers it acts as a source of narrowband thermal light. It is also shown that broadening of the laser optical spectrum at relatively high laser powers occurs, basically due to self-phase modulation (SPM) of laser noise pulses. These features are confirmed by the statistical study of the laser instantaneous power and its derivative and also by the analysis of the radio-frequency spectra of the laser.

Our study of the laser signals using the electronic autocorrelation analysis shows that their coherence time drops from 2 ns at 1 W to ~70–80 ps at 22 W; the latter is in accordance with the value found from the laser linewidth (~30 pm), measured at very low laser power, not from the real spectrum that is approximately three times broader. This effect is explained by the fact that, despite the strong spectral broadening that occurs at high laser powers due to SPM, the noise pulses do not change their statistics after traveling along the laser fiber because of the very long fiber dispersion length (~250 km). Thus the widths of noise pulses, inherent to the narrowband LR-FBG spectrum, do not change along the fibers of which the laser scheme consists and so do not affect the laser coherence properties under examination.

In addition, the autocorrelation function, found for the long oscilloscope traces of the laser signal, demonstrates a series of lateral peaks, or lobes, separated from each other by the round-trip (RT) interval. As is shown, this effect is related to the periodicity of the laser coherence at delays equal to an integer number of the photon RT times: at low laser powers, this number lies within the range of tens of RT intervals, whilst at high laser powers it tends to be equal to one RT. The latter is related to the periodicity of the noise pulsing pattern, established during one RT and preserved for the power-dependent number of subsequent RTs.

Acknowledgements

The Mexican co-authors acknowledge financial assistance from the Consejo Nacional de Ciencia y Tecnología (CONACyT), Goverment of Mexico, and the Centro de Investigaciones en Óptica, A. C. The Spanish co-authors acknowledge financial assistance from the Ministerio de Ciencia e Innovación of Spain—Fondo Europeo de Desarrollo Regional (Ref.: PDI2019-104276RB-I00) and the Generalitat Valenciana of Spain (Ref.: PROMETEO/2019/048).

References

[1] Jeong Y, Sahu J K, Payne D N and Nilsson J 2004 *Opt. Express* **12** 6088–92

[2] Zheng J *et al* 2017 *Opt. Mater. Express* **7** 1259–66

[3] Kir'yanov A V, Barmenkov Y O, Lucio Martinez I, Kurkov A S and Dianov E M 2006 *Opt. Express* **14** 3981–92

[4] Gebavi H, Taccheo S, Milanese D, Monteville A, Le Goffic O, Landais D, Mechin D, Tregoat D, Cadier B and Robin T 2011 *Opt. Express* **19** 25077–83

[5] Kir'yanov A V, Paul M, Barmenkov Y O, Das S, Pal M and Escalante-Zarate L 2013 *IEEE J. Quantum Electron.* **49** 528–44

[6] Guzman Chavez A D, Kir'yanov A V, Barmenkov Y O and Il'ichev N N 2007 *Laser Phys. Lett.* **4** 734–9

[7] Jetschke S, Schwuchow A, Unger S, Leich M, Jäger M and Kirchhof J 2013 *Opt. Mater. Express* **3** 452–8

[8] Xie F, Shao C, Wang M, Zhou Q, Lou F, Xu W, Yu C, Feng S and Hu L 2018 *Opt. Express* **26** 28506–16

[9] Wang Y *et al* 2018 *J. Lightw. Technol.* **36** 3396–402

[10] Ferin A, Gapontsev V, Fomin V, Abramov A, Abramov M and Mochalov D 2012 *Proc. 15th Int. Conf. Laser Optics 2012 (St Petersburg, Russia)* pp 26–8 paper TuSY1-1.4

[11] Shcherbakov E A, Fomin V V, Abramov A A, Ferin A A, Mochalov D V and Gapontsev V P 2013 *Proc. Advanced Solid-State Lasers Congress Technical Digest (San Diego, CA, USA)* paper ATh4A.2

[12] Zervas M N 2014 *Int. J. Modern Phys.* B **28** 1442009

[13] KablukovS I, Zlobina E A, Podivilov E V and Babin S A 2012 *Opt. Lett.* **37** 2508–10

[14] Ying H, Yu Y, Cao J, Huang Z, Pan Z, Wang Z and Chen J 2017 *Laser Phys. Lett.* **14** 065102

[15] Li Q, Zhang H, Shen X, Hao H and Gong M 2016 *J. Opt.* **18** 115503

[16] Bednyakova A E, Gorbunov O A, Politko M O, Kablukov S I, Smirnov S V, Churkin D V, Fedoruk M P and Babin S A 2013 *Opt. Express* **21** 8177–82

[17] Muniz-Cánovas P, Barmenkov Y O, Kir'yanov A V, Cruz J L and Andrés M V 2019 *Opt. Express* **27** 8520–8

[18] Muniz-Cánovas P, Barmenkov Y O, Kir'yanov A V, Cruz J L and Andrés M V 2019 *Sci. Rep.* **9** 1307

[19] Minguela-Gallardo J A, Barmenkov Y O, Kir'yanov A V and Beltrán-Pérez G 2017 *J. Opt. Soc. Am.* B **34** 1407–14

[20] Reddy P H, Kir'yanov A V, Dhar A, Das S, Dutta D, Pal M, Barmenkov Y O, Minguella-Gallardo J A, Bhadra S K and Paul M C 2017 *Appl. Opt.* **56** 9315–24

[21] Mussot A, Kudlinski A, Kolobov M, Louvergneaux E, Douay M and Taki M 2009 *Opt. Express* **17** 17010–15

[22] Larsen C, Noordegraaf D, Skovgaard P M W, Hansen K P, Mattsson K E and Bang O 2011 *Opt. Express* **19** 14883–9

[23] Hejaz K, Shayganmanesh M, Roohforouz A, Rezaei-Nasirabad R, Heidariazar A, Vatani V and Hassan-Nabavi S 2016 *Laser Phys. Lett.* 13035107

[24] Delgado-Pinar M, Villegas I L, Díez A, Cruz J L and Andrés M V 2014 *Opt. Lett.* **39** 6277–80

[25] Roselló-Mechó X, Delgado-Pinar M, Cruz J L, Díez A and Andrés M V 2018 *Opt. Lett.* **43** 2897–900

[26] Agrawal G P 2001 *Nonlinear Fiber Optics* 3rd edn (San Diego: Academic) pp 98–100

[27] Cimek J, Liaros N, Couris S, Stępień R, Klimczak M and Buczyński R 2017 *Opt. Mater. Express* **73** 471–83

[28] Yariv A 1988 *Quantum Electronics* 3rd edn (New York: Wiley) pp 183–6

[29] Diels J-C and Rudolph W 2006 *Ultrashort Laser Pulse Phenomena* 2nd edn (Amsterdam: Elsevier) pp 457–86

[30] Kuusela T A 2017 *Am. J. Phys.* **85** 289–94
[31] Loudon R 2000 *The Quantum Theory of Light* 3rd edn (New York: Oxford University Press) pp 82–124
[32] Goodman J W 2000 *Statistical Optics* (New York: Wiley) pp 239–56
[33] ten Bosch J J and de Voigt M J A 1966 *Am. J. Phys.* **34** 479–82
[34] Pietralunga S M, Martelli P and Martinelli M 2003 *Opt. Lett.* **28** 152–54
[35] Solli D R, Ropers C, Koonath P and Jalali B 2007 *Nature* **450** 1054–7
[36] Akhmediev N and Pelinovsky E 2010 *Eur. Phys. J.* **18** 51–4
[37] Lecaplain C and Grelu P 2014 *Phys. Rev.* A **90** 013805
[38] Akhmediev N *et al* 2016 *J. Opt.* **18** 063001
[39] Smirnov S V, Sugavanam S, Gorbunov O A and Churkin D V 2017 *Opt. Express* **25** 23122–27

Chapter 4

Development of borosilicate optical fibres for particle ionizing radiation dosimetry

H T Zubair, Amal Alqahtani, S F Abdul Sani, Hairul A Abdul-Rashid and D A Bradley

Advanced radiation applications require detectors of high sensitivity and often also of high spatial resolution, and thermoluminescent dosimeters (TLDs) offer a particular solution. Borosilicate glass, which is inexpensive, readily available, suitable for photonic fibres, moisture resistant, and reusable, is investigated here for the first time. The results show that fading is minimal, making this material competitive with commercial LiF TLDs. In the use of megavolt linacs, dose linearity is seen to range from sub-Gy through to MGy. The multiple peaks of glow curve deconvolution relate to silicon oxygen defect centres (Si-ODC), B–O centres, and B structural defects. The activation energies of less than 1.2 eV relate to O^+ and O^- in the silica network. Borosilicate glass is also neutron sensitive. Photoneutron contamination, absent at 6 MV, is appreciable at 10 MV and above. The observed markedly dose-dependent non-linearity is suggested to arise from the $^{10}B(n,\alpha)^7Li$ reaction, with the alpha irradiation creating dose-dependent defect damage in the glass. This chapter describes the development of borosilicate optical fibres for effective passive dosimetry, with neutron detection capabilities.

4.1 Introduction

Radiation dosimetry in health physics and radiation protection refers to the measurement and assessment of the ionizing radiation dose to the human body. This chapter addresses what promises to be a highly effective yet economic means of dosimetry, both for radiotherapy applications and also for product irradiation dose verification. Borosilicate, of which Pyrex is an example, offers a considerable range of characteristics in support of thermoluminescence dosimetry (TLD). Hitherto remaining unexplored for such a purpose, the field has yet to benefit from the commercial borosilicate forms for which current investigations show promise. The potential arises in large part from the fact that borosilicate, in forms such as Pyrex,

doi:10.1088/978-0-7503-3243-9ch4

is widely available and relatively cost effective compared to other types of radiation detectors. At a basic level the utility of borosilicate is based on its amorphous constitution and its ability to withstand the temperatures of everyday laboratory and domestic applications, but with softening at temperatures of around 700 °C, allowing ease of fabrication into a great variety of shapes, some familiar to us in the form of heat resistant laboratory and kitchenware. The particulars of its TL behaviour arise from the mix of elements in the medium (boron included), producing a high level of defects, not only making it quite cheap in many circumstances but also of high potential utility.

A high level of defects in a high-temperature resistant, highly transparent, and colourless insulating medium offers exactly what is needed in providing for a sensitive means of TLD at the levels of interest herein. As an introduction, TLD is a passive form of dosimetry, in other words not offering live-time analysis but instead providing post-irradiation analysis in a two stage process, namely irradiation and subsequent readout (discussed further below). The TL sensitivity of borosilicates, widely available in convenient millimetre to sub-millimetre commercial forms, renders the medium of particular interest for radiotherapy applications, with the spatial resolution needed to define with precision the dose received at a specific point (not least in small field dosimetry). Specifically, it has been recognized that optical fibres and microscope slides are two particular forms of borosilicate glass that are widely available. That the glass is colourless and of high optical quality is a matter of considerable importance, ensuring the avoidance of spectral filtration. Seemingly of little consequence, these excellent optical properties nevertheless point to the reproducibility of TLD performance, a macroscopic expression of homogeneity. The question of whether this translates to reproducibility of TL yield at the millimetre to sub-millimetre scale is one of the issues dealt with in the present work. One further factor in favour of borosilicate glass as a dosimetric medium is that its TL yield is sufficiently similar to that of commercial TLDs to allow continued use of existing TLD readers, making it particularly cost effective.

Prior investigations by other reseachers have shown glass in the form of optical fibres and beads to be an effective medium for radiotherapy TL dosimetry, and also for radiation sterilization and processing dosimetry [1–3]. Here it should be mentioned that a radiation dose of 1 Gy refers to an energy absorption to a receiving medium of 1 J kg^{-1}. Associated with the mode of doping is a recognized potential for inhomogeneity, arising in part from the rapid means used in achieving bulk quantity production. Accordingly, a process of initial screening is required, seeking reproducible sensitivity by selecting for use those samples that give rise to a uniform response for a fixed dose.

The fact that borosilicates incorporate boron at a concentration of some several percent (weight percentage) also points to considerable potential for their use in neutron dose evaluation, an issue of importance in accurate radiotherapy dosimetry for both x-ray and electron beam irradiation at around 10 MV and above. Boron-based proportional counters and boron-doped liquid or glass scintillators have already been established as effective neutron detecting media. These real-time detectors are enriched in ^{10}B, providing a large reaction cross-section for neutrons [4–7]. A challenge in the

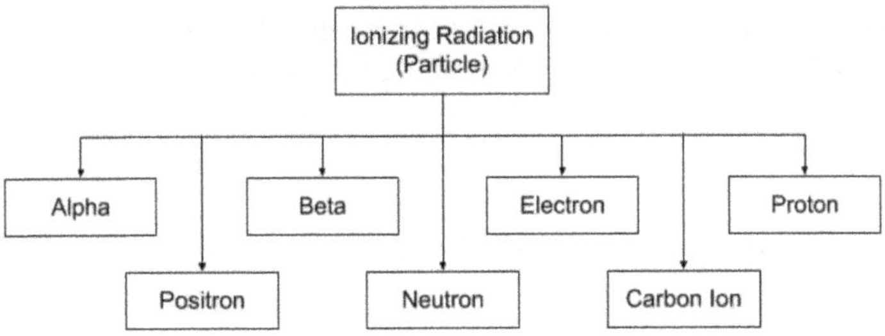

Figure 4.1. The various forms of particulate ionizing radiation.

detection of neutrons is their inherent charge-free nature. As a reminder of the various forms of particulate ionizing radiation, figure 4.1 shows those providing the main coverage. Protons, alpha, carbon, etc, are collectively referred to as heavy ions (the proton is some 2000 times the mass of the electron). Neutrons (also some 2000 times the mass of the electron) are uncharged, giving rise to indirectly ionizing events, for example nuclear reactions.

While in this diagram no hierarchy of energy impartation is intended to be shown, it is nevertheless true that marked differences do arise in how ionization occurs, matters beyond the scope of the current chapter.

The enhanced TL yield of extrinsically doped glass, such as that typically found in the use of photonic fibres, is recognized to be of considerable utility, foremost in the low dose (sub-Gy level) regime where an abundance of defects can be expected to provide an acceptable signal-to-noise ratio. However, excess doping can also restrict the range of doses the dosimeter can accommodate (limiting the so-called dynamic range over which linearity between the TL yield and dose is obtained). Seemingly counter-intuitively, increased sensitivity with dopant concentration is not a reliable expectation, as the issue arises of self-absorption (re-trapping) suppressing TL yield beyond certain limits. Sensitivity can restrict application, in particular in the upper and lower range of dose, matters expounded in greater detail in previous reviews [8, 9]. For borosilicate glass microscope coverslips, restriction at the upper end has been found to be practically absent, with linearity observed over the range from a few kGy to a few hundred kGy, the regime of sterilization and product irradiation.

These opening statements have been aimed at positioning this work within the context of requirements, promise and potential performance, utility, challenges, and scope for exploitation. Accordingly, this chapter is organized into six sections, the first introducing borosilicate media as a viable solution to a number of current challenges and opportunities in passive radiation dosimetry, in particular focusing on heavy particle detections. Section 4.2 describes the physical characterization of borosilicate media and the fabrication techniques for optical fibre as an important structural derivative. Section 4.3 discusses findings from recent radiation sensitivity studies of borosilicate media. Section 4.4 concerns neutron irradiation, use being made of photon beam energies generated from 6 MV through to 15 MV. The particular challenge relates to accounting for closed collimator high-energy irradiations, with beam transmission and photon scattering playing contributory roles.

Section 4.5 presents TL studies of germano-borosilicate flat fibres (FF) subjected to a variety of radiation sources. Section 4.6 concludes with recommendations for further research, focusing on novel borosilicate optical fibre structures for passive radiation dosimetry.

4.2 The synthesis and physical characteristics of borosilicate media

In this work we look at several forms of borosilicate glass, all offering potential utility as radiation sensors, some thin but otherwise of large area (e.g. microscope cover-slips), and some in sub-millimetre fibre form as in commercially available photonic systems or co-doped silica (e.g. GeB). Examples of the utility of borosilicate glass as a radiation sensor is provided in sections 4.4 and 4.5.

4.2.1 The melt quenching technique

The three major broad categories of glass are soda-lime glass (71%–75% silica, 12%–16% sodium oxide, and 10%–15% calcium oxide by weight), lead glass (54%–65% silica, 18%–38% lead oxide, and 13%–15% sodium oxide or potassium oxide by weight), and borosilicate glass, comprising 95% of commercially produced glass. The main glass-forming constituents in borosilicate glass are silica (> 60 wt.%) and boric oxide (> 8 wt.%). For common applications, such as laboratory glassware, cookware, and optical apparatus, borosilicate glass often contains 2–8 wt.% of both alkali and alumina [10], forming a network called alkali aluminosilicate glass. This enables borosilicate glasses to have very low thermal expansion coefficients (some 3×10^{-6} K^{-1} at 20 °C) allowing greater resistance to thermal shock compared other types of glass. Borosilicate glass is synthesized by melting together finely crushed boric oxide, silica sand, soda ash, and alumina at a temperature > 1300 °C, and rapidly cooling down the homogeneous mixture in a process known as melt quenching. Structurally, borosilicate glasses are amorphous, with a variety of networks of borates ($B_3O_6^{3-}$ ring-type groups; BO^{3-} triangles in chain; triborate groups with one non-bridging oxygen, etc) and silicates (SiO_4^- units with one non-bridging oxygen ion, layered with six SiO_4 tetrahedra in a ring; SiO_4^{2-} units with two non-bridging oxygen ions, chains of tetrahedra, etc) depending on the composition of the material. As the proportion of silica increases (> 35 mol%), the tendency of formation of non-bridging oxygen becomes stronger. Many of these networks become precursors to defect centres that form during synthesis. These defect centres enable the glass media to capture electrons and holes in different intermediate energy bands as electron–hole pairs are generated during exposure to ionizing radiation. The electron–hole pairs recombine upon thermal (or optical) stimulation producing thermoluminescence (TL), a process illustrated in figure 4.2.

When ionizing radiation is absorbed in TL materials, electron–hole pairs are generated. Electrons are excited to the conduction band from the valence band. Such electrons may be trapped in the lattice defects or impurities, while others promptly recombine, producing scintillation or radioluminescence. Free holes that travel through the valence band may also become trapped. Many hole centres are thermally unstable and can rapidly decay at normal room temperature. The trapped

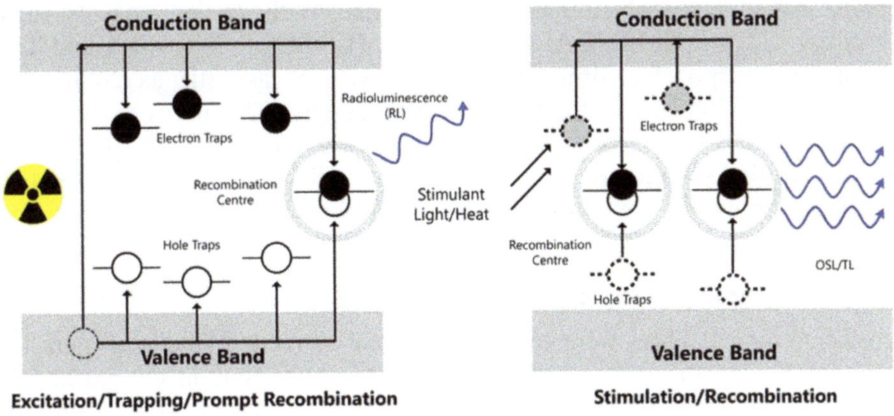

Figure 4.2. A simple energy band model for the TL process.

electrons remain trapped until sufficient energy is provided to escape, the amount of trapped electrons and luminescence intensity being directly related to the applied radiation dose, limited by trap filling saturation.

TL in glass was already being heavily documented since the early–mid 1900s (coinciding with the development of photodetectors). Early studies primarily utilized TL in investigating the colouration of glass. Nyswander and Cohn [11] describe an early apparatus comprising a polarizing photometer to detect TL in doped zinc borate glass (45.5% zinc oxide to 54.5% boric acid) having been exposed to natural sunlight for a prolonged period. Certain specimens, such as thorium doped glasses, produce TL from as little as twenty minutes' exposure. Northup and Lee [12] document an extensive experimental plan investigating TL in 112 mineral samples (32 species) collected from all over America. By the 1960s, as evident in the study of Spurný and Novotný [13], TL in glass was investigated in light of the electronic band theory.

Recent investigations have incorporated rare-earth (RE) ions during the synthesis of borosilicate glass to increase TL sensitivity. During the usual melt quenching technique, samarium (III) oxide in various concentrations (0.2, 0.5, 1, 2, 3, and 5 wt.% of Sm_2O_3) is added to alumina borosilicate glass with a composition of 59.13% SiO_2, 6.38% Al_2O_3, 18.24% B_2O_3, 12.82% Na_2O, and 3.5% ZrO_2 in wt.%. The constituents are first crushed into powder, homogeneously mixed (thus evenly distributing the Sm_2O_3 across the glass mixture), and placed in a platinum crucible. The crucible is gradually heated to 1500 °C in an electric furnace over 14 h, followed by a melting process at the same temperature for a further 2 h. The melt is then poured on a copper plate for quenching and further annealed to 500 °C for removing mechanical or internal strains [14]. X-ray diffraction techniques are used to verify the amorphous nature of the mixture [15]. A similar process can be repeated with the other RE reagents Gd_2O_3, Eu_2O_3, CeO_2, and NdO_2. Depending on the reagent used, the heating temperature and duration may need to be varied between 750 °C and 1500 °C and 10 to 14 h, respectively [16]. The melting time may be reduced to below 1 h if all the reagents are already obtained in the same physical form. Terbium (3+) doped alkali alumina borate glass ($35CaO–15Al_2O_3–50B_2O_3{:}x\mathrm{Tb}$) has been prepared using the same melt quenching technique. With starting reagents, such as high-purity $CaCO_3$ (99.99%), Al_2O_3 (99.99%,), $H_3{}^{10}BO_3$ ($^{10}B > 96\%$), and Tb_4O_7 (99.9%), in fine powder form, the mixture requires as little as 20 min at a temperature of 1100 °C to attain a homogeneous mix. This needs to be followed by annealing at 500 °C, pressurization, and quenching [17]. The glass samples prepared in this technique require polishing with silicon carbide sandpaper (grain size ~10 μm) to ensure a uniform and smooth surface and edges.

4.2.2 The vapour deposition technique

Borosilicate optical fibres were earlier motivated by related studies of pure fused silica core, having the lowest optical attenuation at the time in the red and near infrared portion of the spectrum [18]. Borosilicate glass became a promising cladding material for pure fused silica core fibres, with a slightly lower refractive index that allowed total internal reflection in such optical fibres. Optical waveguides consisting of a pure fused silica core and borosilicate cladding were produced using chemical vapour deposition techniques [19]. In 2008 commercially available borosilicate glass substrates were reported to be used in rod and tube stacking arrangements for photonic crystal fibre fabrication [20].

The modified chemical vapour deposition (MCVD) technique is used to produce optical fibres in a high-temperature environment using $GeCl_4$ and $SiCl_4$ as the precursor [21]. A schematic of an MCVD system is illustrated in figure 4.3, comprising the gas and material delivery systems and the lathe. The gas deposition system is used to transport the glass-forming reagent to the reaction zone. Glass-forming reagents such as $SiCl_4$, $GeCl_4$, $POCl_3$, and BBr_3 are typically used in MCVD systems. A high-purity silica substrate tube is placed on the lathe and heated to the reaction temperature, connected to the bubbler system. The lathe rotates to ensure a uniform deposition on the inner wall of the tube. The burner moves along the tube axis to allow deposition along its length [9].

Once the preforms are fabricated, they are placed in a fibre drawing tower, one at a time, and pulled into an optical fibre diameter. The preform being pulled is first loaded into the top-feed of a two-tier pulling tower. The feed lowers the preform to the furnace in the floor of the top tier to begin the heating process. At about ~2000 °C the lower end of the preform begins to melt, forming a droplet, which falls under

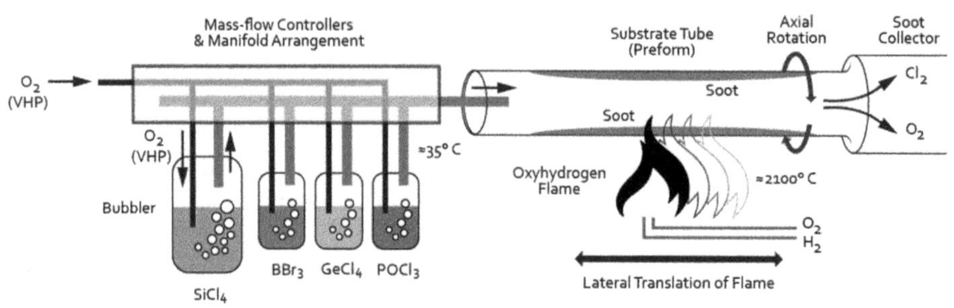

Figure 4.3. A schematic diagram of an MCVD system.

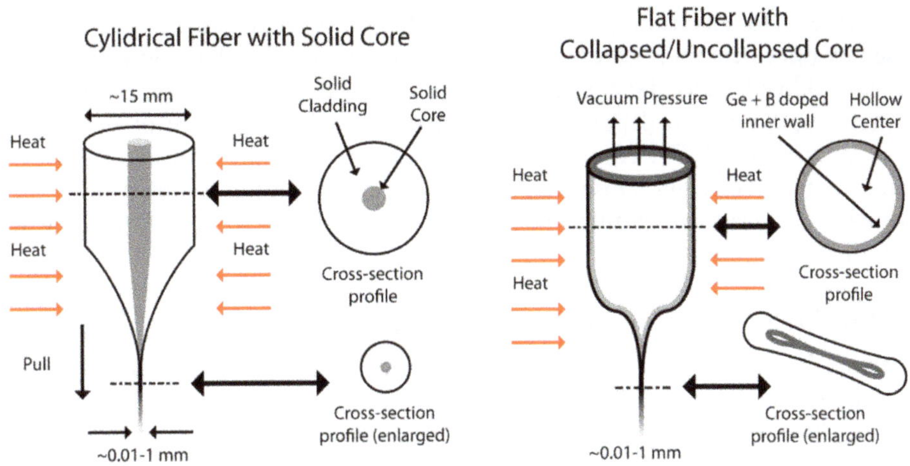

Figure 4.4. Drawing of optical fibre from a preform.

gravity forming a thin thread behind it, illustrated in figure 4.4. A cooling chamber maintains the temperature of the formed thread at room temperature. The droplet is detached from the thread and the thread is passed through a capstan. The rotating speed of the capstan determines the diameter of the fibre. The fibre diameter is monitored continuously to adjust the speed of the capstan and maintain the fibre diameter. Once a sufficient length of optical fibres is obtained (about a few tens of metres), the melting process is stopped.

During drawing, when the vacuum pressure is applied inside a hollow preform, the air inside the hollow preform is sucked out. This causes the pressure inside the hollow preform to drop compared to the atmospheric pressure outside the preform. Inside the neck-down region, a fibre that is pulled down becomes collapsed or flattened due to this low pressure. This produces the optical fibre structure with a planar profile or cross-section, known as a flat fibre. Specialty optical fibres have also been fabricated using the MCVD technique, including hollow cylindrical fibres, collapsed fibres, photonic crystal fibres, and flat fibres, a detailed discussion being available in Ghomeishi *et al* [22]. Dambul *et al* [23] detail a fabrication process of a Ge- and B-doped silica flat optical fibre. A simple arrangement for pulling a Ge- and B-doped flat fibre is shown in figure 4.4. The process starts with depositing Ge and B dopants using the MCVD technique. Unlike a solid core cylindrical fibre, the centre of a flat fibre preform is left hollow. During pulling, vacuum pressure is applied to this hollow region, which contracts the walls of the fibre to form the planar structure. Depending on the pressure, feed speed, drawing speed, and temperature, remnants of the hollow region may be preserved as two small air holes towards the extreme edges of the fibre cross-section, referred as uncollapsed core.

4.2.3 Heavy metal coating

An additional process of coating optical fibre samples with heavy metal ions, such as with platinum and gold, has been demonstrated to be effective for dose enhancements for electromagnetic ionizing radiation sources (gamma or x-rays). It is caused by photo-electron generation in the heave metal layer that contributes to the locally modified secondary electron dose. The laboratory method involves a sputtering technique to deposit Pt or Au particles on the surface of optical fibre samples (strands of 0.5–1 cm with a diameter of a few hundreds of micrometres) to form an external coating. Dose enhancement factors (DEF) of $0.0160 \pm 0.001\ \text{nm}^{-1}$ have been achieved for coating thicknesses of up to 80 nm in optical fibres, with a total DEF of ~1.30 [24, 25].

4.3 Outlook for borosilicates for radiation dosimetry

In this section a brief review will first be provided of some of the previous photonic fibre neutron and alpha-particle irradiation studies conducted by members of this group, subsequently proceeding to GeB fibres. While most of our work on doped glass has been carried out using silica fibres for photons and electrons, members of this group did make relatively early studies of neutron, proton, and alpha irradiated responses. The majority of such studies were on silica fibres of core diameter $125.0 \pm 0.1\ \mu$m, manufactured by INOCORP (Canada). Fast-neutron studies were performed by Hashim *et al* [26], investigating the TL response of Al- and Ge-doped fibres subjected to ^{241}AmBe neutrons in a fixed geometry, with the results showing proportionality between the duration of irradiation and the response. Outside the resonance region the cross-sections for ^{70}Ge are markedly greater than that for ^{28}Si or ^{27}Al, with total neutron absorption cross-sections of 2.18, 0.16, and 0.23 b, respectively [27]. A comparison of the thermal neutron absorption cross-section of various elemental constituents of optical fibres are presented in table 4.2. While these cross-sections are minor (two orders of magnitude less for ^{70}Ge, and three orders of magnitude less for ^{28}Si and ^{27}Al) compared to the boron-10 (^{10}B) thermal neutron cross-sections to be discussed later in this chapter, the results nevertheless showed the particular fibres to be neutron sensitive. For linacs operating in the 6 and 15 MV photon mode, Noor *et al* [28] investigated the use of Ge-doped optical fibre for *in vitro* IMRT prostate dosimetry, good agreement being found between the calculated and measured photon dose values for both 6 and 15 MV irradiations. While the thermal neutrons provided an insignificant contribution to the in-field total dose, conversely for out-of-field doses TLD-100 and Ge-doped fibres were

Table 4.1. Elemental composition of GeB-doped flat silica optical fibre fabricated in Dambul *et al* [23].

Element	Weight%
Si	54.98
O	38.45
Ge	0.34
B	6.23
Total	100

Table 4.2. Thermal neutron cross-section for common optical fibre constituent elements.

Element	Thermal neutron cross-section (barn)
Si	0.17100
O	0.00019
Cl	35.5000
H	0.33260
Ge	2.18000
P	0.17200
B	767.000
Al	0.23300
F	0.00960
Ca	0.43000
Ce	0.60000
Yb	35.0000
Sc	27.2000

shown to be sensitive to the thermal neutron component for an incident 15 MV nominal beam energy. Compared to calculated dose values, the average discrepancy was found to be 2.0%, 2.1%, and 2.5% for Ge-doped optical fibres, TLD-700, and TLD-100, respectively.

With respect to alpha-particle irradiations and TL yield, reference is first made to the nascent doped SiO_2 optical fibres studies of Ramli *et al* [29], as context acknowledging that heavy ion sources and ion beams are now of particular vogue in radiotherapy development, with the ability to deliver a more precise dose to the target volume while minimizing the dose to the surrounding healthy tissue. This particular study investigated the TL characteristics of variously doped SiO_2 optical fibres irradiated with ^{241}Am alpha particles, favourable comparison being made against that of TLD-100 rods. More recently, Bradley *et al* [30] studied the TL response of tailor-made sub-millimetre doped silica fibres exposed to the α-emitter $^{223}RaCl_2$. This is used in the palliative treatment of bone metastases, benefitting from a high linear energy transfer (LET) and associated short path length (<100 μm) of the α-particles. Various forms of Ge-doped SiO_2 fibres were used, including photonic crystal fibre (PCF) collapsed, PCF uncollapsed, flat, and single-mode fibres.

With regard to borosilicates, the work herein is intended to indicate high potential for the use of fibres made from this medium, most particularly with respect to neutron irradiation. It has already been established that the medium works well as a dosimeter of photons and electrons [31, 32], also showing linearity of response from sub-Gy to several hundred kGy. The thermoluminescence glow curve of borosilicate microscope coverslips (under ^{60}Co gamma-ray irradiation, delivering a dose rate of 1.61 Gy min^{-1}) is presented in figure 4.5, marking the glow curve peak at a significantly lower temperature, at some 150 °C to 175 °C for 0.13 mm thick coverslips, compared to the 200 °C–220 °C range for that of 1.00 mm thick samples.

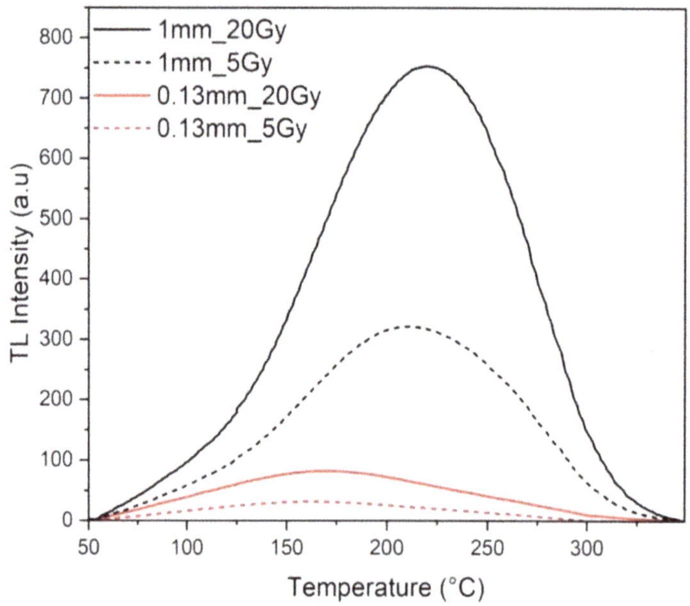

Figure 4.5. Glow curves for 0.13 and 1.00 mm coverslips, extracted from measured data using ^{60}Co irradiations. Reproduced with permission from [76]. Copyright 2020 Elsevier.

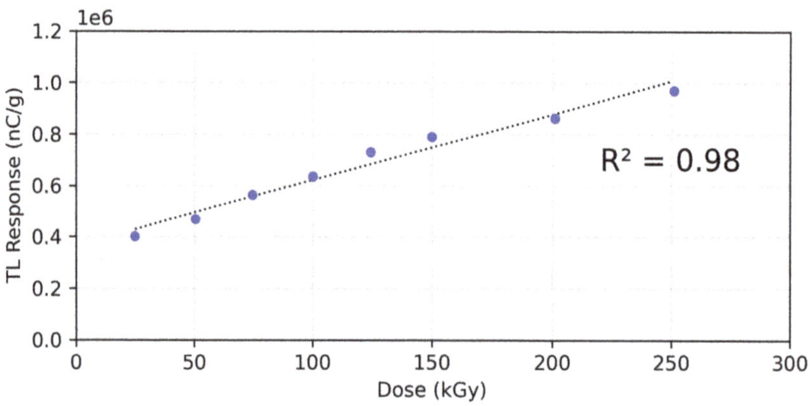

Figure 4.6. TL response of the coverslips for 1 MeV electron doses from 25 to 250 kGy.

This is suggestive of the changes in thermoluminescence characteristics that come about with variation of sample geometry. Figure 4.6 shows the borosilicate microscope slides responding linearly to electron beam industrial product irradiation, up to at least 250 kGy. Compared to many other forms of passive dosimetry, the result is dramatic, with no practical indication of saturation. The interpretation is that defect density is such that up to the investigated dose level full occupancy has not been approached. This is sufficient to serve most product irradiation needs. While clinical linacs are not suited to delivering doses of much beyond 0.1 kGy (in the usual set-up the typical dose rate for the clinical accelerator is some 2 Gy min^{-1}), with electron irradiators such as that used herein, the dose-rates are enormous, including use of the current dose rate of 1.3 kGy s^{-1}.

4.4 Neutron detection with borosilicate glass

4.4.1 Motivation

Pure and applied neutron physics encompass a great many areas of endeavour (including from space travel through to medical and security scanning applications). Herein we confine the research question to whether borosilicate glass can be effective in sensing neutron emission, even in the presence of a much larger fluence of photons. With this interest in mind, we have investigated the possibilities for neutrons delivered by clinical accelerators operating at several megavolts. As an example 15 MV operations give rise to an estimated x-ray to neutron ratio of 10^5 [33]. Our work to date concerns proof of capability rather than seeking to obtain full calibration of the glass. The latter would need to accommodate a range of complexities, beyond the present scope, including use of Bonner spheres for neutron moderation and access to a secondary standards dosimetry lab (SSDL) with neutron capabilities. The focus of investigation has been neutron contamination to the dose delivered to the patient from use of linacs operating in the photon mode. In the electron mode neutron production is suggested to be at least two orders of magnitude below that obtained in the photon mode [34].

For the energies of interest the principal mode of neutron production is photoneutron generation, which is exothermic and thus with a photoneutron threshold. For linacs working in the photon mode, the lowest bound for photoneutron production has been suggested to be some 6 MV [35, 36]; 7 to 10 MV is more commonly quoted, the threshold depending on the materials involved [33, 37]. Exceptions include the target nuclides and corresponding reactions $^{2}H(\gamma,n)^{1}H$ and $^{9}Be(\gamma,n)^{8}Be$, with respective photoneutron reaction thresholds of 2.226 and 1.666 MeV, representing the lowest neutron separation energy of all nuclides. This said, deuterium and beryllium are not typical constituents of borosilicate glass and if present would almost certainly not exceed ultra-trace levels (ppb), accordingly challenging the detection limits of routine elemental analysis methods (as an example, wavelength-dispersive x-ray spectroscopy). ^{13}C is yet another exception, the associated photoneutron production threshold being 4.9 MeV for the reaction $^{13}C(\gamma,n)^{12}C$. The presence of natural carbon in the current borosilicate samples is 1.3%. Added to this is that the abundance of ^{13}C in natural carbon is just 1.109%, the reaction $^{13}C(\gamma,n)^{12}C$ accordingly being of little consequence.

4.4.2 The case for borosilicate glass as a potential neutron detection media

The use of borosilicate glass to sense neutrons emitted from the linac head of a clinical accelerator is a matter of particular interest, as current MV phototherapy makes no account of the neutron contribution to patient dose. Of note is that for neutrons around 1 MeV the biological damage is some 20 times that of x-rays. Of course in all such considerations the low cost of borosilicate glass and its availability, homogeneity, and potential for high spatial resolution studies all point to particular potential for radiation therapy. Conversely, in surveys of dose from linac-based product irradiators the neutron component is very much less interesting; such irradiators typically operate in the direct electron mode and generally provide beams of some 1 to 2 MeV.

Thermalized neutron interactions are a further important factor in providing for detection of the photoneutron component, the mean free path (mfp) of fast neutrons being some few centimetres in low Z materials, effectively yielding a thermalized field. At thermal energies the $^{10}B(n,\alpha)^{7}Li$ interaction cross-section can achieve values as large as 3837 b. Particular use of this is made in various forms of neutron detector. With boron in borosilicate glass at the several percent level and with ^{10}B making up 19.8% (^{11}B accounting for the remaining 80.2%) the detection capability of this medium is of particular interest. While many existing detectors are competitive, for example with TLD-100 being similarly passive, there is a considered potential for borosilicate glass applied to in- and out-of-field collimated beam neutron surveys.

Turning to active detectors, particular difficulties arise in neutron monitoring in and around the primary beam [34]. First, given their bulk, they are very much less able to offer point-by-point mapping capability across a beam. Furthermore, photon interference and the intensity of the photon pulse will typically overwhelm such detectors. In such circumstances it is therefore common to only make use of passive detectors. Conversely, in conducting measurements in outer maze areas, the passive devices are of very much more limited utility, requiring lengthy duration measurements given their relatively low sensitivity. Finally, given that neutron detection needs are spread over many decades of energy (from thermal energy, $k_T = 0.025$ eV, through to several tens of MeV), accordingly no single detector is able to accurately measure the fluence or dose equivalent over the entire range.

Of the existing passive neutron detectors, the most common of these include TLDs with lithium additives (^{6}Li and ^{7}Li), Brunckhorst *et al* [38] detailing the use of these for photoneutron detection. Other passive forms are bubble detectors, solid-state track detectors, and activation foils (gold and indium for thermal neutrons). Varying detection capabilities are available, leading to particular choices for use either in the primary beam or outside the immediate treatment area (e.g. within the maze) or both in some cases. Of the active types, one can mention moderated boron trifluoride (BF_3) detectors and Rem-meters (both used predominantly for outer maze surveys and outside the room). The BF_3 fluence detecting proportional counter depends on the $^{10}B\ (n_{thermal}, \alpha)^{7}Li$ reaction, as already mentioned above, with a threshold of 2.31 MeV and a cross-section σ of 3837 b. The α and recoil ^{7}Li nucleus produce large pulses, orders of magnitude greater than the photon pulse, accordingly with good photon rejection capability. A comprehensive account of the various neutron detectors has been provided by Keehan [39].

4.4.3 Photoneutron cross-section

With the patient 100 cm from the photoneutron source (accelerator head components), the incident photoneutrons are typically fast, with mfp in air > 100 m for

2 MeV neutrons [40]. Reactions such as $^{10}O(n,\alpha)^{13}C$ can occur in borosilicate glass, the fast-neutron cross-sections of these being small compared to that for $^{10}B(n,\gamma)^{9}B$ photoneutrons. The photonuclear reaction cross-sections exhibit giant dipole resonances (GDR), photon absorption being essentially the converse of the electric dipole gamma-decay process. In ^{208}Pb these resonances occur at a photon energy of 15 MeV, the resonance energy being proportional to $1/R^{1/2}$, also $1/A^{1/6}$, with R the radius of the nucleus and A its mass number [41]. As a result of absorbed photons, these vibrational resonances within the nucleus are sufficient to excite the nucleus, causing emission of neutrons, protons, etc.

4.4.4 Photoneutron production

Since photoneutrons are produced by above-threshold photons interacting with linac components (target, collimator, beam flattening filter, shielding, etc), the chief materials are chosen to be Cu, Fe, W, Au, and Pb, with minimal neutron attenuation occurring in these. The maximum number of neutrons is produced inside the head when the collimators and jaws are completely closed, Brunckhorst [33] providing an indication of this. For high atomic number nuclei the threshold is around 8 MeV. For even–even low atomic number nuclides the threshold is greater (16 MeV for oxygen and 18 MeV for carbon) as shown by Brunckhorst [33]. In photoneutron production two processes occur [34]. The first is direct strongly forward emission, with a $\sin^2\theta$ angular distribution and an average neutron energy of a few MeV, peaking at energies > 2 MeV, contributing some 10% to 20% of the neutron yield from bremsstrahlung photons. The second is evaporation neutrons, dominant for heavy nuclei and isotropically emitted, with an average energy of the order of 1–2 MeV and with the spectra peaking at the order of 200–700 keV. The overall situation is near isotropic neutron emission, penetrating the linac head in all directions [34].

4.4.5 Neutron detection factors

As previously mentioned, the boron content comprises ^{10}B (19.8%) and ^{11}B (80.2%). The incident fast-neutron cross-section for $^{10}B(n,\alpha)^{7}Li$ has been discussed [42]. Howell *et al* [43] have shown secondary neutron spectrum measurements for three types of 15 MV accelerator, the cross-section increasing between 200 keV and 1 MeV. Important in interpreting the measurements is that the photon beam penetration through the collimators and jaws is expected to be less than 1% [44].

4.4.6 Materials and methods

The work presented here used an Elekta Infinity™ accelerator at the Oncology Centre, Medical City, Riyadh. The borosilicate detectors were cut from microscope coverslips that were 1.0 mm thick, 0.6 × 0.6 cm^2 segments fitting within the TLD planchette. The detectors were placed at 3 cm depth within a 30 × 30 cm Perspex phantom formed of 1 cm thick slabs, with 5 cm of Perspex below the detector. The surface of the Perspex phantom was fixed at an SSD of 100 cm. Doses from 50 to 1000 cGy were used, defined for photons, the neutron contribution being unknown.

Out-of-field measurements were taken 2 cm outside the edge of the collimated beam, the penumbra accounting for some 1% to 3% of the central-axis dose at this location [44]. Evaluations at 3 cm depth equate with the point of maximum dose build-up, the medium also allowing for a considerable degree of neutron moderation, from fast, to epithermal, to thermal. The primary set of measurements concerned comparison between open and closed collimator conditions. Use was made of a Harshaw 600 TLD reader, with each measurement made within seven days of exposure, corrected for fading.

4.5 Results and discussion

Using a closed collimator only a small fraction of the response is expected to be due to photon beam penetration, for multi-leaf collimators (MLCs) some 3% to 5% of isocentre dose [45] but less than 1% for the Elekta Infinity™ linac low leakage design [46]. Additionally, clinical accelerators operating at high MV give rise to very high x-ray to neutron ratios, 10^5 in the 15 MV case. Thus, below 15 MV the weaker neutron component may well be masked.

Figure 4.7, for 6, 10, and 15 MV beams, shows results all obtained from the same accelerator, with isocentre evaluations for both closed jaw/collimators and open jaw/collimator conditions. Observed is a quasi-linear TL yield for doses from 100 cGy onwards, proportional changes being noted at 6 and 10 MV. The absence of neutron contamination at 6 MV aligns with the consensus that the spectral distribution is below the photoneutron threshold conditions. At 10 MV there is also an apparent absence of neutron contamination, differences in measured output for each paired condition remaining within the same range over the entire dose range. This apparent result will be treated cautiously and is revisited below.

The results for 15 MV, shown in figures 4.7 and 4.8, provide strong evidence for enhancement of TL yield as a result of a photoneutron component, for both closed and open beam situations. For the closed beam, in changing from 6 to 10 MV the response at 1 Gy increases by 3% and by an additional 11% in changing from 10 MV to 15 MV. Conversely, for the closed beam situation the measured output at 1 Gy increases by 5.2% from 6 to 10 MV and by an additional 202% in changing from 10 to 15 MV. Difficulty arises in apportioning the contributions to dose, predominantly as a result of the relative efficiencies of the photon and neutron dose deposition mechanisms in the borosilicate medium. This situation has previously been noted by Kry *et al* [47] in making use of TLD-100 for out-of-field measurements; measurements were accurate at 6 and 10 MV but a substantial over-response was observed at 15 and 18 MV (by up to 1063% at 18 MV) relative to response in the absence of a neutron component. Suppression of the neutron response was obtained by wrapping the TLD-100 in up to 0.75 mm of Cd (making use of its high neutron capture cross-section), reducing the response to 39% on average.

Finally, figure 4.9 provides comparison results from the use of an optically stimulated luminescence (OSL) detector based on the reference OSL medium carbon-doped aluminium oxide, Al_2O_3:C [48]. In the conventional form this is used for

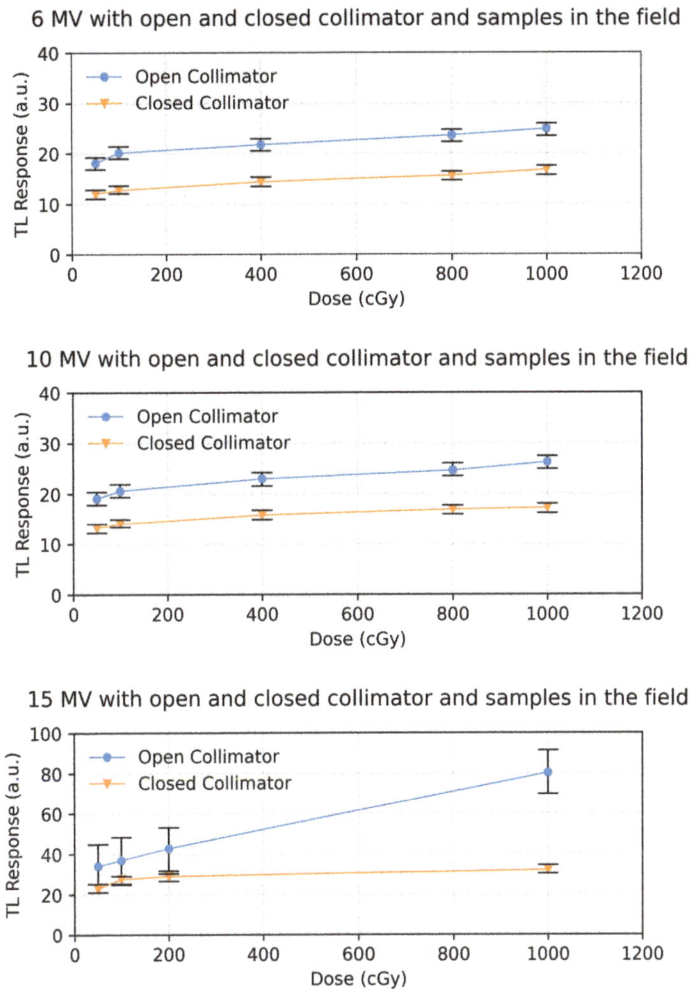

Figure 4.7. TL yields from use of a 6 MV beam (upper panel), a 10 MV beam (centre panel), and a 15 MV beam (lower panel), SSD 100 cm. The results are for both open and closed collimator conditions.

photon and electron dosimetry, offering high sensitivity, favourable optical properties, and a relatively low neutron sensitivity. Conversely, for enhanced neutron sensitivity an adaptation has been made to the Al_2O_3:C dosimeter using a proprietary $^6Li_2CO_3$ coating thickness, forming a so-called OSLN detector, a type produced by Landauer as the Albedo Neutron Detector. The neutron response is dependent on the $^6Li(n,\alpha)^3H$ reaction. The current irradiations were made at 15 MV with the detectors located at the isocentre, measurements being made with open jaw/collimator as well as closed jaw/collimator conditions for comparison. The differences between open and close conditions increase with dose, by a factor of up to 10 at 1 Gy.

To understand such differences, reference is first made to the neutron sensitivity of the LiF based TLD-100 (Harshaw Chemical Company, Solon, OH) [47] using a

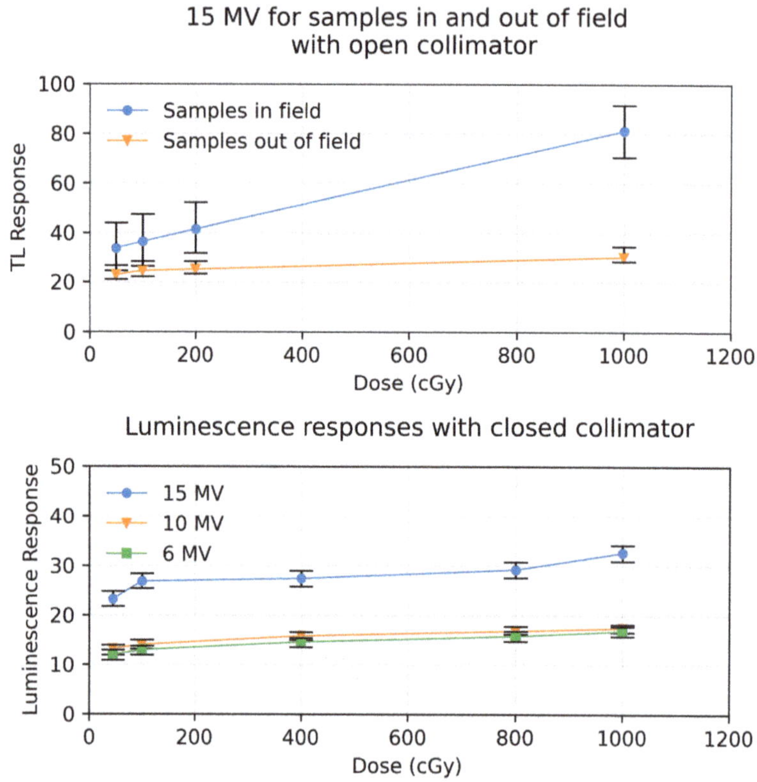

Figure 4.8. TL yield at an SSD of 100 cm, for a 15 MV beam, open jaw/collimator condition at isocentre and at 2 cm beyond the collimator edge (upper panel) and a closed jaw/collimator condition comparison for 6, 10, and 15 MV closed beams with measurements made at 2 cm beyond the collimator edge (lower panel).

15 MV accelerator, comparison being made with TLD-700 (also from Harshaw). The latter is neutron insensitive, TLD-100 containing a natural abundance of ^{6}Li (7.5%) and ^{7}Li (92.5%), TLD-700 containing 99.99% ^{7}Li, devoid of any significant (n,α) cross-section. The relative difference between the two ranged from 85% to up to 650%, depending in good part on the concentration of ^{6}Li. Yukihara *et al* [49] examined the neutron response of Al_2O_3:C OSL coated with differing thicknesses of ^{6}Li as well as other neutron convertors, clear differences being seen.

A proposition made herein is that radiation damage to the borosilicate glass produces additional TL yield, giving rise to a second-order (non-linear) response. Borosilicate radiation damage investigations date back to work at Brookhaven National Laboratory [50] concerning reactor irradiations of Corning glass containing 8% boron. Slow neutron reactions in ^{10}B were found to cause cracking of the glass, a result of induced strains from the alpha-particle effects. Further to this, at the Institut Laue-Langevin, Boffy *et al* [51] observed neutron guides formed of polished flat borosilicate glass to be suffering alpha-particle damage,

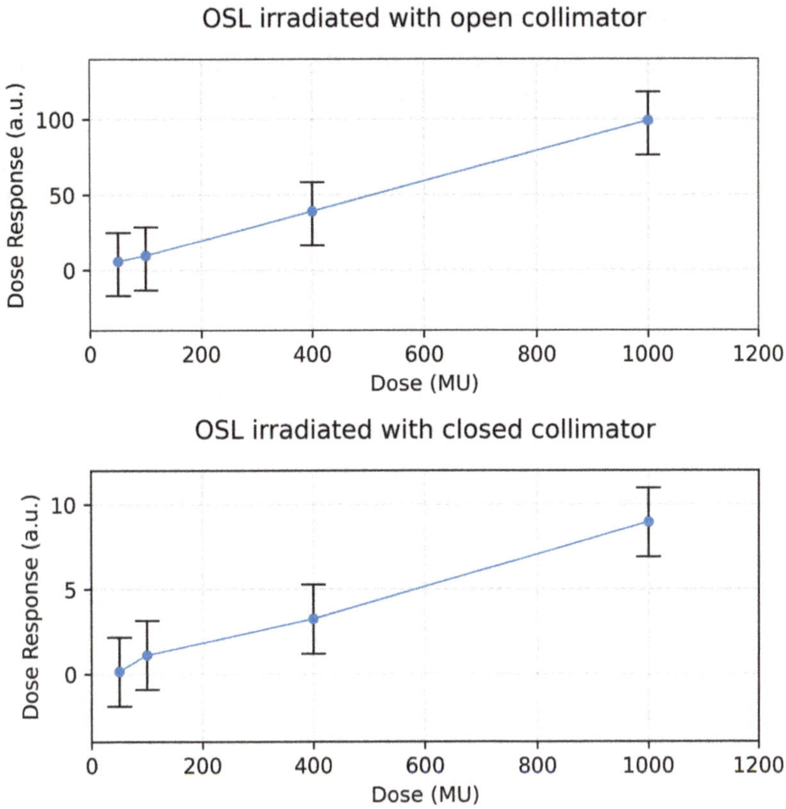

Figure 4.9. OSL results for irradiations at 15 MV, with detectors at the isocentre. Upper panel: open jaws and collimator; lower panel: closed jaws and collimator.

with notable shrinkage for neutron doses below 0.5×10^{18} n cm^{-2}, and beyond this by swelling. For borosilicate glass encapsulating radioactive waste Karakurt *et al* [52] reported alpha damage causing volume changes of up to −0.7%, depending on the composition of the glass. While these might be considered to be studies of extremes, Bradley *et al* [53] found the deliberate mechanical deformation of silica fibres to provide major TL yield improvement in the use of radiotherapy photon doses, TL yields per unit mass being enhanced by some 6.0 to 17.5 times, depending on the extent of induced strain. The over-response reported here and by Kry *et al* [47] for photoneutron emissions at 15 MV and 18 MV are suggested to be a non-linear effect resulting from the high linear energy transfer alpha-particle irradiation. These arise from the presence of ^{6}Li in TLD-100 and ^{10}B in borosilicate with consequent energy deposition and damage from the alpha recoil reactions. Kry *et al* [47], rather than seeking a mechanism for the over-response, focused on empirical evidence for its existence. The non-linearity creates a sensitive means of detection of neutrons but otherwise presents particular difficulty in evaluating neutron dose.

4.5.1 Activation products in borosilicate glass

In certain circumstances neutrons generated in the operation of linacs can necessitate consideration of neutron activation [54]. A well know example is the reaction $^{27}Al(n,\gamma)^{28}Al$, producing ^{28}Al with a half-life of 2.24 min, β^- decaying with a gamma emission of 1178.8 keV, a particular concern given the natural abundance of ^{27}Al of close to 100%. Accordingly, the presence of Al in treatment rooms should be minimized, potentially needing a short cooling off period before room entry. Thomadsen *et al* [55] have provided a summary of common radionuclides produced through photo-activation processes. Further concerns included the concrete bunker, with potential activation of ^{56}Mn and ^{24}Na (both 100% natural abundant), emitting photons of 846.8 and 1368.6 keV, respectively, with a half-life in each case of 2.58 and 14.96 h. As a result, barite is the favoured material, with very low presence (<0.5%) of Al_2O_3 and no practical presence of Na or Mn. In the case of borosilicate glass the presence of Na at several % indicates potential for activation, an internal source of irradiation that can be a further contribution to readout values.

4.6 A germano-borosilicate flat fibre dosimeter

Harnessing the dosimetry capabilities of optical fibres is based on their material composition, comprising a silica host and a variety of dopants in their core and cladding, namely, Ge, P, F, B, etc. As mentioned earlier, the presence of these dopants creates defect centres that facilitate the processes needed for passive TLD. Early experimental investigations on deliberate radiation exposure and readout of commercial optical fibres were carried out by Abdulla *et al* [56] and Espinosa *et al* [57]. These studies made use of commercially available optical fibres with a Ge-doped core (containing the majority proportion of defect centres) and silica host. The dose range obtained with these optical fibres was between 0.01 Gy to 5 kGy, irradiated by a Co-60 gamma source. The effectiveness of silica-based optical fibres as TLD was extended to alpha sources [29], beta sources [58], protons [59], electrons [60], and fast neutrons [26] in subsequent studies.

While the response of silica optical fibres has proven effective for applications in dosimetry under different irradiation conditions, a parallel development of tailoring the structure of optical fibres by adjusting the fabrication parameters has appeared to influence and enhance sensitivity. Starting with a base silica material, the two main techniques of increasing sensitivity are (i) incorporating dopants during fabrication and (ii) manipulating the structure of an optical fibre from the conventional cylindrical core/cladding geometry. During the pulling of flat fibres, the deformation of the cylindrical cross-section into an elongated cross-section produces stresses on the circumference of the neck-down region of the preform that generate additional structural defects, in the form of non-bridging oxygen hole centres (NBOHCs) and breakage of the strained Si–O bonds in the continuous random silica network of the amorphous base material [61], thus producing potential trap centres for radiation induced electron–hole pairs.

Following this principle, Abdul Sani *et al* [24] and Mahdiraji *et al* [62] verified the increase in sensitivity of optical fibres as they are drawn into a flat fibre structure, followed by incorporation of Ge and then B (enhancing the sensitivity at each stage of doping). This fortuitous development came about as a photonics research group had been investigating novel optical fibre structures for ultraviolet (UV) waveguiding [23], enabling potential use of the produced samples for dosimetry studies. An SEM image of the highly sensitive Ge–B silica flat fibre cross-section and its TL glow curves obtained alongside that of undoped and Ge-doped silica optical fibre samples in cylindrical and flat structures after exposure to 8 Gy of 6 MeV electron radiation in clinical linear accelerators are shown in figure 4.10 and figure 4.11,

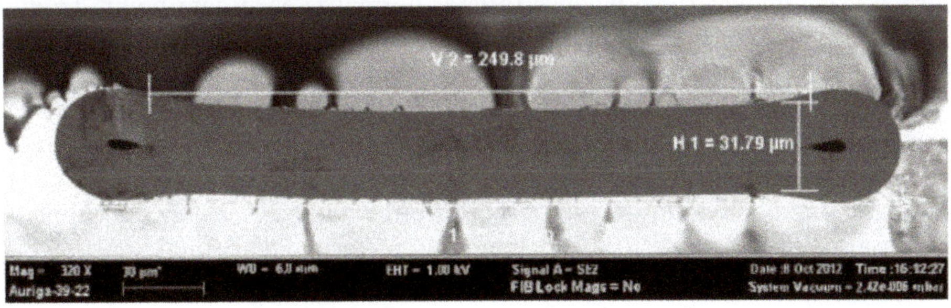

Figure 4.10. FE-SEM image of a boron and germanium co-doped flat fibre with dimensions of 320 μm × 32 μm and a core length of 250 μm. Two small air holes are visible at the sides of the sample. The core layer is not visible in this image. Reproduced with permission from [23]. Copyright 2012 IEEE.

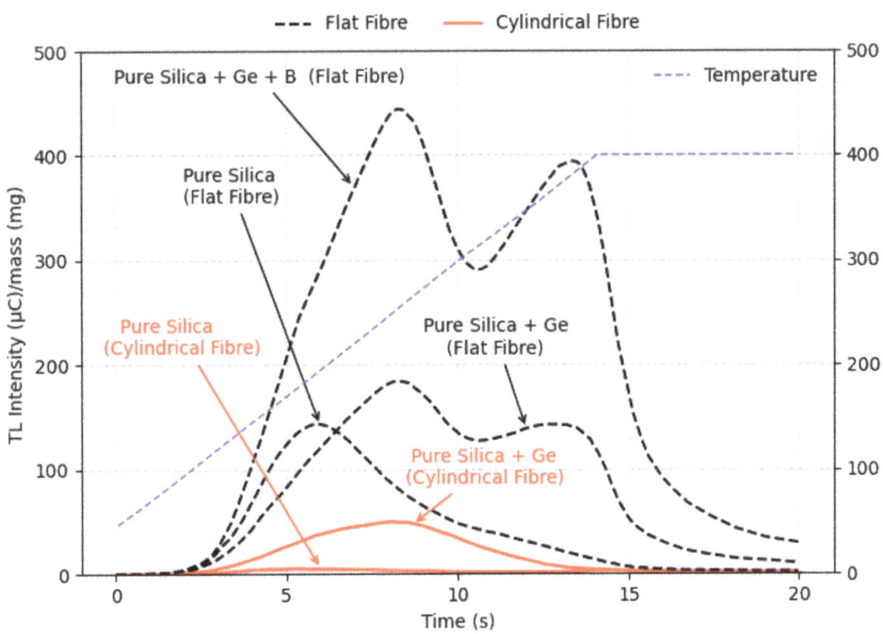

Figure 4.11. Glow curves obtained for a dose of 8 Gy using 6 MeV electrons from five optical fibre structures: pure silica (cylindrical), silica + Ge (cylindrical), pure silica (flat), silica + Ge (flat), and silica + Ge + B (flat).

respectively. It is specifically noted that manipulation of the structure of the optical fibre (pure silica composition, cylindrical to flat structure) is more effective in the increase of TL yield compared to the addition of dopants (cylindrical structure, pure silica to Ge-doped silica composition).

4.6.1 Experimental procedures

The sample preparation for TL investigations of optical fibres is well outlined by Begum *et al* [63]. The first step is to remove any polymer coating and dust using acetone. Samples are then cut into approximately 3–5 mm lengths (in line with the TLD reader planchet) using an optical fibre cleaver. The mass of each sample is then recorded using an electronic balance to later normalize each data point. To avoid surface contamination, vacuum tweezers are usually used for handling TL materials. As a standard operating procedure, before irradiation the samples first need to be annealed to remove any residual TL resulting from the various potential background sources of TL excitation including environmental exposures. In studies with optical fibres, samples are commonly found to be oven annealed at a temperature of 400 °C for one hour, following the procedure for TLD-100. After annealing, the samples are slowly cooled to room temperature to avoid thermal stress. Recent studies have shown that raising the temperature to 500 °C during annealing of optical fibres is more effective for dose history erasure [64].

4.6.2 Experimental studies

Investigations with the highly sensitive GeB FF have been performed in a wide range of radiation dose environments with dose exposures from mGy through kGy. Alawiah *et al* [65] utilize a synchrotron microbeam facility to expose GeB FF samples to hard x-ray doses up to 1000 kGy at ~3000 kGy h^{-1} (calculated at the surface of the sample). The TL glow curve is shown in figure 4.12—note the prominent TL peak around 250 °C. The peak integral of the GeB FF at 1000 kGy is 2.7 times that of 500 kGy, suggesting formation of radiation induced traps at higher doses. Kinetic parameter evaluation, using the TL glow curve analysis, identifies three constituent peaks at 178 °C, 245 °C, and 330 °C, as shown in figure 4.13. During electron (6, 15, and 21 MeV) detection with pure silica flat fibres, five TL glow peaks have been observed at 219.5 °C, 279.7 °C, 328.8 °C, 372.0 °C, and 400.0 °C [66]. Table 4.3 lists the maximum temperatures (T_{max}) for glow curve peaks as observed for different varieties of the doped silica media. The glow peaks have been grouped into nine groups based on the range of values. As per this taxonomy, the pure silica flat optical fibres exhibited glow peaks in the D, F, H, and I groups when subjected to electron radiation. In contrast, the GeB FF was deconvolved into three glow peaks in the B. E, and H regions for synchrotron hard x-ray exposures.

Initial investigations of pure silica FF also revealed a minimum detectable dose of <10 mGy under 9 MeV electron irradiation conditions [67]. In a recent related development, Alyahyawi *et al* [68] made use of high spatial resolution GeB FF in investigating clinical radiosurgery (at the Gamma Knife Perfexion facility) for *in vivo* dose verification measurements, use of the well-established LiF (Mg,Ti)

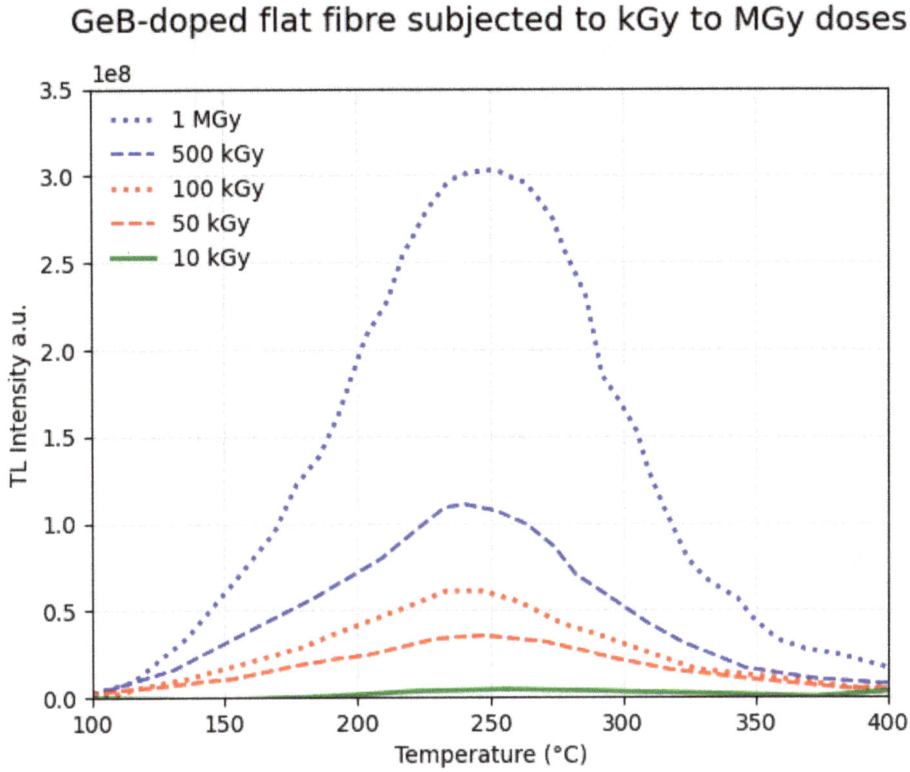

Figure 4.12. TL glow curve for GeB-doped flat optical fibre subjected to synchrotron irradiation giving doses of 10 kGy to 1 MGy with a maximum peak temperature at 245 °C.

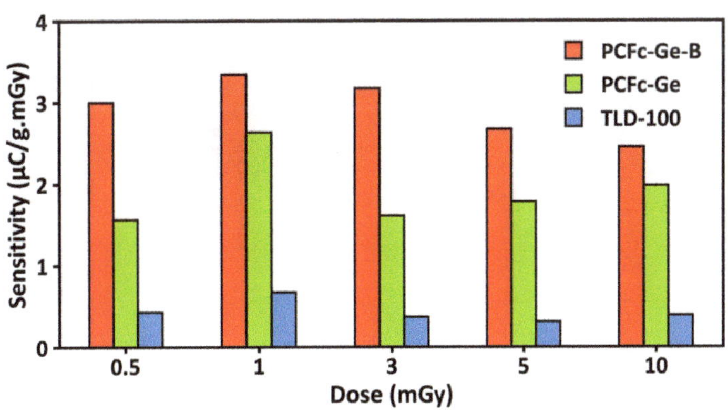

Figure 4.13. Sensitivities of collapsed PCFs and TLD-100. Reproduced with permission from [69]. Copyright 2016 Elsevier.

phosphor TLD-100 being limited in performance at the elevated doses seen in radiotherapy. The intent was to compare the performance of the two forms of TL dosimeter, GeB FF and TLD-100, measuring scattered radiation resulting from cranial cavity radiosurgery procedures. The highest readings for the absorbed dose

Table 4.3. Variation of maximum temperature (T_{max}) in celcius for glow curve peaks of germano-borosilicate media as observed in experimental studies.

Author	Sani *et al* [32]		Alawiah *et al* [74]	Fadzil *et al* [72]	Ghomeishi *et al* [22]	Alawiah *et al* [65]
Material	Borosilicate glass coverslips		Ge-doped SiO_2 flat optical fibres	Ge-doped SiO_2 flat optical fibres	Ge-doped SiO_2 flat optical fibres	GeB-doped SiO_2 flat optical fibres
Radiation type	Electron 9 MeV (Gy min^{-1})	Electron 2 MeV (kGy min^{-1})	Electron 2.5 MeV (kGy min^{-1})	Photon 6 MV/10 MV (Gy min^{-1})	Photon 6 MV/10 MV (Gy min^{-1})	Photon 8.9 keV (MGy h^{-1})
Peak A	84.2–91.5					
Peak B	170.3–189.9		166.17–181.02	170.07–179.00	185	178
Peak C		188.9–200	199.27–219.68			
Peak D	209–231.2			203.52–219.71		
Peak E	227.9–265.8	226.2–241	234.53–251.85	230.75–243.57	235	245
Peak F	263–294.2	266.9–278.9	261.74–278.75	264.50–279.43	283	
Peak G	287.5–320	304.9–310.3	300.09–311.85	286.62–309.41		
Peak H		333.8–334.3			330	330
Peak I					380	

from scattering were obtained for the thyroid, followed by the chest and pelvis. Both dosimeter responses showed general accord (±3 mGy at 1σ), covering mean doses ranging from some 10 mGy down to less than 1 mGy, representing doses of <1% to <0.1% of the prescribed dose at the treatment site. Observed readings obtained from use of the GeB FF are 1.4 times larger than those of the TLD-100 on average, being attributed to photoelectric absorption interactions taking place more predominantly in the silica media than in the LiF based TLD-100, thus requiring corrections for energy dependencies. Rozaila *et al* [69] made use of Ge and GeB photonic crystal fibre (collapsed) samples [70] along with TLD-100 for dose measurements in the range 1–10 mGy, irradiated by an x-ray tube with accelerating potential 80 kVp. The GeB-doped photonic crystal fibres were verified to be ~9 times more sensitive than conventional TLD-100s in the given energy range, as shown in figure 4.13.

As noted previously in figure 4.11, the TL glow curves of Ge FF and GeB FF are visually similar, with GeB FF exhibiting much greater TL intensity overall. In various studies concerning glow curve convolution, the familiar shape with two large peaks around 245 °C and 380 °C has appeared consistently. This consistency has also been maintained for a variety of radiation sources, such as 150 MeV protons [71], 6 MeV electrons [62], 6/10 MV photons [72], and 100–150 kVp photons [73].

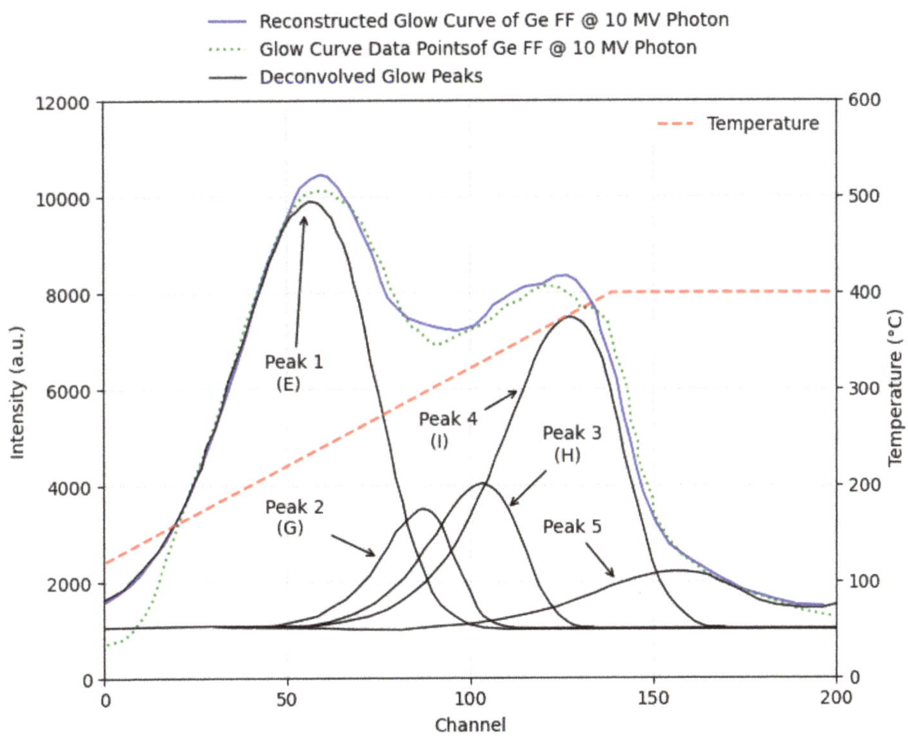

Figure 4.14. Glow curve deconvolution of a Ge-doped flat fibre subjected to 2 Gy of 10 MV photon irradiation consisting of five glow peaks (a heating rate of 30 °C s^{-1}).

Hassan *et al* [71] and Rais *et al* [73] note a different glow curve shape when the concentration of Ge is varied in the optical fibres. As the concentration of Ge content is lowered from 6 mol% to 2.3 mol% (the content composition used in these studies), the shape changes from a double-peaked distribution to a broad single-peaked distribution. As Ghomeishi *et al* [22] have suggested, the second peak, commonly referred to as the higher temperature glow peak (HTP; channel 101–200), is a result of Ge structural defects in the fibre. Figure 4.14 presents a deconvolved TL glow curve of a Ge FF subjected to 10 MV photons, showing five peaks, where peak 3 (H), peak 4 (I), and peak 5 are HTPs. Figure 4.15 presents the glow curve for a GeB FF subjected to ultra-high dose radiation (kGy–MGy). The deconvolved glow peaks consist of three components, with the third component being in the H region, the dominant peak in Ge-doped samples. A significant observation of this glow curve is the dominance of peak 2 (E), suggestive of a saturation of Ge-induced defects as doses extend beyond hundreds of kGy. Glow peaks in the region E (centred around 245 °C) are commonly found in all silica-based optical fibres, regardless of the structure and dopant composition, as shown in table 4.3. These are attributed to silicon oxygen defect centres (Si-ODCs), and O^+ and O^- defect centres.

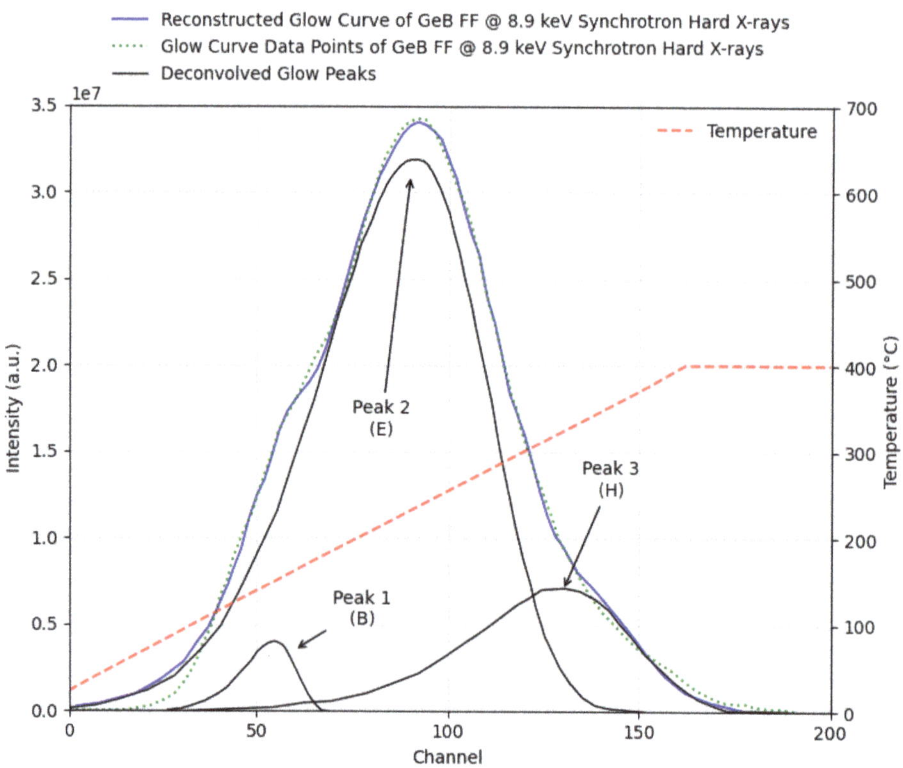

Figure 4.15. The deconvolved TL glow curve of a GeB-doped flat fibre following 1 kGy to 1 MGy of synchrotron (hard x-ray) irradiation at 8.9 keV, consisting of three glow peaks (a heating rate of 35 °C s^{-1}).

4.7 Conclusions

In single-mode communication optical fibres, phosphorus (P_2O_5) has long replaced boron (B_2O_3) as a choice of dopant for lowering the process temperature and tailoring the refractive index [75], leading to a decline in commercially available boron-doped silica optical fibres. However, as the discussion has progressed in this chapter, a promising picture of borosilicate media as dosimeter materials has clearly emerged. Considering the recent investigations on silica-based TL dosimeters, the optical fibre structure doped with boron appears to have dosimetry capabilities for both particle and non-particle ionizing radiation detection. The fabrication and study of novel borosilicate optical fibre structures is an interesting avenue, yet to be explored, promising both scientific and commercial value.

The work described herein has been intended to showcase the ability of borosilicate media to act as a dosimeter of energetic radiations, neutron sensing being used as a particular exemplar of the versatility of the medium, which is also shown to offer linear sensitivity over a range of extremely high electron doses, in the kGy to several hundred kGy range. This is a range covering radiation sterilization and radiation processing needs. The borosilicates provide the capability to validate the doses that are given to such products, in principle also being able to be implemented at very little cost, also challenging the capability of the existing dosimetric technology applied within this range.

To date the majority of work has been carried out on borosilicates in the form of microscope coverslips, which are cheap, abundant, available with relatively large areas, and of various millimetre and sub-millimetre thicknesses. Also investigated to date are GeB optical fibres in various forms, shown to be good in detecting photons, electrons, protons, and alpha-particle irradiation. Indeed, there is every indication that the existing borosilicate forms are capable of covering the entire radiotherapy dose range, and also for the various radiations to which the patient is subjected. Yet to be explored is their use in ion beam radiotherapy for ions heavier than the alpha particle, for example carbon ions. In principle it would appear that borosilicate fibres might be expected to offer similar performance, albeit this remains to be demonstrated, as they do not differ substantially in composition from microscope coverslips. What can be expected to be offered is tremendous scope for positional and spatial delineation.

References

[1] Jafari S M, Bradley D A, Gouldstone C A, Sharpe P H G, Alalawi A, Jordan T J, Clark C H, Nisbet A and Spyrou N M 2014 *Radiat. Phys. Chem.* **97** 95–101

[2] Ley K, Jafari S M, Lohstroh A and Bradley D A 2019 *Radiat. Phys. Chem.* **155** 178–83

[3] Ley K 2019 Thermoluminescence of silica beads for dosimetry up to 250 kGy *PhD Thesis* University of Surrey, Guildford

[4] Bollinger L M, Thomas G E and Ginther R J 1962 *Nucl. Instrum. Methods* **17** 97–116

[5] Yamashita M, Stephens L D, Smith A R and Patterson H W 1966 *J. Nucl. Sci. Technol.* **3** 343–53

[6] Pino F, Stevanato L, Cester D, Nebbia G, Sajo-Bohus L and Viesti G 2014 *Appl. Radiat. Isot.* **92** 6–11
[7] Hunt S, Iliadis C and Longland R 2016 *Nucl. Instrum. Methods Phys. Res.* A **811** 108–14
[8] Bradley D A, Khandaker M U and Alanazi A 2020 *Radiat. Phys. Chem.* **170** 108680
[9] Zubair H T *et al* 2020 *IEEE Photonics J.* **12** 1–25
[10] Pfaender H G 1996 *Schott Guide to Glass* (Dordrecht: Springer)
[11] Nyswander R E and Cohn B E 1930 *J. Opt. Soc. Am.* **20** 131
[12] Northup M A and Lee O I 1940 *J. Opt. Soc. Am.* **30** 206
[13] Spurný Z and Novotný J 1965 *J. Phys. Chem. Solids* **26** 1107–10
[14] Malchukova E, Boizot B, Ghaleb D and Petite G 2005 *Nucl. Instrum. Methods Phys. Res.* A **537** 411–4
[15] Malchukova E and Boizot B 2010 *Mater. Res. Bull.* **45** 1299–303
[16] Malchukova E V, Nepomnyashchikh A I, Boizot B and Terukov E I 2018 *Glas. Phys. Chem.* **44** 356–63
[17] Kawamura I, Kawamoto H, Fujimoto Y, Koshimizu M, Okada G, Koba Y, Ogawara R, Yanagida T and Asai K 2019 *J. Ceram. Soc. Japan* **127** 663–8
[18] Van Uitert L G, Pinnow D A, Williams J C, Rich T C, Jaeger R E and Grodkiewicz W H 1973 *Mater. Res. Bull.* **8** 469–76
[19] French W G, Pearson A D, Tasker G W and MacChesney J B 1973 *Appl. Phys. Lett.* **23** 338–9
[20] Carvalho I C S, Fokine M, Cordeiro C M B, Carvalho H and Kashyap R 2008 *Opt. Mater. (Amst).* **30** 1816–21 Online
[21] Mat-Sharif K A, Zulkifli M I, Muhamad-Yassin S Z, Tamchek N, Aljamimi S M, Yusoff A, Amin Y M, Shafiqah S A S and Abdul-Rashid H A 2013 *4th Int. Conf. Photonics, ICP 2013—Conf. Proc.* pp 284–7
[22] Ghomeishi M, Mahdiraji G A, Adikan F R M, Ung N M and Bradley D A 2015 *Sci. Rep.* **5** 1–10
[23] Dambul K D *et al* 2012 *Photonics Glob. Conf. PGC* **2012** 49–51
[24] Abdul Sani S F, Alalawi A I, Azhar A R H, Amouzad Mahdiraji G, Tamchek N, Nisbet A, Maah M J and Bradley D A 2014 *Radiat. Phys. Chem.* **104** 134–8
[25] Alalawi A I, Hugtenburg R P, Abdul Rahman A T, Barry M A, Nisbet A, Alzimami K S and Bradley D A 2014 *Radiat. Phys. Chem.* **95** 145–7
[26] Hashim S, Bradley D A, Saripan M I, Ramli A T and Wagiran H 2010 *Appl. Radiat. Isot.* **68** 700–3
[27] Baum E M, Knox H D and Miller T R 2002 *Nuclides and Isotopes Chart of the Nuclides* 16th edn (New York: KAPL: Lockheed Martin)
[28] Noor N M, Hussein M, Bradley D A and Nisbet A 2011 *Nucl. Instrum. Methods Phys. Res.* A **652** 819–23
[29] Ramli A T, Bradley D A, Hashim S and Wagiran H 2009 *Appl. Radiat. Isot.* **67** 428–32
[30] Bradley D A, Sani S F A, Shafiqah A S S, Collins S M, Hugtenburg R P, Rashid H A A, Najua Zulkepely N and Maah M J 2018 *Appl. Radiat. Isot.* **138** 65–72
[31] Alqahtani A, Bradley D A, Alanizi A and Nisbet A 2020 *Radiat. Phys. Chem.* **168** 108630
[32] Sani S F A, Othman M H U, Alqahtani A, Nazeri A A Z A, Almugren K S, Ung N M, Hashim S A, Alkallas F H and Bradley D A 2020 *Radiat. Phys. Chem.* 108903
[33] Brunckhorst E 2009 Experimental investigations of the neutron contamination in high-energy photon fields at medical linear accelerators *PhD Thesis* (Universität Hamburg)

[34] Ipe N E 2007 *Neutron Shielding Design and Evaluation AAPM 2007 Summer School* (Collegeville, MN)
[35] Hall E J, Martin S G, Amols H and Hei T K 1995 *Int. J. Radiat. Oncol.* **33** 225–30
[36] Anon 1984 *Neutron Contamination from Medical Electron Accelerators* (Bethesda, MD: National Council on Radiation Protection and Measurements)
[37] Salehi D, Sardari D and Jozani M S 2013 *Chinese Phys.* C **37** 078201
[38] Brunckhorst E, Sheng X, Todorovic M, Becker J and Cremers F 2008 *Radiat. Prot. Dosimetry* **131** 513–20
[39] Keehan S 2017 Risks posed by neutron contamination in high energy radiotherapy *PhD Thesis* RMIT University, Melbourne
[40] Forman L, Vanier P E and Welsh K 2004 Fast neutron source detection at long distances using double-scatter spectrometry *Proceedings Volume 5198, Hard X-Ray and Gamma-Ray Detector Physics V* ed L A Franks, A Burger, R B James and P L Hink (Bellingham, WA: SPIE) 217
[41] Hussein E M A 2007 *Radiation Mechanics* (Amsterdam: Elsevier) pp 153–245
[42] Bevilacqua R, Hambsch F-J, Vidali M, Ruskov I and Lamia L 2017 ^{10}B(n,α)^{7}Li and ^{10}B(n, $\alpha_1\gamma$)^{7}Li cross section data up to 3 MeV incident neutron energy *EPJ Web Conf.* **146** 11010
[43] Howell R M, Kry S F, Burgett E, Hertel N E and Followill D S 2009 *Med. Phys.* **36** 4027–38
[44] Podgorsak E B 2005 *Radiation Oncology Physics: A Handbook for Teachers and Students* (Vienna: IAEA)
[45] IAEA 2005 Dosimetry and medical radiation physics (DMRP) (Slides to *Radiation Oncology Physics* handbook) www-naweb.iaea.org/nahu/DMRP/slides.html
[46] Prabagaran C, Prabhu R, Vyankatesh S and Manigandan D 2014 MLC transmission measurement for Elekta-infinity integrated agility160 leaf with IEC standard *35th National Annual Conf. of Medical Physicist of India—AMPICON 2014*
[47] Kry S F, Price M, Followill D, Mourtada F and Salehpour M 2007 *J. Appl. Clin. Med. Phys.* **8** 169–75
[48] Klemic G A, Azziz N and Marino S A 1996 *Radiat. Prot. Dosimetry* **65** 221–6
[49] Yukihara E G, Mittani J C, Vanhavere F and Akselrod M S 2008 *Radiat. Meas.* **43** 309–14
[50] Truell R, DeKlerk J and Levy P W 1958 *J. Appl. Phys.* **29** 225–6
[51] Boffy R, Peuget S, Schweins R, Beaucour J and Bermejo F J 2016 *Nucl. Instrum. Methods Phys. Res.* B **374** 14–9
[52] Karakurt G, Abdelouas A, Guin J-P, Nivard M, Sauvage T, Paris M and Bardeau J-F 2016 *J. Nucl. Mater.* **475** 243–54
[53] Bradley D A, Mahdiraji G A, Ghomeishi M, Dermosesian E, Adikan F R M, Rashid H A A and Maah M J 2015 *Appl. Radiat. Isot.* **100** 43–9
[54] Fischer H W, Tabot B E and Poppe B 2006 *Phys. Med. Biol.* **51** N461–6
[55] Thomadsen B, Nath R, Bateman F B, Farr J, Glisson C, Islam M K, LaFrance T, Moore M E, George Xu X and Yudelev M 2014 *Health Phys.* **107** 442–60
[56] Abdulla Y A, Amin Y M and Bradley D A 2001 *Radiat. Phys. Chem.* **61** 409–10
[57] Espinosa G, Golzarri J I, Bogard J and García-Macedo J 2006 *Radiat. Prot. Dosimetry* **119** 197–200
[58] Bajuri F, Mustafa S, Hassan M F, Dollah T, Kadir A B A, Tamchek N, Saad F F A, Bradley D A and Noor N M 2018 *Int. J. Nanoelectron. Mater.* **11** 169–78
[59] Hassan M F, Abdul Rahman W N W, Fadzil M S A, Tominaga T, Geso M, Akasaka H, Bradley D A and Noor N M 2017 *J. Phys. Conf. Ser.* **851**

[60] Yaakob N H, Wagiran H, Hossain M I, Ramli A T, Bradley D A, Hashim S and Ali H 2011 *J. Nucl. Sci. Technol.* **48** 1115–7

[61] Hibino Y and Hanafusa H 1988 *J. Light. Technol.* **6** 172–8

[62] Mahdiraji G A, Adikan F R M and Bradley D A 2015 *J. Lumin.* **161** 442–7

[63] Begum M *et al 4th Int. Conf. Photonics, ICP 2013—Conf. Proc.* pp 291–3

[64] Moradi F, Mahdiraji G A, Dermosesian E, Khandaker M U, Ung N M, Mahamd Adikan F R and Amin Y M 2017 *Radiat. Phys. Chem.* **134** 62–70

[65] Alawiah A, Fadhli M M, Bauk S, Abdul-Rashid H A and Maah M J 2013 *Proc. SPIE* **89235** 89235F

[66] Alawiah A *et al* 2013 *Proc. SPIE* **8775** 87750S

[67] Hashim S, Omar S S C, Ibrahim S A, Hassan W M S W, Ung N M, Mahdiraji G A, Bradley D A and Alzimami K 2015 *Radiat. Phys. Chem.* **106** 46–9

[68] Alyahyawi A, Dimitriadis A, Nisbet A and Bradley D A 2020 *Radiat. Phys. Chem.* 108973

[69] Rozaila Z S, Alyahyawi A, Khandaker M U, Amin Y M, Bradley D A and Maah M J 2016 *Radiat. Phys. Chem.* **126** 9–13

[70] Amouzad Mahdiraji G, Dermosesian E, Safari M J, Mahamd Adikan F R and Bradley D A 2015 *J. Light. Technol.* **33** 3439–45

[71] Hassan M F, Rahman W N W A, Kadir A B A, Isa N M, Tominaga T, Geso M, Akasaka H, Bradley D A and Noor N M 2018 *AIP Conf. Proc.* **2045** 020084

[72] Fadzil M S A, Tamchek N, Ung N M, Alawiah A, Abdullah N, Bradley D A and Noor N M 2018 *J. Sains Nukl. Malaysia* **30** 1–14

[73] Rais N N M, Bradley D A, Hashim A, Isa N M, Osman N D, Ismail I, Hassan H A and Noor N M 2020 *Radiat. Phys. Chem.* 108935

[74] Alawiah A *et al* 2015 *Radiat. Phys. Chem.* **113** 53–8

[75] Nagel S, MacChesney J and Walker K 1982 *IEEE J. Quantum Electron.* **18** 459–76

[76] Alqahtani A *et al* 2020 Microscope cover-slip glass for TLD applications *Appl. Radiat. Isot.* **160** 109132

Chapter 5

The ionizing radiation effect of erbium, ytterbium, or bismuth doped/co-doped optical fibres

Yanhua Luo, Desheng Fan, Mingjie Ding, Qiancheng Zhao, Binbin Yan, Jianxiang Wen and Gang-Ding Peng

Targeting harsh radiation environment applications, i.e. inter-satellite communication, lidar, nuclear plants, etc, the ionizing radiation effects of erbium, ytterbium, or bismuth doped/co-doped optical fibres have been investigated. First, the radiation-induced photodarkening in erbium/ytterbium co-doped optical fibre (EYDF) and the thermal effect in these irradiated EYDFs have been studied, which evidently display a reversible thermal effect. Then the radiation-induced activation of a bismuth active centre (BAC) from inactive bismuth has been assessed through the analysis of the absorption, emission, and lifetime of bismuth-doped optical fibre (BDF) and bismuth/erbium co-doped optical fibre (BEDF). The factors, such as Yb co-doping and radiation type, that influence the radiation effect have also been inspected. In addition, thermal treatment has been implemented to improve the performance of BEDF after gamma/electron irradiation. To improve the radiation resistance of the fibres, it has been attempted to investigate radiation hardening by combining thermal treatment and hydrogen loading. Finally, a demonstration of a radiation-independent temperature sensor based on the fluorescence intensity ratio (FIR) of EYDF has been carried out, indicating great potential for the sensing field in the radiation environment.

5.1 Introduction

With the development of modern optical fibre technology, optical fibres, in particular doped fibres, have been widely used in various fields, including inter-satellite communications, LiDAR, fibre-optic gyroscopes, laser altimeters, nuclear plants, radiation therapy, etc [1–4]. Therefore it is necessary to study the radiation effect on these doped fibres for future applications in harsh environments.

doi:10.1088/978-0-7503-3243-9ch5

It is well known that ionizing radiation from high-energy rays often produces changes in material structure and characteristics. For example, in silica optical fibres ionizing radiation can easily break the bond of Si–O in the silica matrices, thereby generating many Si-related defects, such as peroxy radicals (PORs), oxygen deficient centres (ODCs), non-bridging oxygen hole centres (NBOHCs), self-trapped holes (STHs), self-trapped excitons (STEs), etc [5, 6]. If other doping elements, such as P, Al, Er, etc, are introduced, more defects, e.g. aluminium oxygen hole centres (AlOHCs), phosphorus oxygen hole centres (POHCs), etc, will be produced due to the radiation-induced break of the chemical bonds. These defects produce a series of absorption bands in the range of 0.5–8.5 eV, as shown in figure 5.1 [6–10], thereby evidently increasing the fibre loss. This excess loss is often called radiation-induced attenuation (RIA) [11]. In addition, the reduction of doped ions can happen when the doped fibres are exposed to ionizing radiation. For example, under ionizing radiation Yb^{3+} can be reduced to Yb^{2+} and Bi^{n+} to $Bi^{(n-1)+}$ [12].

At present, there are many kinds of functional doped optical fibres, mainly including rare-earth and Bi-doped fibres, which cover different spectral regions ranging from 950 to 2100 nm [13]. Different doped optical fibres target different operation bands or wavelength ranges. For example, Er-doped fibre (EDF) operates

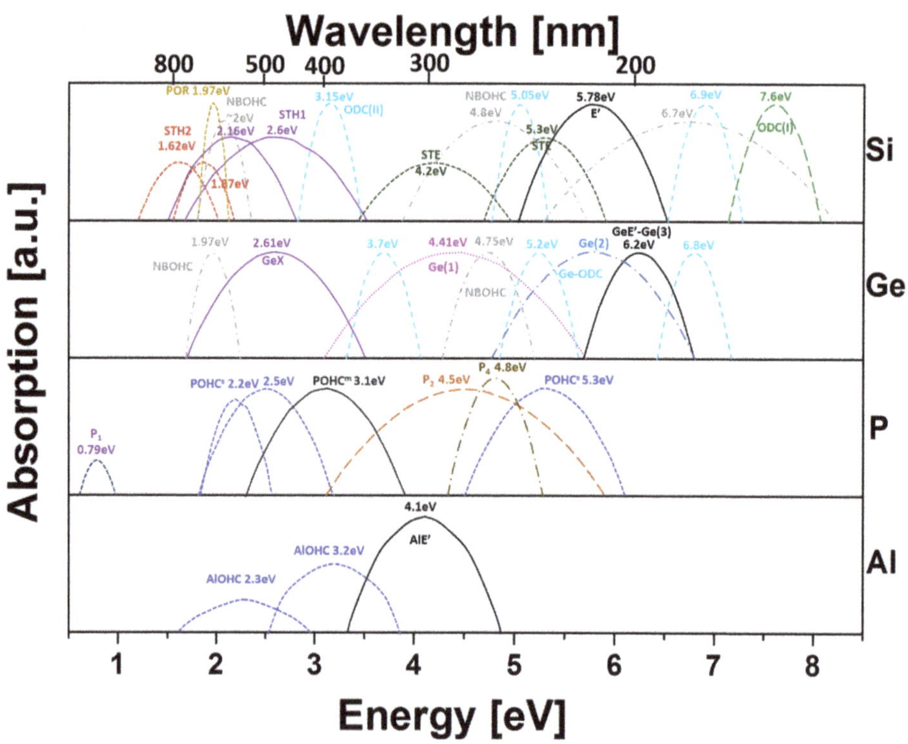

Figure 5.1. Absorption bands associated with Si, Ge, P, or Al-related point defects [6–10]. POR: peroxy radical; ODC: oxygen deficient centre; NBOHC: non-bridging oxygen hole centre; STH: self-trapped hole; STE: self-trapped exciton; Ge-ODC: Ge-oxygen deficient centre; POHC: P-oxygen hole centre; AlOHC: Al-oxygen hole centre; and GeE′: Ge related E′ centre.

in the C- and L -bands (1520–1620 nm), Tm-doped fibre (TDF) in the S-band (1450–1490 nm), Yb-doped fibre (YDF) in a wavelength range of around 1055–1070 nm, Pr-doped fibre around the 1300 nm region, and the recently developed Bi-doped optical fibre (BDF) in the broadband ranges of 1100–1500 nm and 1600–1800 nm.

In our previous work [5, 11, 14–26] much attention was paid to the development of erbium, ytterbium, or bismuth-doped/co-doped optical fibres for photonic applications in radiation environments. Thus in this chapter the published works on the ionizing radiation effects on erbium-, ytterbium-, or bismuth-doped/co-doped optical fibres fabricated using modified chemical vapour deposition (MCVD) are first reviewed. Then radiation hardening using combined thermal treatment and H_2 loading was introduced to improve the radiation resistance of the fibres. Finally, the demonstration of a radiation-independent temperature sensor based on the fluorescence intensity ratio (FIR) has been carried out, indicating great potential for radiation environment applications.

5.2 Ionizing radiation effects on Er/Yb co-doped optical fibre

EDF is mainly based on the emission of erbium ions at 1530 nm under 980 or 1480 nm pumping [27]. At present, EDF is widely used in fibre amplifiers, fibre lasers, broadband light sources, and fibre sensors. Taking advantage of the broad absorption band and large pumping absorption coefficient of Yb^{3+}, it is often co-doped into EDF, obtaining a Er/Yb co-doped optical fibre (EYDF) for the realization of the luminescence sensitization of Er^{3+}.

5.2.1 Radiation-induced photodarkening

When the EYDF is exposed to Co^{60} γ-radiation, a significant photodarkening effect can be observed due to the RIA [14]. By comparing the change in the transmission spectra before and after radiation, RIA spectra were obtained, as shown in figure 5.2(a). Figure 5.2(a) clearly demonstrates that the higher the radiation dosage, the stronger the RIA induced. Further, it is found that the RIA spectrum (in units of photon energy (eV)) can be fitted well by a Gaussian function. In particular, the

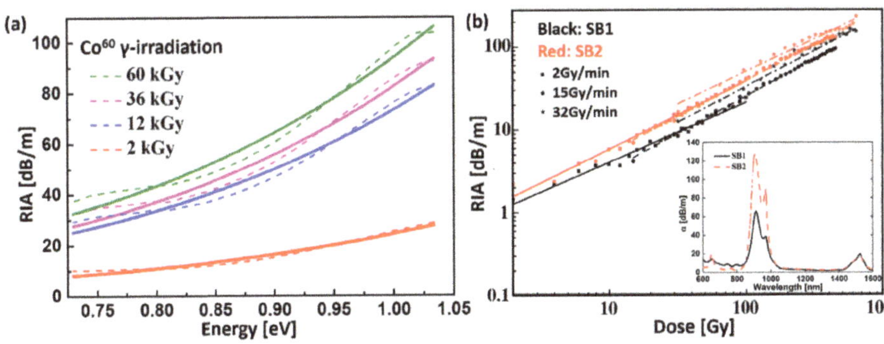

Figure 5.2. (a) RIA spectra (solid line) and their Gaussian fittings (dashed line) [14]. (b) RIA responses for two EYDFs, SB1 and SB2, with high and low Yb-doping concentrations, respectively, reflected in their absorption spectra in the inset.

centre position is located at 3.2 eV with a full width at half maximum (FWHM) of 1.08 eV [14], which is properly matched with the absorption band of AlOHC, as shown in figure 5.1. Such a result indicates that the RIA is mainly derived from the radiation-induced AlOHC defect centre.

Furthermore, the growth of RIA with dose can further be fitted by the power law [28]

$$\mathrm{RIA} = c \times D^f, \tag{5.1}$$

as shown in figure 5.2(b), where D is the radiation dose, and c and f are constants. As can be seen from figure 5.2(b), RIA is almost dose-rate independent within the low dose-rate range (2–15 Gy min^{-1}), but a higher dose rate (>=32 Gy min^{-1}) tends to produce higher RIA. In particular, for the sample with high Yb doping concentration (reflected by higher absorption at 980 nm as shown by the inset of figure 5.2(b)), RIA is evidently stronger than that with low Yb-doping.

5.2.2 A reversible thermal effect in irradiated EYDF

Subsequently, the thermal effect of EYDFs after 2 kGy gamma irradiation was studied [11]. Interestingly, the induced thermal effect is reversible, as shown in figure 5.3. It shows that when the sample is heated up to 300 °C the transmittance at 830 nm decreases sharply. After the heating stops the transmittance basically returns to its original position. This thermally controllable recovery process of transmittance can be repeated for at least three cycles. Such a reversible thermal effect indicates that EYDFs can be applied to design a novel fibre-optic temperature sensor with tailored properties [11].

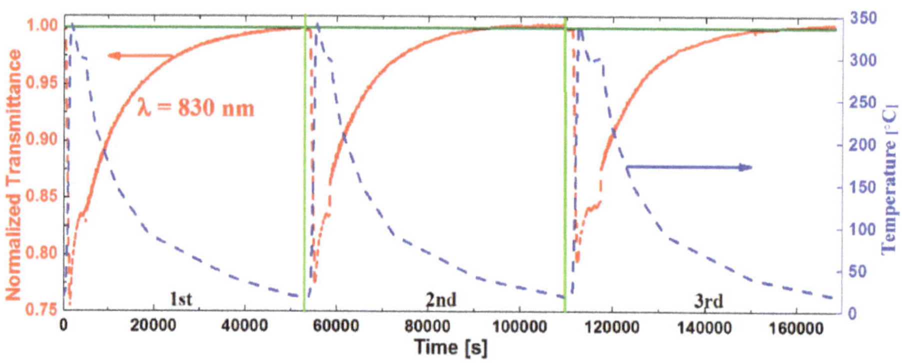

Figure 5.3. Repeatability of the induced thermal effect (normalized transmittance) between room temperature and 300 °C.

5.3 Ionizing radiation effects in Bi-doped optical fibres

Unlike EDFs, the emission spectra of BDFs are varied and much more sensitive to the microenvironment. This is mainly because Bi is a transition metal element and its outer electron shell is fully open, which makes it extremely susceptible to the influence of the external environment. According to its surrounding environments of

SiO_2, GeO_2, Al_2O_3, or P_2O_5, different bismuth active centres (BACs), such as BAC-Si, BAC-Ge, BAC-Al, and BAC-P, will be formed, respectively [13]. These BACs often have different emission profiles. In addition, despite the great progress in BDF and its devices, the configuration of the near-infrared (NIR) luminescence centre is still controversial. Previously, Bi^{+}, Bi^{5+}, BiO, Bi clusters, etc, have been assumed to be the origin of the NIR emission. Recent reports suggest that both Bi with low valence and defect centres work together for NIR emission [29], but the specific valence state and interaction are still not clear, which greatly affects the practical applications of BDF and BDF based devices. In addition, ionizing radiation has demonstrated the capability of regulating of both the valence of Bi ions and defects in the glass [5, 6, 12]. Hence, ionizing radiation is uniquely suited for the creation of and tailoring the properties of bismuth-doped materials, in particular BDFs. In the following, the effect of radiation on BDF, bismuth/erbium co-doped optical fibre (BEDF), and bismuth/erbium/ytterbium co-doped optical fibre (BEYDF) is described along with the influential factors.

5.3.1 Activation of BAC from inactive bismuth

5.3.1.1 BDF

After gamma irradiation, the absorption of Bi-doped silica fibre at 520, 700, and 800 nm increases. Absorption of BDF is clearly increased with the increase of the radiation dose. In particular, a new peak appears at ~580 nm associated with AlOHCs [16]. The luminescence evidently increases under 980 nm pumping, as shown in figure 5.4 [16], but decreases slightly under 532 nm pumping [15]. Seen from figure 5.4, when the BDF was excited at 980 nm, the luminescence intensity increases significantly with the increase of radiation doses up to 2.0 kGy. Then the luminescence decreases when the radiation doses are increased up to 3.0 kGy.

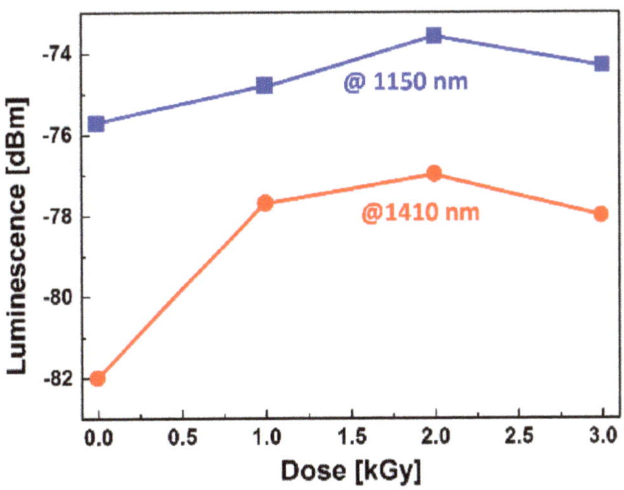

Figure 5.4. Luminescence intensity of irradiated BDF with different dosages under 980 nm 0.025 mW pumping.

In addition, the luminescence of BAC-Si at 1410 nm increases much more than that of BAC-Al at 1150 nm [16]. Furthermore, the microstructural characteristics using electron spin resonance (ESR) indicate the generation of many radiation-induced defect centres by the radiation, where the ESR signals increase with the increase of dose [16]. The increase of fluorescence under 980 nm pumping is possibly linked to the formation of the radiation-induced subvalence Bi ions [16].

5.3.1.2 BEDF

The effects of γ irradiation on BACs and the related photoluminescence properties of BEDF, fabricated by *in situ* solution doping and MCVD, have also been investigated. The samples were irradiated to 1, 5, 15, 30, and 50 kGy doses at a dose rate of 5.5 kGy h^{-1} at room temperature. The optical properties of the BEDF samples were measured before and after γ irradiation [18]. The optical transmission over the spectral range of 900–1650 nm under γ irradiation increases overall. In particular, figure 5.5(a) shows a relative change of the absorption at 1400 nm referred to a value at 1530 nm for the five values of the irradiation dose [18]. As can be seen from figure 5.5(a), such a relative change increases with the increase of radiation dose. The attenuation spectrum in [18] also indicates that the RIA has lower values at 1400 nm than that at 1530 nm after irradiation, demonstrating potential application at the shorter wavelength. In particular, it appears to be beneficial even for the increase of the dose. Thus the BEDF proves to be more reliable at 1400 nm and tolerates high radiation exposure. The photoluminescence evidently degrades gradually, as shown in figure 5.5(b), while the corresponding emission at 1100 nm remains almost unchanged after irradiation. Further investigation indicates that BEDF has good reliability under low and medium irradiation doses at quite high dose rates. Moreover, it is possible that emission and pump efficiency could be increased by irradiation if the BEDF parameters and pump conditions are optimized [17, 18].

In addition, the fluorescence lifetime of BAC-Al and Er^{3+} in BEDF was studied after gamma irradiation, as shown in figure 5.6 [19]. As can be seen from figure 5.6,

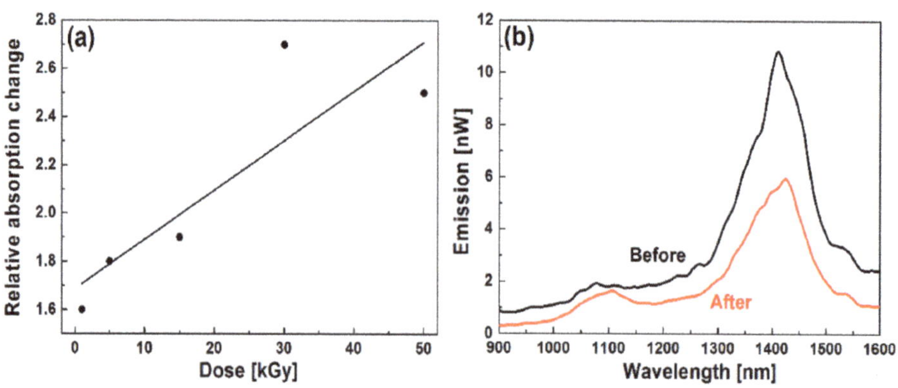

Figure 5.5. (a) The absorption change at ~1400 nm referred to that at ~1530 nm of BEDF versus dose. (b) The spectral emission of BEDF before and after 50 kGy gamma irradiation under 44.1 mW 830 nm pumping. Reprinted by permission from Macmillan Publishers Ltd [18], Copyright (2016) CCBY 4.0.

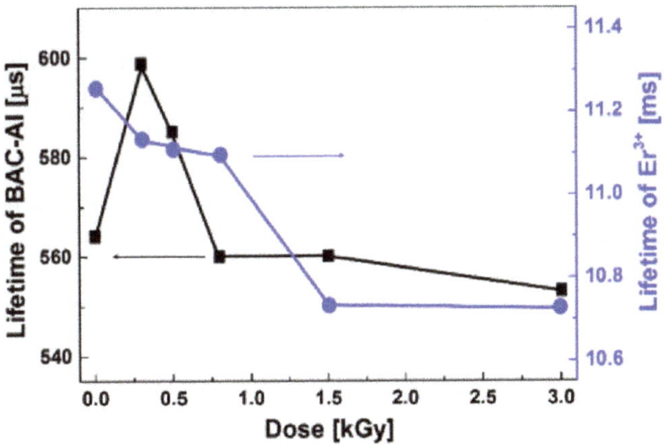

Figure 5.6. The fluorescence lifetime of the BAC-Al at 1138 nm and Er^{3+} at 1534 nm as a function of irradiated dose.

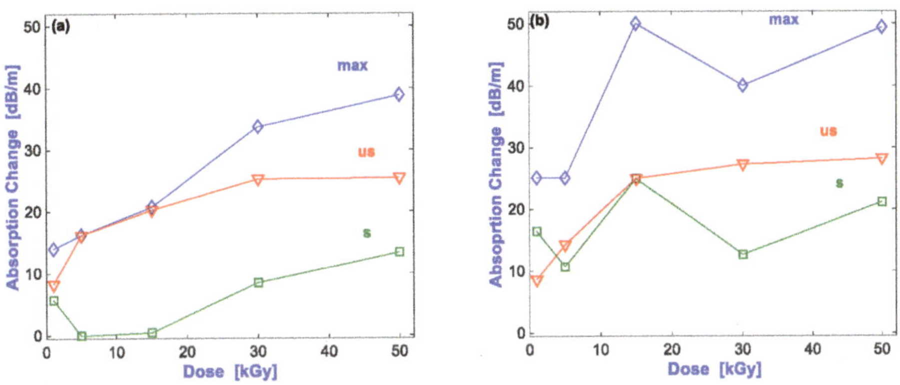

Figure 5.7. Radiation-induced absorption change of (a) BEDF and (b) BEYDF for the pump at 830 nm. Note: s, us, and max are the saturable absorption, the unsaturable absorption, and the maximum absorption. Adapted with permission from [21]. Copyright 2016 The Optical Society.

the lifetime of BAC-Al at 1138 nm is increased by 35 μs after irradiation at 0.3 kGy, possibly due to the increment of the BACs. The increment of the BACs may be caused by the change of valence state of Bi or the generation of more colour centres due to the gamma irradiation. However, this does not occur in erbium ions as the lifetime of Er^{3+} at 1534 nm slowly decreases with increasing dose.

5.3.2 Factors that influence the radiation effect

5.3.2.1 Yb^{3+} co-doping

The results above have demonstrated that the gamma irradiation can enhance the NIR luminescence of BDF and BEDF through the activation of inactive bismuth into BAC. A study of the pump absorption of BEDF at 830 nm found that after radiation the pump absorption was significantly enhanced, as shown in figure 5.7(a). With the increment of radiation dose, saturable absorption and non-saturable absorption also increase to different degrees. In particular, the increase of saturable absorption, which is the effective absorption to some extent, indicates the increase of BAC. Although more BACs are generated, the luminescence is still suppressed due to the strong RIA [20]. Furthermore, through a comparison of pump absorption, saturable absorption, and non-saturable absorption of BEDF and BEYDF (in figure 5.7) [21], it is found that for both BEDF and BEYDF gamma irradiation contributes to the conversion of inactive Bi to BAC. The increment of saturable absorption for BEYDF is higher than that for BEDF, indicating that the addition of Yb will strengthen this radiation effect. The enhanced radiation effect may be due to the conversion of Yb^{3+} to Yb^{2+} [12].

5.3.2.2 Radiation type

The radiation effect is also related to the radiation type. To see their differences, the radiation effects on BEDF induced by γ, electron, and mixed neutron/γ irradiation have been studied and compared, as shown in figure 5.8 [24]. Seen from figure 5.8, RIA and pump RIA are clearly generated, and the emission is greatly restrained. The effects produced by γ rays and electron radiation are comparable to some degree, while that induced by the mixed neutron and γ rays is relatively weak. Such results hint that the neutron radiation may have a certain inhibitory effect compared to γ rays [24].

5.3.3 Thermally induced effects in irradiated BEDF

5.3.3.1 Gamma irradiation

The thermal stability of BACs in BEDF after γ-irradiation has also been investigated through the study of their spectral characteristics [22], where the experimental results of the thermal effect on the absorption and luminescence properties of pristine BEDF and irradiated BEDFs have been obtained, as shown in figure 5.9. As can be seen from figure 5.9(a), the heat-induced loss of the pristine BEDF at 1300 nm increases slowly before the temperature is increased to 400 °C. However, a sharp increase of 48 dB m^{-1} is observed when the temperature increases to 500 °C, indicating a strong thermal darkening effect at high temperatures [30, 31]. In the cooling process, the induced loss decreases with the decrease of temperature and

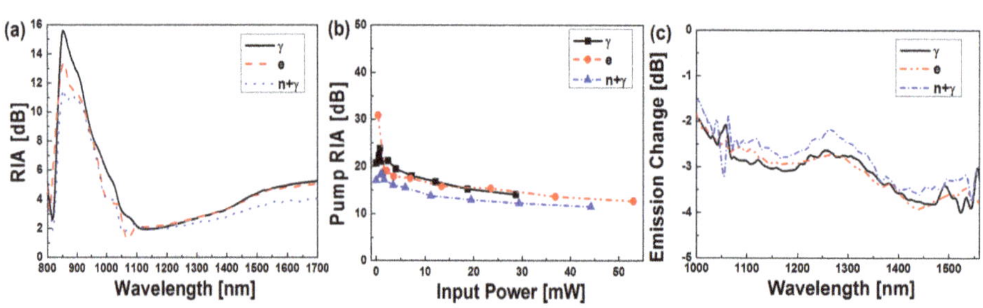

Figure 5.8. (a) RIA spectra, (b) pump RIA, and (c) emission change of BEDF with the gamma irradiation of 60 kGy, electron beam irradiation of 50 kGy, and the mixed radiation of neutron and gamma equivalent to 74 kGy. (γ: gamma irradiation; e: electron irradiation; n+γ: mixed neutron/γ irradiation).

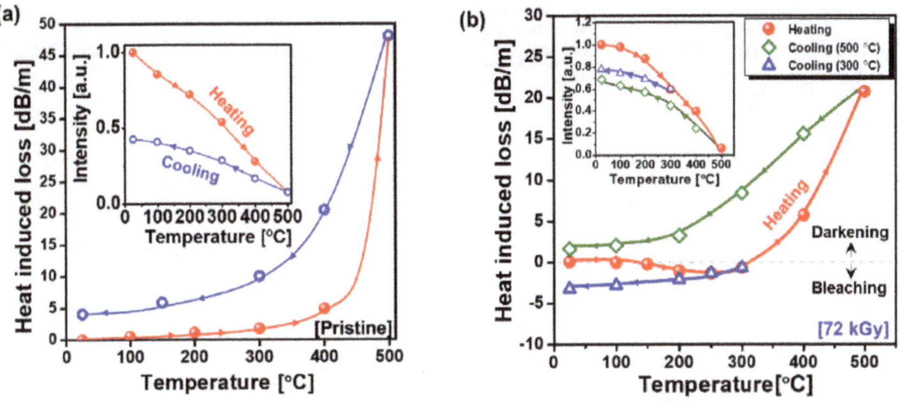

Figure 5.9. (a) Evolution of heat-induced loss at 1300 nm of the pristine BEDF during the heating (~500 °C, red dot) and cooling process (blue dot). The inset shows the evolution of luminescence at 1180 nm under 980 nm pumping. (b) Evolution of heat-induced loss at 1300 nm of the gamma irradiated BEDF (~72 kGy) heated to 300 °C and 500 °C. The inset shows the evolution of luminescence at 1180 nm corresponding to different temperatures. Adapted with permission from [22]. Copyright 2019 The Optical Society.

finally reaches 5 dB m^{-1} at room temperature. In addition, the emission of BAC-Al at 1180 nm follows a trend similar to the loss change. In contrast, as can be seen from figure 5.9(b), remarkable thermally induced bleaching of irradiated BEDF has already been observed after heating to 300 °C, possibly associated with the degradation of AlOHC. However, after further heating, induced loss at 1300 nm displays the opposite trend and thermal darkening dominates until the temperature is increased to ~500 °C. After cooling to room temperature, a smaller residual of heat-induced loss (~1.6 dB m^{-1}) is retained. Through comparison of their loss and emission performance upon heating to 500 °C, it can be seen that ~70% of the emission of BAC-Al for irradiated BEDF recovered. This is evidently higher than that (less than 50%) for pristine BEDF, which may be attributed to the annihilation of AlOHC at low temperatures and the following alleviation of the radiation-induced darkening effect to some degree. These thermal treatment results further suggest that BEDF features higher thermal stability upon γ irradiation [22].

5.3.3.2 Electron irradiation

Later on, the thermal annealing effect on the spectroscopic properties of electron beam irradiated BEDFs (140 kGy) from 500 °C was also investigated, as shown in figure 5.10 [23]. Seen from figure 5.10(a), the luminescence of BAC-Al at 1190 nm in both pristine and irradiated BEDFs reduces significantly with the heating, and is hardly observed at 500 °C. In particular, the luminescence of the pristine BEDF degrades faster than that of the irradiated one. With sample cooling, the luminescence at 1190 nm for both samples recovers slowly. But the recovery amount for irradiated BEDF (~71.4%) is larger than that of the pristine one (~42.4%). Such observation is in agreement with the evolution of the heat-induced loss as shown by the inset of figure 5.10(b), where induced loss of irradiated BEDF is almost two times higher than that of the pristine one.

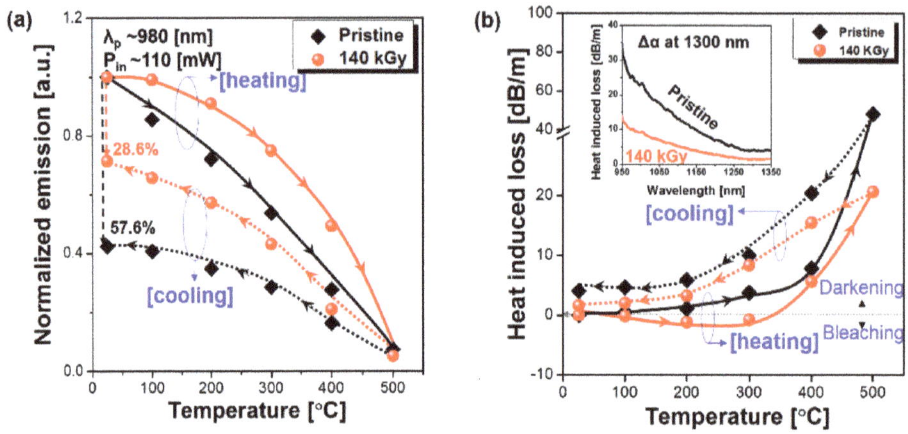

Figure 5.10. (a) Evolution of luminescence at 1190 nm and (b) heat-induced loss at 1300 nm of pristine and electron beam irradiated (140 kGy) BEDFs during heating and cooling. Inset: heat-induced loss spectra of pristine and irradiated BEDFs after thermal treatment. Adapted with permission from [23]. Copyright 2019 The Optical Society.

As can be seen from figure 5.10(b), noticeable thermal bleaching has been observed in electron irradiated BEDF at ~300 °C, while the thermal darkening in pristine BEDF is aggravated by further heating. The thermal bleaching evidently confirms the annihilation of AlOHC in irradiated BEDF, similar to the observation in gamma irradiated BEDF [22]. In particular, due to the alleviation of thermal bleaching of AlOHC in the irradiated BEDF, it features a slower decrease rate and a higher recovery ratio compared to the pristine case. All these results demonstrate that electron irradiation can also be used to improve the thermal stability of BEDF, similar to gamma irradiation.

Moreover, through comparison of figure 5.9 and figure 5.10, it can be seen that the luminescence of electron irradiated BEDF recovered more than that of gamma irradiated BEDF, although their residual of thermally induced loss is almost equivalent to each other. Such an observation suggests that the irradiation of BEDF with an electron beam should be more beneficial for thermal stability than that with gamma radiation.

5.4 Radiation hardening

In order to eliminate and suppress RIA, combined thermal treatment and hydrogen loading methods have been employed to try to improve the radiation resistance of fibres [5, 25]. In addition, it is worth mentioning that there are many radiation hardening methods, which include Ce co-doping [32–35], H_2 or D_2 loading [2, 35–38], hermetic coatings [2, 37], hole-assisted carbon-coatings [39], bleaching by lasers or heating [11, 40–42], etc [43]. In the following, combined thermal treatment and H_2 loading will be described.

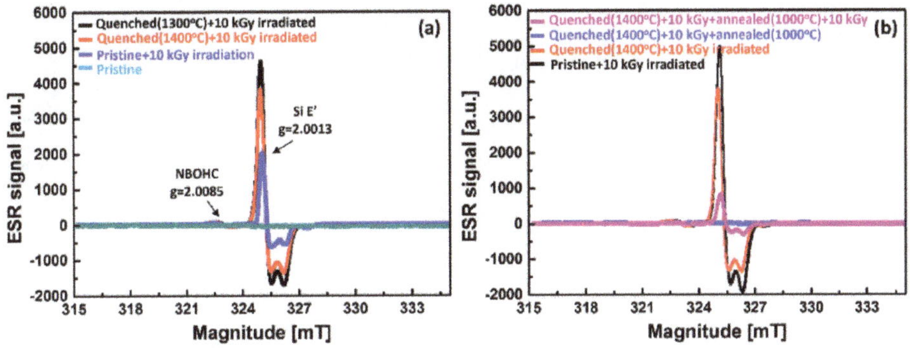

Figure 5.11. The ESR spectra of silica fibre cladding materials (a) after quenching and irradiation processing and (b) with further annealing and re-irradiation processing. Adapted with permission from [5]. Copyright 2016 Elsevier.

5.4.1 Combined thermal treatment of silica fibre cladding materials

To improve radiation hardening, combined thermal treatments, including thermal quenching, irradiation, then annealing and subsequent re-irradiation, have been applied to the cladding materials of silica fibre preforms [5]. Through the investigation of the microstructural properties after treatment, such as ESR, the radiation effect, and resistance, have been assessed. Figure 5.11(a) and (b) show the ESR spectra of silica fibre cladding materials after quenching and irradiation processing and with further annealing and re-irradiation processing, respectively [5]. As can be seen from figure 5.11(a), the ESR signal is greatly enhanced after radiation due to the breakage of Si–O bonds by the gamma irradiation. In particular, for samples after quenching from high temperature, the ESR signal becomes stronger due to the modification of the microstructure by thermal quenching. Interestingly, the ESR signal of a sample quenched from 1400 °C is weaker than that from 1300 °C, indicating that higher temperature quenching will increase the radiation stability. As can be seen from figure 5.11(b), subsequent annealing at 1000 °C will further improve the radiation hardening of the materials, as the ESR signal with the full combined treatment becomes significantly weaker than that without thermal annealing and quenching. All these results demonstrate that combined thermal treatment can provide an alternative route to fabricate silica optical fibre materials with improved radiation hardness [5].

5.4.2 H_2 loading of BEDF

The radiation hardening of BEDF through H_2 loading has been investigated, as shown in figure 5.12 [25]. As can be seen from figure 5.12(a), the RIA at 710 nm corresponding to BAC-Al absorption in the BEDF without H_2 loading increases quickly and reaches 48.0 dB m^{-1} when the radiation dose is 6.7 kGy. When the radiation dose is up to 35.6 kGy, the RIA arrives at a maximum of 63.6 dB m^{-1}. However, after H_2 loading the increase of RIA is significantly restrained with only 11.5 dB m^{-1} at the maximum radiation dose. A similar variation trend has been

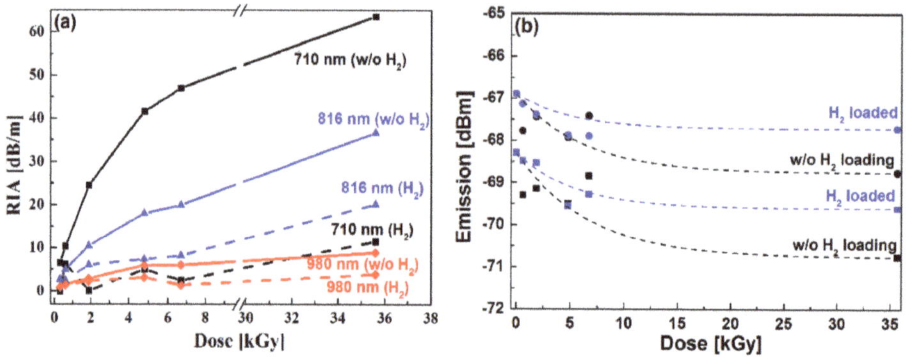

Figure 5.12. (a) RIA at 710 nm, 816 nm, and 980 nm for BEDF with and without H_2 loading as a function of radiation dose. (b) The emission intensity of BAC-Al at 1130 nm (square symbols) and BAC-Si at 1420 nm (round symbols) of BEDF with and without H_2 loading as a function of radiation dose.

observed for the RIA at the resonant absorption of BAC-Si at 816 nm and Er^{3+} at 980 nm. All these results clearly demonstrate that after H_2 loading the RIA is significantly decreased. In addition, as can be seen from figure 5.12(b), the emission of BACs in irradiated BEDFs was decreased, in particular the decrement of H_2-loaded BEDF is ~ 1 dB less than that of the counterpart. All these results indicate that H_2 loading is an effective method to achieve radiation resistance of BEDF [25], similar to the reports for EDF [44–47].

5.5 Applications

As mentioned above, there are many applications for the radiation effect of optical fibres, such as inter-satellite communications, lidar, fibre-optic gyroscopes, nuclear plants, and radiation therapy, etc [1–4]. Here, a very promising fibre-optic temperature sensor based on the fluorescence intensity ratio (FIR) in EYDF is demonstrated as an example.

5.5.1 A radiation-independent fibre-optic temperature sensor

This temperature sensor will use the FIR of the emission transitions of $^2H_{11/2} \rightarrow {}^4H_{15/2}$ and $^4S_{3/2} \rightarrow {}^4H_{15/2}$ of Er^{3+}, which are centred at 524 nm (I_1) and 545 nm (I_2), as shown in figure 5.13(a) [26]. As can be seen from figure 5.13(a), the emission (I_1) increases monotonically with the increment of the temperature from 300 K to 460 K, while that of I_2 varied a little. As a result, the FIR increases almost monotonically with the increment of the temperature (T). In particular, it can be fitted well with the equation

$$\mathrm{FIR} = A\exp\left[\frac{-\Delta E}{k_B T}\right] + B, \tag{5.2}$$

as shown in figure 5.13(b) [26], where ΔE is the difference of the corresponding two energy levels, k_B is the Boltzmann constant, and A and B are the fitting parameters.

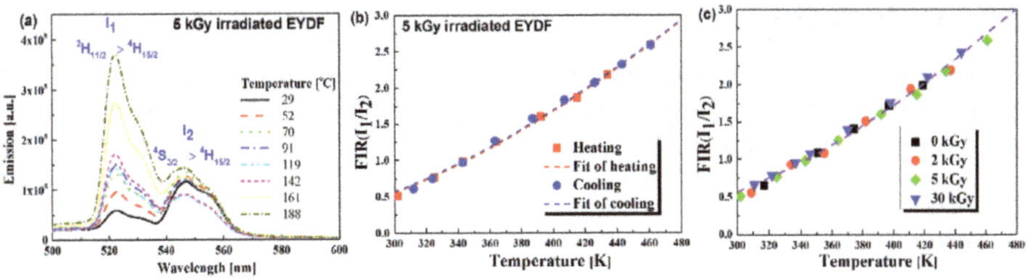

Figure 5.13. (a) Emission spectra of Er^{3+} for 5 kGy irradiated EYDF at different temperatures. (b) The corresponding FIR values versus temperature in the heating and cooling process. (c) FIR versus temperature for EYDFs irradiated with different doses.

As can be seen from figure 5.13(b), the fitting curves of FIRs in the heating and cooling processes overlap well with each other, demonstrating good repeatability. Furthermore, the FIR of EYDFs irradiated with different gamma radiation doses are measured and are almost the same, as shown in figure 5.13(c) [26]. Such results demonstrate that the temperature response of the FIR temperature sensors are largely not influenced by the radiation, showing great potential for temperature sensing in a gamma radiation environment [26].

5.6 Conclusion

For future radiation applications, the radiation effects on optical fibres, including EYDF, BDF, BEDF, etc, were systematically studied. These investigations indicate that strong RIA is commonly observed in all these doped fibres, which often results in a decrease of the luminescence. RIA in EYDF is mainly attributed to the defect centre of AlOHC. In particular, irradiated EYDF has a reversible thermal effect, which can lay the foundation of the temperature sensing in the radiation environment. In BDFs the radiation can activate some inactive Bi ions into BACs, which sometimes enhances the luminescence. However, due to the strong RIA the emission of BAC is often restrained. In addition, Yb co-doping can effectively enhance such a radiation effect. A comparative study found that the effects of gamma ray and electron beam irradiation are comparable. Combined thermal treatment or H_2 loading can effectively enhance the radiation hardening of the optical fibres. Finally, based on the reversible thermal effect a very promising radiation-independent FIR based fibre-optic temperature sensor has been demonstrated with EYDF, which exhibits great application potential in the field of temperature sensing in a radiation environment.

Funding

National Natural Science Foundation of China (NSFC) (61520106014 and 61675032), Science and Technology Commission of Shanghai Municipality, China (SKLSFO2018-02), and 111 Project (D20031).

Acknowledgement

The authors acknowledge many colleagues and partners who are involved with this work including Tingyun Wang, Dan Sporea, L Mihai, D Negut, Justin Davies, Shuen Wei, Bowen Zhang, Jiajun Ma, Edward Obbard, Yuan Tian, Xinghu Fu, etc.

References

[1] Brichard B, Fernandez A F, Ooms H and Berghmans F 2003 Gamma dose rate effect in erbium-doped fibers for space gyroscopes *Int. Conf. on Optical Fiber Sensors* p 336

[2] Zotov K V *et al* 2008 Radiation resistant Er-doped fibers: optimization of pump wavelength *IEEE Photon. Technol. Lett.* **20** 1476–8

[3] Haddad E H E *et al* 2016 Comparison of gamma radiation effect on erbium doped fiber amplifiers *Int. Conf. on Space Optics, ICSO 2016 (Biarritz, France, 17–21 October 2016)*

[4] Alam M, Abramczyk J, Madasamy P, Torruellas W and Sanchez A 2007 Fiber amplifier performance in gamma radiation environment *Optical Fiber Communication Conf. and Exposition and The National Fiber Optic Engineers Conf. (Anaheim, CA)* (Washington, DC: Optical Society of America) OMF4

[5] Wen J *et al* 2016 Effects of quenching, irradiation, and annealing processes on the radiation hardness of silica fiber cladding materials (I) *Opt. Fiber Technol.* **30** 95–9

[6] Skuja L 1998 Optically active oxygen-deficiency-related centers in amorphous silicon dioxide *J. Non-Cryst. Solids* **239** 16–48

[7] Griscom D L, Friebele E J, Long K J and Fleming J W 1983 Fundamental defect centers in glass: electron spin resonance and optical absorption studies of irradiated phosphorus-doped silica glass and optical fibers *J. Appl. Phys.* **54** 3743–62

[8] Girard S *et al* 2019 Overview of radiation induced point defects in silica-based optical fibers *Rev. Phys.* **4** 100032

[9] Girard S *et al* 2013 Radiation effects on silica-based optical fibers: recent advances and future challenges *IEEE T. Nucl. Sci.* **60** 2015–36

[10] Origlio G, Girard S, Cannas M, Ouerdane Y, Boscaino R and Boukenter A 2009 Paramagnetic germanium-related centers induced by energetic radiation in optical fibers and preforms *J. Non-Cryst. Solids* **355** 1054–6

[11] Fan D *et al* 2018 Radiation-induced reversible thermal effect in Er^{3+}/Yb^{3+}-codoped silica fibers *Opt. Lett.* **43** 3385–8

[12] Wei S, Ding M, Fan D, Luo Y, Wen J and Peng G-D 2018 Effects of post treatments on bismuth-doped and bismuth/erbium co-doped optical fibres *Bismuth—Advanced Applications and Defects Characterization* (Rijeka: IntechOpen)

[13] Luo Y, Yan B, Zhang J, Wen J, He J and Peng G-D 2018 Development of Bi/Er co-doped optical fibers for ultra-broadband photonic applications *Front. Optoelectron.* **11** 37–52

[14] Fan D, Sporea D, Xiao G, Luo Y, Wei S and Peng G-D 2017 γ-radiation-induced photodarkening in Er/Yb co-doped fibre and its thermal recovery *ANZCOP2017 (New Zealand)*

[15] Wang J *et al* 2014 Influence of gamma-ray irradiation on the spectral properties of Bi-doped silica fibers *Asia Communications and Photonics Conf. 2014 (Shanghai)* (Washington, DC: Optical Society of America) ATh4C.4

[16] Wen J *et al* 2015 Radiation-induced photoluminescence enhancement of Bi/Al-codoped silica optical fibers via atomic layer deposition *Opt. Express* **23** 29004–13

[17] Sporea D *et al* 2015 Effects of gamma rays on rare Bi/Er co-doped optical fibers *11th Int. Conf. on Optics 'Micro- to Nano-Photonics IV' (Bucharest, Romania)*

[18] Sporea D *et al* 2016 γ irradiation induced effects on bismuth active centres and related photoluminescence properties of Bi/Er co-doped optical fibres *Sci. Rep.* **6** 29827

[19] Chen L *et al* 2018 Influence of gamma-ray irradiation on the fluorescence lifetime of Bi/Er co-doped fibers *2018 Asia Communications and Photonics Conf. (ACP) (26–29 October 2018)* pp 1–3

[20] Yan B *et al* 2015 Gamma radiation-induced formation of bismuth related active centre in Bi/Er/Yb co-doped fibre *Asia Communications and Photonics Conf. 2015 (Hong Kong)* (Washington, DC: Optical Society of America) ASu2A.56

[21] Yan B *et al* 2016 Enhanced gamma radiation effect in Bi/Er co-doped optical fibre by co-doping Yb *Asia Communications and Photonics Conf. 2016 (Wuhan)* (Washington, DC: Optical Society of America) AF2A.135

[22] Zhao Q C, Zhang J Z, Sporea D, Luo Y H, Wen J X and Peng G D 2019 Gamma radiation and thermal-induced effects on the spectral performance of BACs in Bi/Er codoped aluminosilicate fibers *Opt. Express* **27** 9955–64

[23] Zhao Q, Luo Y, Hao Q and Peng G-D 2019 Electron beam irradiation and thermal-induced effects on the spectral properties of BAC-Al in Bi/Er codoped aluminosilicate fibers *Opt. Mater. Express* **9** 4287–94

[24] Luo Y *et al* 2018 Ionising radiation induced effects on bismuth/erbium co-doped optical fibres *2018 Asia Communications and Photonics Conf. (ACP) (26–29 October 2018)* pp 1–3

[25] Ding M *et al* 2019 Spectroscopic study of the radiation hardening of bismuth/erbium co-doped optical fiber (BEDF) by hydrogen loading *Opt. Mater.* **95** 109246

[26] Fan D *et al* 2019 Gamma radiation effects on the performance of FIR-based fiber-optic temperature sensors *the Asia Communications and Photonics Conf. (Chengdu, Sichuan, China)* M4A.145

[27] Luo Y, Yan B, Wen J, Zhang J and Peng G-D 2019 Measurement of optical fiber amplifier *Handbook of Optical Fibers* ed G D Peng (Singapore: Springer Nature)

[28] Griscom D L, Gingerich M E and Friebele E J 1993 Radiation-induced defects in glasses: origin of power-law dependence of concentration on dose *Phys. Rev. Lett.* **71** 1019–22

[29] Dianov E M 2015 Nature of Bi-related near IR active centers in glasses: state of the art and first reliable results *Laser Phys. Lett.* **12** 095106

[30] Zlenko A S *et al* 2012 Mechanisms of optical losses in Bi:SiO_2 glass fibers *Opt. Express* **20** 23186–200

[31] Wei S *et al* 2017 Thermal effect on attenuation and luminescence of Bi/Er co-doped fiber *IEEE Photon. Technol. Lett.* **29** 43–6

[32] Vivona M *et al* 2011 Influence of Ce codoping and H_2 pre-loading on Er/Yb-doped fiber: radiation response characterized by confocal micro-luminescence *J. Non-Cryst. Solids* **357** 1963–5

[33] Thomas J *et al* 2012 Radiation-resistant erbium-doped-nanoparticles optical fiber for space applications *Opt. Express* **20** 2435–44

[34] Ladaci A *et al* 2018 X-rays, γ-rays, electrons and protons radiation-induced changes on the lifetimes of Er^{3+} and Yb^{3+} ions in silica-based optical fibers *J. Lumin.* **195** 402–7

[35] Girard S *et al* 2012 Radiation hardening techniques for rare-earth-based optical fibers and amplifiers *Proc. SPIE* **8246** 82460Y

[36] Zotov K V, Likhachev M E, Tomashuk A L, Bubnov M M, Yashkov M V and Gur'yanov A N 2007 Radiation-resistant erbium-doped silica fibre *Quan. Electron.* **37** 946–9

[37] Zotov K V *et al* 2008 Radiation-resistant erbium-doped fiber for spacecraft applications *IEEE T. Nucl. Sci.* **55** 2213–5

[38] Girard S *et al* 2014 Radiation-hardened erbium-doped optical fibers and amplifiers for future high-dose space missions *Proc. SPIE* **8971** 89710E

[39] Girard S *et al* 2014 Radiation-hard erbium optical fiber and fiber amplifier for both low- and high-dose space missions *Opt. Lett.* **39** 2541–4

[40] Ladaci A *et al* 2018 Radiation hardened high-power Er^{3+}/Yb^{3+}-codoped fiber amplifiers for free-space optical communications *Opt. Lett.* **43** 3049–52

[41] Sheng Hsiung C, Ren-Young L, Chu-En L, Fong-In C, Chao-Yi T and Chii-Chang C 2013 Photo-annealing effect of gamma-irradiated erbium-doped fibre by femtosecond pulsed laser *J. Phys. D: Appl. Phys.* **46** 495113

[42] Mady F, Duchez J, Mebrouk Y and Benabdesselam M 2015 Equilibrium degradation levels in irradiated and pumped erbium-doped optical fibers *IEEE T. Nucl. Sci.* **62** 2948–55

[43] Ladaci A *et al* 2017 Optimized radiation-hardened erbium doped fiber amplifiers for long space missions *J. Appl. Phys.* **121** 163104

[44] Likhachev M E *et al* 2013 Radiation resistance of Er-doped silica fibers: effect of host glass composition *J. Lightwave Technol.* **31** 749–55

[45] Sheng Y *et al* 2012 Improvement of radiation resistance by introducing CeO_2 in Yb-doped silicate glasses *J. Nucl. Mater.* **427** 58–61

[46] Girard S *et al* 2012 Radiation hardening techniques for Er/Yb doped optical fibers and amplifiers for space application *Opt. Express* **20** 8457–65

[47] Zotov K *et al* 2007 Radiation-resistant erbium-doped fiber for spacecraft applications *9th European Conf. on Radiation and Its Effects on Components and Systems* (Piscataway, NJ: IEEE) pp 1–4

IOP Publishing

Optical Fiber Technology and Applications
Recent advances
Mário F S Ferreira and Mukul Chandra Paul

Chapter 6

Optical fiber sensors

Daniele Tosi

Optical fiber sensors are an emerging technology for the detection of physical, electrical, and biochemical parameters. Optical fibers are compact, immune to electromagnetic fields, and can implement several high-performance sensing systems that outperform many of their electronic or mechanical counterparts. The key feature of optical fiber sensors is the possibility to stack several sensors within the same fibers, using advanced multiplexing methods that can detect hundreds of sensors up to distributed sensors with their hundreds of thousands of sensing points. In this chapter, an overview of the main optical fiber sensor architectures and applications is provided, defining the basic principle and simple methods of detection up to the most advanced distributed sensors and biosensors.

6.1 Introduction

Optical fiber sensors (OFSs) make use of optical fibers, or optical fiber devices, to detect the change of physical, chemical, or biological parameters such as temperature, pressure, strain, or concentration of specific analytes (Udd and Spillman 2011). Compared to other traditional sensing technologies, such as piezoelectric sensors, electrochemical sensors, or micro/nano-electromechanical systems, OFSs have specific advantages that are of extreme importance in many applications. The first key advantage is related to the fiber properties, as the fiber is made of silica and is designed to both transport light to the sensing points and as a sensing mechanism itself. OFSs are passive networks, based on a safe and biocompatible design, with a miniature footprint (usually 0.1–1 mm thickness), and a lightweight and compact form factor. As fibers have loss around 0.2 dB km^{-1}, OFSs can perform long-range remote sensing over tens, or hundreds, of kilometers. For applications in medicine, fibers are compatible with the ISO10993 standard and are magnetic resonance imaging (MRI)-compatible (Tosi *et al* 2018). The other key properties rely on the electromagnetic properties of the sensing devices—methods based on filters (such as the fiber Bragg gratings), or on optical reflectometry in the time or frequency

doi:10.1088/978-0-7503-3243-9ch6

domain, are able to resolve physical parameters in space. Systems based on time-domain reflectometry can resolve strain and/or temperature over hundreds of kilometers with a resolution of 10–100 cm (Baldwin 2018) and are widely used in pipeline monitoring and geotechnical applications, whereas optical frequency-domain reflectometry can resolve below the millimeter and plays a significant role in modern healthcare sensors (Tosi *et al* 2018). Through proper design it is possible to detect multiple parameters with a single sensor. For example, the dual detection of pressure and temperature with a single probe is a key asset in high-pressure high-temperature detection (Poeggel *et al* 2015). As a further property, fiber optic sensors have a well-known temperature sensitivity, suitable for compensation, which limits the thermal fluctuations characterizing many electronic or piezoelectric sensors. Finally, OFSs are suitable for harsh environments, thanks to the radiation-compatibility of fibers and their ability to survive and detect in high temperature, high pressure, and high radiation environments (Mihailov 2012).

Thanks to their properties OFSs have found applications in many contexts where traditional sensors cannot approach the same combination of spatial resolution, size, and performance. In civil engineering OFSs can be integrated into composites (Leng and Asundi 2003) and implement structural health monitoring with non-invasive structures, such as long-term monitoring of bridges and infrastructure (Rodrigues *et al* 2010). Long-haul temperature and acoustic sensors can perform monitoring of pipelines over several kilometers, detecting early cracks and providing an essential tool for downstream and upstream oil and gas applications (Baldwin 2018). Biomedical sensors instead take advantage of the miniature form factor of the fibers and are currently applied as a method to monitor mini-invasive catheters, to detect torsion and shape, as well as to map temperatures in 1D, 2D, and 3D configurations with resolution even better than imaging (Tosi *et al* 2018). OFSs can also be used to design biosensors, which selectively detect the presence and concentration of specific analytes such as proteins, cancer biomarkers, or the antibody–antigen interaction. Compared to traditional technologies such as immunoassays, OFSs are label-free and do not need any activation or labeling procedure but instead work as an instantaneous detector (Leung *et al* 2007).

In this chapter an overview of the main OFS technologies is provided, presenting how the fiber optic properties can be used to turn a fiber device into a sensor. In the first part, the basic definitions and schemes, as well as simple sensors are presented. Then the focus is moved to gratings and interferometers, which constitute the backbone of current sensing technologies. Subsequently, the emerging technologies in distributed sensors and in biosensors are presented. Finally, conclusions are drawn, highlighting the current trends and challenges.

6.2 Overview of fiber optic sensor techniques

The key principles of OFSs make use of the well-known electromagnetic properties to convert the changes of the mechanical, thermal, or biological parameters under test into a change of the light properties travelling through the sensing region. Light travels through the fiber following the basic electromagnetic properties of guided propagation (Agrawal 2012); as such, light transports an optical power and carries a spectral envelope, a phase, and a polarization, and propagates through well-defined modes. Each of these parameters can be exploited for sensing by modulating the light parameter through the measurand. Intensity-based sensors make use of the changes of the light intensity, which can be measured through a photodiode or a camera, and constitute one of the simplest and most inexpensive formats of OFSs (Tosi *et al* 2008). Spectral detection is a common method of detection and it applies to gratings, interferometers, surface plasmon resonance, and many other methods; it involves the detection of wavelength shifts on the reflection or transmission spectrum of resonant devices (Tosi 2017). Phase changes of light travelling into multiple optical paths, or into a fiber span, are at the basis of optical interferometers (Lee *et al* 2012) and distributed sensors (Lu *et al* 2010), respectively. The possibility to engineer the fiber modes is at the basis of the emerging shape-sensing technologies, which exploit multi-core fibers to track 3D strain vectors (Gasulla *et al* 2017).

Among the numerous configurations, we can classify different typologies of fiber optic sensors, as outlined in figure 6.1. The first distinction is between sensors working in transmission and in reflection. In the first case, the light source transmits the light through the sensing regions and the detector is placed on the opposite side. This configuration is often used in time-domain sensing, as it allows higher power to be delivered through the photodetector and is compatible with all sensors. On the other hand, for some applications, particularly in medical devices (Tosi *et al* 2018), it is not practical to work on transmission sensors as we cannot place the source and detector at opposite sides of the link. Reflection-based sensors instead collect the light reflected by the sensor(s); the systems make use of a coupler, or circulator, to

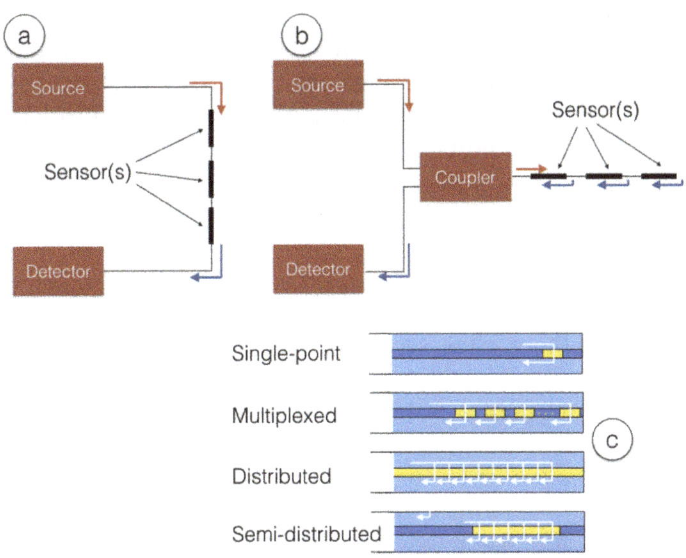

Figure 6.1. Fiber optic sensing systems. (a) Schematic of a system working in transmission. (b) Schematic of a system working in reflection. (c) Typologies of fiber optic sensors, in terms of the design of the sensing points: single-point; multiplexed; distributed; and semi-distributed.

route the light from the source to the sensing fiber and collect the reflection element with a detector. In this system the light source and detector, as well as the coupler and additional circuitry, can be embedded in the same device which works as both a transmitter and detector. The drawback is that some sensors (long period gratings, tilted fiber Bragg gratings, surface plasmon resonance) are transmission devices and do not encode the sensing information into a reflection component. To circumvent this stalemate it is possible to terminate the fiber with a mirror and collect the so-called transmission–reflection spectrum, i.e. the transmission spectrum backreflected by the mirror.

As previously anticipated, OFSs have the unique advantage of implementing sensing systems with different designs of multiple sensing points, as outlined in figure 6.1(c). In addition to standard single point sensors, there are methods to interrogate multiple sensors in the same fiber. The first method is to multiplex an array of sensors. This approach is very common with fiber Bragg gratings, which have spectra that can be disambiguated in the wavelength domain through the wavelength-division multiplexing (WDM) approach (Tosi *et al* 2018). Distributed sensors turn the whole fiber into a sensor by measuring the backscattering components generated in every section of the fiber (Bao and Chen 2012). In this case, rather than having a plurality of sensing points placed in different parts of the fiber, the whole fiber is treated as a series of sensors, with a spatial resolution that depends on the interrogation system. Finally, semi-distributed sensors are characterized by the presence of an active region where the sensing mechanism is encoded, but within the sensing region the different parts of the sensors are resolved in space. This approach is used in chirped fiber Bragg gratings (Korganbayev *et al* 2018), whereas the spectrum of the device is interpreted to measure temperature or strain distributions over a few centimeters of length. The possibility to have a multiplexed, distributed, or semi-distributed sensor allows spatially resolving parameters in space, hence moving from the standard measurement of a physical parameter to resolving a parameter along the fiber direction. The placement of the fiber, as well as the simultaneous detection of multiple fibers, allows measuring along 1D, 2D, and 3D structures.

6.3 Intensity-based sensors

The simplest OFS architectures rely on the change of intensity of the light propagating through the sensing area—the change of a physical parameter, such as displacement or pressure, results on a variation of the power at the photodetector. These methods are inherently low-cost, as the detector is based on a simple photodiode rather than on a spectrometer or optical spectrum analyzer. Intensity-based sensing methods are best compliant with multi-mode fibers (MMFs), which give specific advantages in terms of sensitivity and simplicity. MMFs, such as glass fibers with a 400 μm core diameter or polymer optical fibers (POFs) with a 1 mm diameter, have a large numerical aperture, which allows an easy coupling even from LED sources, hence simplifying the optical circuitry. Modern systems, such as the strain sensor presented in (Aitkulov and Tosi 2019) or the refractometer presented in

(Bremer and Roth 2015), make use of smartphones as light inputs, as the flashlight of the phone can be coupled into the fiber link and from the fiber to the camera. Visible light LEDs are also heavily used in intensity-based sensors and multiple wavelengths can be operated. For example, the system of Vallan *et al* (2012) makes use of a two-wavelength system (red/blue) and a network of filters to provide a reference arm to the sensor. The other key advantage is that multimode fibers have a better capacity to modulate the optical power in lossy structures, such as fiber bends, thanks to the fact that there are plenty of weakly guided modes that can easily be converted into evanescent waves. For example, a standard POF operating in the green wavelength propagates around 3.6 million modes, hence small perturbations of the fiber (e.g. micro-holes, fiber bends, fiber thinning) can easily couple the light out of the fiber, changing the intensity in a measurable way.

6.3.1 Fiber bundles

The fiber bundle system makes use of a bundle of at least two fibers, transmitting light into a movable target. The most common configuration is the bifurcated fiber bundle (BFB), which transmits light from a transmitting fiber (TX) into a target having a set displacement, and collects the output with a secondary receiving fiber (RX) as well as on the TX fiber. This configuration can be used as a displacement sensor, as shown in figure 6.2 (Tosi *et al* 2013). When the light from the TX fiber is reflected from the target, depending on the displacement value a different amount of fiber is collected by the TX and RX fibers. The system reported in Tosi *et al* (2013)

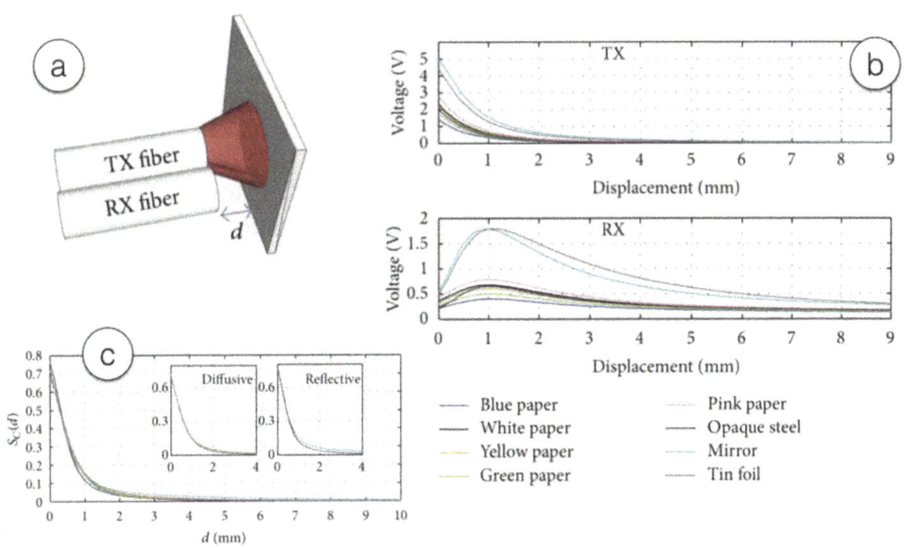

Figure 6.2. Fiber bundle based on a pair of POFs for displacement (d) detection. (a) Schematic of the system based on a TX and one RX. (b) Output voltage detected from the TX and RX fibers for different values of displacement and different types of targets. (c) Compensated output function S_C, calculated as the ratio between the RX and TX fiber outputs, and showing the reflectivity-compensated displacement dependence. Reproduced from Tosi *et al* (2013) with permission from Hindawi.

makes use of a bundle of two POFs made of polymethyl methacrylate, with a 1 mm cladding size, 20 μm perfluorinated cladding thickness, and 2.2 mm protective jacket, with fibers placed in contact.

The amount of light collected by each fiber is shown in figure 6.2(b), which shows the output voltage at the photodetector as a function of displacement. When the target is in contact with the fiber, the TX fiber receives the maximum power, which tends to decrease as the displacement increases. The RX fiber instead has a peak in correspondence to 1 mm, while for lower values of displacement only a portion of the fiber is illuminated. A simple measurement of the displacement based on the TX (or RX) fiber would be dependent on the reflectivity of the target. The results obtained for different diffusive targets (such as paper or opaque metals) and reflective targets (such as mirrors or bright metal) show that the set of curves has a trend proportional to the target reflectivity. By normalizing the curves into a compensation factor S_C, computed calculating the weighted ratio of TX and RX elements, we can normalize away the reflectivity term and provided a readout that is independent of the reflectivity of the target in a range up to 2 mm. As shown in Vallan *et al* (2012), the system can be further improved through the presence of wavelength filters, which allow or block the transmission to the mirror.

The fiber bundle is a simple yet effective configuration for the measurement of a movable target or relative displacement (for example, to evaluate the thickness of layers or mechanical components) as it can work with any reflectivity value. If the target is mounted on a pressure-sensing diaphragm, converting a change of environmental pressure into a displacement, the fiber bundle implements pressure detection; this is the principle of Camino pressure sensors. Instead, by moving the RX fiber onto the same plane as the TX fiber, it is possible to detect the breathing rate pattern (Aitkulov and Tosi 2019), measuring its frequency and shape.

6.3.2 Bent fiber sensors

Another possibility for simple and high-sensitivity systems is to use bending as a way to modulate the losses of an optical fiber. When a multi-mode optical fiber is bent around a tight corner, with a bending radius that is comparable with the radius of the fiber (up to 30 times), each mode experiences a different loss that strongly depends on the confinement. Lower order modes with a higher confinement factor experience lower loss, while higher order weakly confined modes experience higher loss, which is also dependent on the geometry of the sensor (bending radius, refractive index around the fiber, and change of temperature) (Sai *et al* 2009). In comparison to sensors based on the fiber coupling, such as fiber bundles, in which the relative movements of the fiber(s) with respect to a reference is exploited, sensors based on bending effects aim to detect the changes in the environment around the fiber.

A single fiber bent in a U-shaped pattern is used for the measurement of the refractive index (RI) of the environment. This system has been reported by Sai *et al* (2009), for the measurement of RI in the range 1.33–1.35, while Gowri and Sai (2016) reported the dependence of U-bent sensors for different bending radii and fiber geometries. POFs, with 1 mm diameter, experience the highest sensitivity to RI, as the presence of a higher number of modes increases the sensitivity. In this fiber rays input to the bent region can be either totally internally reflected and reach the detector, or be partially reflected into the fiber and partially transmitted into the outer medium. This latter contribution converts to fiber losses, which are therefore RI-dependent, and provide a simple and yet effective way to measure the changes of RI around the fiber. For a POF with 1 mm diameter the power detected in air is 42% higher than the power detected in water, corresponding to a change of ~127%/RIU (refractive index unit) of the output power in this range, while the sensitivity is higher for values of RI > 1.4.

Another popular approach is macro-bending, where a fiber (in general a POF or a glass MMF with core radius $\geqslant 400\ \mu$m) is bent along a patterned shape, such as a continuous alternated 'snake'-shaped pattern as proposed by Massaroni *et al* (2015). In this configuration the fiber is bent at a larger radius, but as the light travels through a longer lossy region the intensity modulation has a high efficiency. When the light travels through a patterned bending section losses are modulated by the strain acting on the shape. Extensive movements loosen the curve pattern reducing the losses, while compressive motions shrink the bending corners, which result in larger attenuation. Since the fiber has a fast response, macrobending sensors are commonly used in the detection of alternated signals in the time/frequency domain, such as the detection of cardiocirculatory and breathing patterns. The fiber can be incorporated in a flexible element, which transfers the strain from the outer body to the sensor as such; this approach is commonly employed in medical smart textiles (Massaroni *et al* 2015), which incorporate the sensors into a wearable device for real-time analysis.

6.4 Fiber Bragg grating sensors

Fiber Bragg gratings (FBGs) are one of the most popular and versatile sensing devices based on optical fibers, thanks to their straightforward sensing principle, ease of fabrication, and possibility of multiplexing (Kersey *et al* 1997). An FBG is a periodic modulation of the refractive index of an optical fiber along the propagation direction z. The modulation is impressed in the fiber by means of a direct inscription with a femtosecond laser, or through an interferometer set-up exploiting the photosensitivity of the fiber (Othonos and Kalli 1999). The strength and the pattern of the modulation define the strength of the grating, which affects the reflectivity per unit length. The period of modulation Λ, instead, defines the Bragg wavelength λ_B, as $\lambda_B = 2n_{\text{eff}}\Lambda$ (where n_{eff} is the fiber effective refractive index).

The concept of an FBG is sketched in figure 6.3, which reports different types of gratings, each corresponding to a pattern of modulation that highlights different spectral characteristics. In a uniform, or standard, FBG the modulation pattern has a uniform period and it results in the reflection of a narrow bandwidth, centered around the Bragg wavelength. All the other wavelengths are transmitted through the FBG. The bandwidth of the FBG usually ranges between 0.2 nm to 0.4 nm, depending on the strength of the grating.

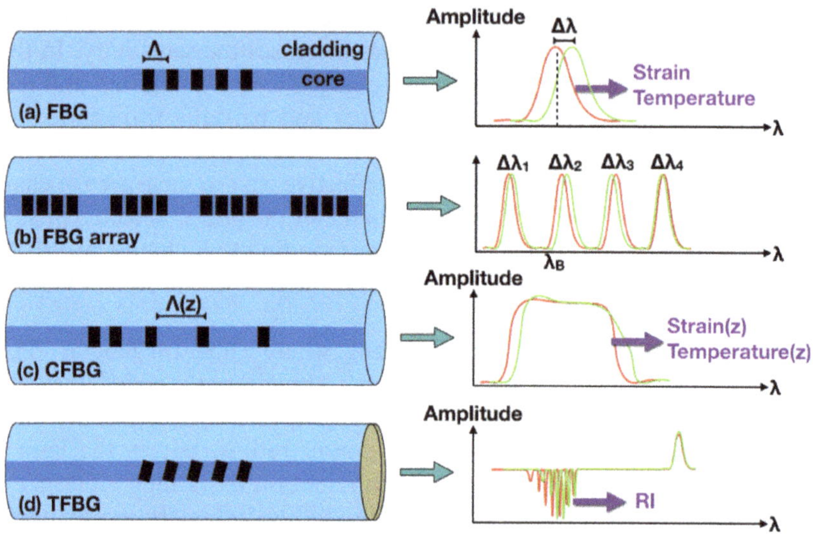

Figure 6.3. Schematic of a fiber Bragg grating sensor. Left: refractive index modulation patterns of each type of grating; right: corresponding reflection spectrum. (a) Uniform FBG; (b) FBG array; (c) chirped FBG; and (d) tilted FBG.

The sensing properties of the grating are encoded in the Bragg wavelength, as it depends on both the grating period and on the fiber refractive index. When an FBG experiences a temperature variation both of these terms are subjected to a linear variation due to the thermal expansion of the material and its thermo-optic coefficient. In a similar fashion, when the FBG experiences a longitudinal strain the material expansion and the strain-optic coefficient also result in a change of the resonant wavelength. By detecting the reflection spectrum of the FBG, we detect a wavelength shift $\Delta\lambda$ with respect to the reference condition that is a function of both temperature and strain. The sensitivity is linear, with coefficients equal to 10 pm $°C^{-1}$ and 1 pm/$\mu\varepsilon$, respectively. Hence, by detecting the FBG spectrum and estimating the wavelength shift with respect to the reference conditions using the methods described in temperature (Tosi 2017), it is possible to obtain a precise estimation of strain or temperature.

6.4.1 FBG arrays

The key advantage of FBGs, as they outperform traditional technologies for strain or temperature sensors (such as strain gauges or thermocouples), is the possibility to work with arrays of FBGs and separate each contribution, as outlined in figure 6.3(b). An array of FBGs is composed of multiple grating elements, each located in a different position within the fiber. Each grating is designed in order to have a different Bragg wavelength, with sufficient separation between adjacent Bragg wavelengths of each FBG. With this arrangement, rather than observing a single reflection peak, the spectrum comprises a comb of resonance conditions and each Bragg wavelength appears as an individual peak in the spectrum.

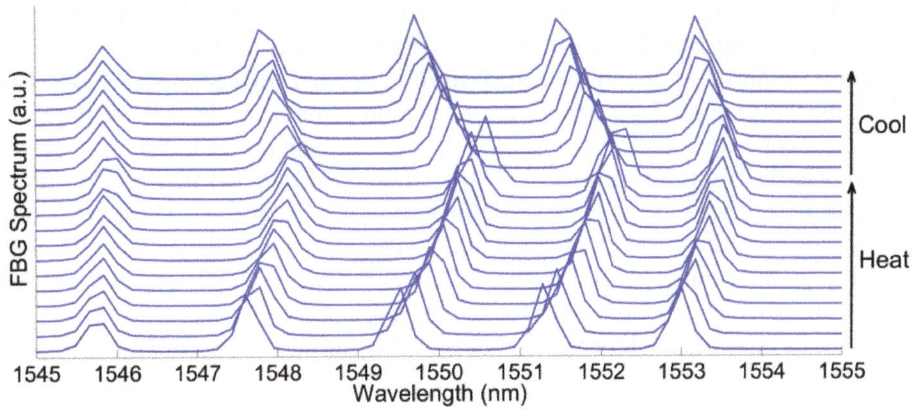

Figure 6.4. Example of WDM applied to FBG arrays. The image shows the evolution of a five-element FBG array during a thermo-therapy (the heating and then the cooling stages). Reproduced from Schena *et al* (2016) with permission of MDPI.

By engineering the system in order to separate each spectral contribution by a sufficient amount, it is possible to extend the spectral estimation to each grating element. In this way we can estimate multiple parameters in different parts of the fiber, measuring the spatial distributions of temperature and/or strain (Tosi *et al* 2018). This principle is known as WDM and it applies to FBG sensing networks. Most systems operate with FBGs having a ~2 nm separation between adjacent Bragg wavelengths. Infrared detectors have bandwidths ranging from 80 to 140 nm for most systems, hence it is possible to interrogate 40–70 FBGs for each channel, which allows a dense monitoring of parameters.

In figure 6.4 (Schena *et al* 2016) the principle of WDM is shown. The reflection spectrum of a five-element FBG array is shown. Each FBG has 5 mm length and the distance between each FBG is 10 mm, hence the system provides a temperature measurement over 5 cm with 1 cm spatial resolution. The spectra of each FBG, detected with an infrared spectrometer, show an array of Bragg wavelengths, each subjected to a different shift. In this experiment the thermal patterns during thermo-therapy are recorded and each FBG detects a different temperature, resulting in a different wavelength shift that rises during heating and drops during cooling. Through this set-up it is possible to obtain a centimeter-level spatial resolution over long distances, which is an enabling tool for high-resolution needles and manometers in medicine (Arkwright *et al* 2009)

6.4.2 Chirped FBG

In a chirped FBG the period of modulation is not constant, but varies along the fiber direction, as shown in figure 6.3(c). Hence the Bragg wavelength also changes in each portion of the grating. The most common configuration is a linear chirp, where the Bragg wavelength varies linearly with the grating position. As a result the chirped FBG spectrum appears as a broadband device, in which a range of wavelengths is reflected. The bandwidth of the grating depends on the grating length and the chirp rate coefficient, which defines the rate of variation of Bragg wavelength over space, usually in the range of 0.6–1.8 nm mm^{-1} (Korganbayev *et al* 2018).

When a strain or temperature distribution is applied to the chirped FBG, each local Bragg wavelength shifts by a different amount. For example, if the temperature variation is spatially constant the whole grating spectrum shifts, but in the case of a non-uniform distribution, such as the Gaussian-shaped patterns that are commonly experienced in thermal ablation, the deformation of the chirped FBG spectrum is different, as the spectral envelope appears to vary. In this case the chirped FBG can act as a semi-distributed sensor as the grating spectrum depends on each value of temperature or strain experienced in each location of the grating.

The results reported in (Korganbayev *et al* 2018) show the most advanced method of semi-distributed sensing, as they apply a full reconstruction of the temperature distribution. The process is based on an optimization technique and considers two gratings, the 'real' grating which is the measured device and the 'simulated' grating which is simulated to have an identical spectrum to the real one. At each measurement the real grating spectrum varies. Then, different spatial distributions are applied to the 'simulated' grating until the measured and simulated grating spectra have the best match, estimated as the minimum of the mean square error between each reflectivity value. The algorithm proposed in this work brings the average measurement time to ~1 s and is therefore compatible with real-time operation in thermal sensing.

6.4.3 Tilted FBG

In a tilted FBG, the pattern of modulation of the refractive index is tilted with respect to the fiber plane, at an angle of 5°–15°. The tilt angle does not affect the grating behavior in correspondence to the Bragg wavelength, which appears similar to the standard FBG. However, the tilt angle excites a plurality of cladding modes, which interact at the boundary between the fiber cladding and the outer medium. These modes appear only in the transmission spectrum. They can also be visualized in reflection through applying a mirror at the end of the fiber and collecting the transmission–reflection spectrum (Guo *et al* 2016). Cladding modes appear as a comb of modes at lower wavelength than the Bragg resonance condition and each corresponds to a different effective refractive index. The cladding modes appearing as a spectral dip in the grating spectrum shift when the fiber is exposed to an outer refractive index (RI) change, as shown in figure 6.3(d). Each mode experiences a different shift and, depending on the RI value, in general one mode exhibits a different sensitivity. Hence TFBG sensors can effectively work as refractometers and through proper functionalization they can be used as biosensors (Albert *et al* 2013).

6.5 Fabry–Perot interferometers

The use of interferometers in optical fibers has offered a reliable and compact method to perform sensing of temperature, pressure, and refractive index (Udd and Spillman 2011). In modern OFS implementations, the Fabry–Perot interferometer (FPI) is the most important structure, as it allows a compact design, high sensitivity, and fabrication that can be performed with modern laser-based splicers. The FPI is simply a cavity formed by two mirrors, having fixed spacing and RI, which reflect a variable amount of the input light. The combination of the light that is reflected or transmitted multiple times through the cavity forms standing waves, which have a periodic frequency spectrum. High values of mirror reflectivity form high-finesse interferometers, characterized by sharp transmission peaks, due to the large number of reflections occurring in the cavity. FPIs with low values of mirror reflectivity, instead, lead to low-finesse interferometers in which higher orders of reflection are significantly attenuated. Low-finesse FPI spectra are more similar to sine-wave functions. Fiber-optic FPI sensors utilize cavities that are either fabricated in the fiber or externally mounted to the fiber itself as the sensing element. Most systems utilize the ~4% Fresnel reflectivity that is obtained at the boundary between the glass fiber and the air contained in a cavity to form the mirrors composing an FPI.

The intrinsic FPI (IFPI) is formed within the cavity itself, as an inline device, by forming a cavity that acts as the interferometer. The design by Jáuregui-Vázquez *et al* reported in figure 6.5 (Jáuregui-Vázquez *et al* 2013), reports the efficient design of an IFPI cavity on the tip of the fiber. The cavity is formed by an air gap in proximity to the fiber tip, which is achieved by splicing a hollow-core fiber (characterized by air holes) into a single-mode fiber. The resulting structure, shown in figure 6.5(b), acts as a three-mirror interferometer: the inner cavity formed by the

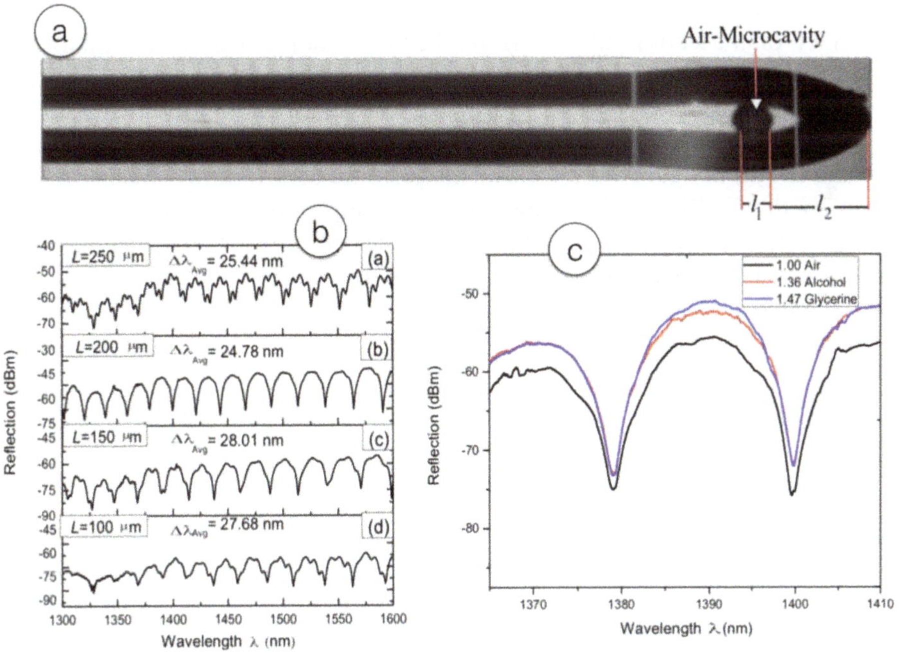

Figure 6.5. Fabrication of an intrinsic FPI on a micro-probe. (a) Photograph of the IFPI probe fabricated as an air cavity within the fiber. (b) Reflection spectrum of IFPI probes with different cavity splice offset lengths (100–250 μm). (c) Spectrum of the IFPI probe for different values of external refractive index. Adapted from Jáuregui-Vázquez *et al* (2013) with permission of MDPI.

two air/fiber interfaces act as the main interferometer, and the outer mirror is formed at the interface between the fiber and the outer medium. The characteristic of this device is that the interferometer is entirely fabricated on the splicer, hence making the procedure fast and repeatable. The IFPI spectrum is shown in figure 6.5(b) for different values of splice offset, which change the properties of the interferometer. The IFPI reflection spectrum appears as a periodic envelope, with a period of 24–28 nm. For the intermediate values of offset, the spectrum has a clear envelope with a ~10 dB fringe. The IFPI finds application as a refractometer, as the spectrum depends on the outer medium. Figure 6.5(c) shows that when the external RI changes, the envelope experiences a variation of peak amplitude due to the change of the reflectivity of the fiber/RI interface.

The extrinsic FPI (EFPI) is instead fabricated using an external element, serving as diaphragm, mounted on the fiber tip and integrated in the interferometer. When exposed to a pressure changes in the environment the diaphragm compresses, changing the cavity length of the interferometer and therefore resulting in a change of the reflection spectrum. Since the pressure is directly encoded in the interferometer sensitivity and the response is linear, the EFPI is a very effective approach for the fabrication of compact pressure probes (Poeggel *et al* 2015), with the key advantage of being all-glass structures that do not require epoxy or sealants and therefore ensure a higher biocompatibility.

Figure 6.6 shows the design of an all-glass EFPI with a tip-based diaphragm (Tosi *et al* 2016). The structure is formed by encapsulating an air gap within two optical fibers (SMF on the lead-in side, MMF on the outer side). The inner air gap forms the body of the EFPI. Subsequently, the MMF is cleaved, polished, and etched until it achieves a small thickness. The EFPI works as a thre-mirror interferometer, in which the outer cavity is shorter and appears in the spectral envelope. The reflection spectrum, shown in figure 6.6(b), shows the presence of two peaks. When pressure is applied to the EFPI the diaphragm bends, compressing the inner air gap and reducing its length, as shown in figure 6.6(c). The compression of the cavity is linear with pressure variations, with a slope of 1.61 nm/kPa. Pressure increases therefore cause a compression of the length, resulting in a wavelength shift in the EFPI spectrum, which can be measured at 1563 nm.

6.6 Distributed sensors

Distributed sensors interrogate the multiple scattering events that occur in the optical fiber as the light propagates through the medium. With respect to the previously described architectures, which host a specific active point at which the sensing mechanism occurs, distributed sensors do not have an ad hoc sensing device, but rather the whole fiber acts as a sensor. The multiple scattering centers present in the fiber act as a continuous reflection, several orders of magnitude below the forward wave, which can be detected and interpreted either in the time domain, through optical time-domain reflectometry (OTDR), or in the frequency (spectral) domain, through optical frequency-domain reflectometry (OFDR) (Bao and Chen 2012).

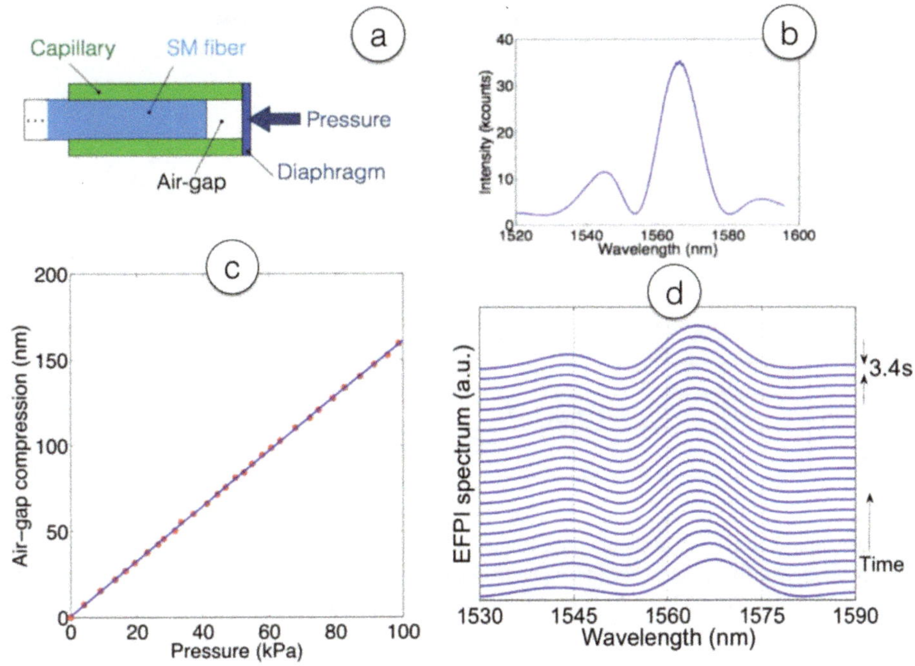

Figure 6.6. An all-glass extrinsic FPI sensor for pressure measurement and its application in measuring intra-tissue pressure during thermotherapy. (a) Schematic of the EFPI. (b) Reflection spectrum of the EFPI. (c) Air-gap compression as a function of pressure. (d) EFPI spectrum measured during a laser ablation procedure, recorded at each measurement time. Reproduced from Tosi *et al* (2016) with permission of MDPI.

The key principle of distributed sensors is the interpretation of the scattering signals that are generated in each fiber location as the light propagates. The scattering components are shown in figure 6.7 (Bao and Chen 2012). Considering a wave with input frequency ω_0. As the fiber is a non-homogeneous medium, characterized by scattering centers, the first component that we observe is the Rayleigh scattering, which occurs at the same input frequency ω_0 and for this reason is defined as elastic scattering. The overall Rayleigh scattering effect is due to the non-propagating density oscillations of the fiber medium, which give rise to a series of reflection events at the same input frequency and have a spectrum that appears as a random signal (Froggatt and Moore 1998). Rayleigh scattering is the basis of optical backscatter reflectometry (OBR), which interprets the signals emitted at each fiber location and estimates the temperature or strain values in each fiber location. The Raman and Brillouin scattering components are, instead, defined as inelastic scattering as they imply a change of energy between the input wave and the scattering components. For each scattering phenomenon we observe a combination of waves with a lower frequency (Stokes) and waves with a higher frequency (anti-Stokes). Clearly the propagation of Stokes and anti-Stokes waves in the fiber is affected by the different attenuations of the fiber at these new wavelengths. Raman scattering appears with a series of spectral lines at wide separation from the input

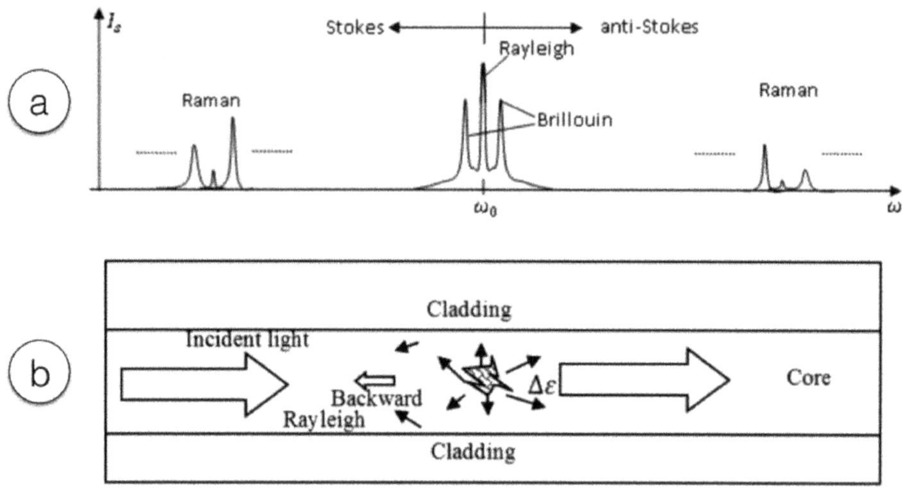

Figure 6.7. Characteristics of the scattering signals underlying distributed optical fiber sensors. (a) Identification of Rayleigh, Brillouin, and Raman scattering components (Stokes and anti-Stokes) in the frequency spectrum, for a signal having input frequency ω_0. (b) Sketch of the Rayleigh scattering effect generated in the fiber. Reproduced from Bao and Chen (2012) with permission of MDPI.

(about 100 nm at 1550 nm input), which have an intensity that has a significant dependence on the temperature. For this reason Raman scattering is the basis of distributed temperature sensors (DTS), used in oil and gas pipeline monitoring and infrastructure monitoring (Dakin and Pratt 1985). Finally, Brillouin scattering appears as the combination of two spectral lines with frequency closer to the input wave. This scattering phenomenon arises from the acoustic oscillations of glass and therefore transduces acoustic fields and vibrations into a local change of the Brillouin frequency shift.

6.6.1 Distributed temperature sensors

DTS systems are based on a pulsed laser source which emits a short pulse into the fiber under test (FUT). As the light propagates into the FUT the Raman scattering components are generated and propagated backwards into the fiber, each arriving at the detector at a different time. The time of arrival of each reflection corresponds to the round trip time of the pulse to each fiber location and back. The OTDR therefore discriminates the location of each event in the time domain by computing the propagation time of the pulse from the source to the scattering center and back along the same fiber. Figure 6.8 shows the basic schematic of a DTS (Schenato 2017). Since the Raman scattering components have different frequencies with respect to the input wave, a WDM filter is used to collect the backreflections and route each scattering component (Stokes and anti-Stokes) to a separate detector. The waveform detected by each receiver in the time domain shows a sawtooth-shaped curve in which the reflected components are attenuated by a different amount, corresponding to the losses in the fiber along the forward and backward path. The spatial resolution of the DTS system, corresponding to the shortest distance between two adjacent measurement points,

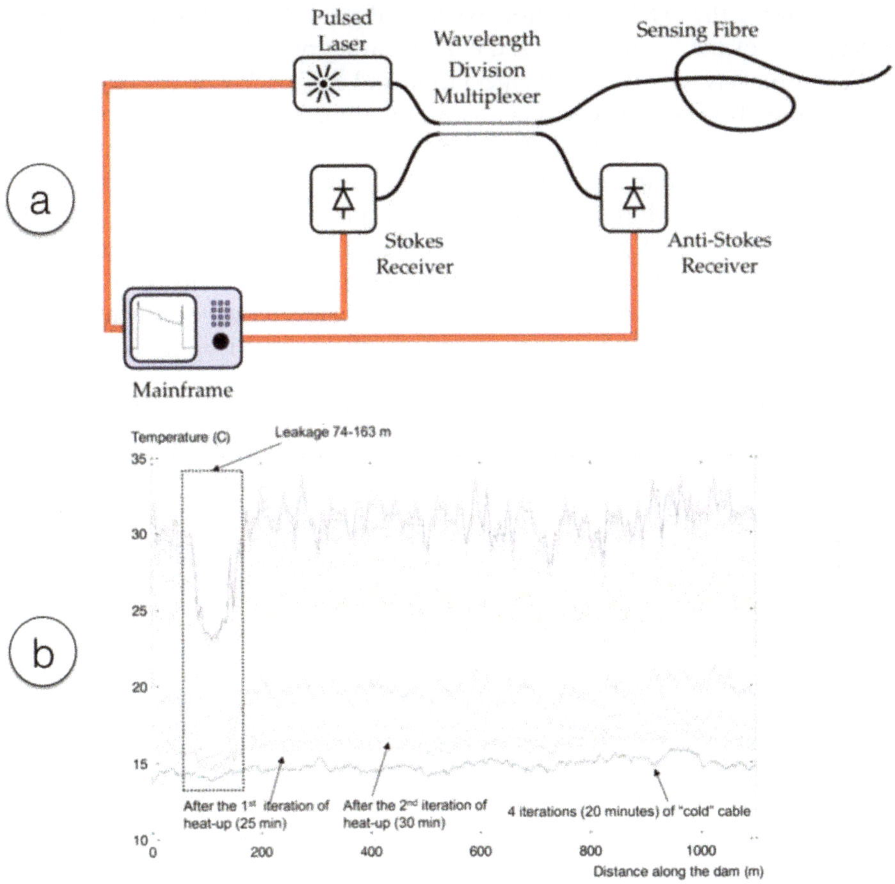

Figure 6.8. Schematic and application of a DTS system based on OTDR. (a) Diagram of a DTS system. (b) Application to leak identification in a dam. Reproduced from Schenato (2017) with permission of MDPI.

corresponds to $c/(2n_{\mathrm{eff}}B)$, where c is the speed of light and B is the bandwidth of the mainframe analyzer, usually a data acquisition card (DAQ) with on-board processing capabilities. A 1 GHz DAQ system achieves a 10 cm spatial resolution, while most commercial models operate within 100–500 MHz to achieve 20–100 cm resolution over tens or hundreds of kilometers. On the other hand, the repetition rate of the pulsed laser corresponds to the maximum length of fiber that can be interrogated—a 1 kHz laser can detect up to 150 km in fiber.

The working principle of a DTS relies on the detection and interpretation of the anti-Stokes and Stokes traces in the time domain. The intensity of the Stokes signal is independent of the temperature. The anti-Stokes signal intensity instead has a temperature dependence and the sensitivity is ~0.8%/°C (Schenato 2017, Bao and Chen, 2012). Hence by compensating the attenuation of the fiber of each component, the ratio between anti-Stokes and Stokes provides a signal that is proportional to the fiber temperature in each fiber location. The fiber attenuation is usually estimated during the DTS calibration, whereas the anti-Stokes/Stokes ratio provides compensation of the fluctuations of the laser during operation.

DTS sensors have a simple implementation and can work with resolution below 1 m over hundreds of kilometers. The main limitation is implicit in the weakness of Raman scattering. As shown in figure 6.7(a), this is the weakest scattering phenomenon and therefore, in order to collect a reliable signal and mitigate the effect of noise, several thousands of waveforms are averaged to obtain each measurement, slowing down the measurement time to several minutes. Hence multimode fibers are typically used for DTS systems in order to increase the amount of light collected by the receivers. Since the power decreases exponentially in the fiber, accuracy is also an exponential function of the fiber length, as the sensor is much more accurate in the early part of the fiber rather than at the far end. Due to the need to collect weak signals with a large bandwidth, avalanche photodiodes are often used in DTS systems (Dakin and Pratt 1985). DTS finds applications mainly in the identification of leaks and cracks over a very long range. A primary application is in oil and gas, in order to detect leaks in oil pipelines of tens of kilometers of length (Bao and Chen 2012). The installation of fiber cables into pipelines and the subsequent interrogation with a network of DTS systems positioned in order to interrogate each pipeline segment provides an early warning of pipeline cracks which might induce leaks by detecting the temperature increase or decrease of each region (depending on the type of fluid contained in the pipeline). Recently, geotechnical applications have made use of DTS systems to detect leaks in specific structures. Figure 6.8(b) (Schenato 2017) shows the application of a DTS to the identification of a leak in a dam. As the fiber cable is heated, the section of the fiber containing the leak (74–163 m) senses a much smaller temperature due to its contact with soil, hence identifying a dam failure.

6.6.2 Optical backscatter reflectometry

The OBR system is a frequency-domain interrogation method for the detection of Rayleigh scattering patterns in the fiber. Since Rayleigh scattering originates from the local fiber defects and impurities located in the core, which provide scattering centers localized in each point of the fiber, the spectral properties of each fiber segment are random signals and identify the fiber section as a fingerprint. For this reason the elastic scattering properties of the fiber are known as the *signature* of the fiber. When a temperature variation or a mechanical strain is applied to the fiber the local length and the effective refractive index of the fiber segment change. The coefficients of sensitivity are the same as for FBG, i.e. 10 pm/°C and 1 pm/$\mu\varepsilon$, respectively. However, different to FBGs which have a well-determined spectrum, the spectral shifts due to physical changes of the fiber apply as a shift of each random fiber signature. In order to estimate the wavelength shift, in each fiber section the correlation between the reference and measured signatures is computed, estimating the wavelength shift as the peak of the mutual correlation (Froggatt and Moore 1998).

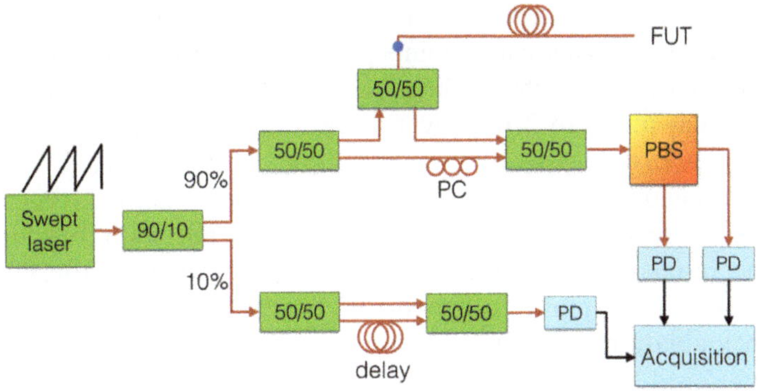

Figure 6.9. Schematic of an OBR system for distributed strain/temperature sensing; PD = photodiode; PBS = polarization beam splitter.

The OBR is a system engineered to detect and process all the fiber signatures due to Rayleigh scattering in the frequency domain (Bao and Chen 2012). The schematic of an OBR system is shown in figure 6.9 and it is based on a swept laser interferometer. The laser source scans a frequency range Δf. Commercial infrared sources scan with speeds up to 200 nm s^{-1} over a ~80 nm range with stable wavelength repeatability. The system is a split interferometer containing a measurement arm, where 90% of the input light is delivered, and a reference arm (10% of the input light) which contains a delay unit that matches the round trip time of the FUT. The light backscattered by the FUT is returned to the interferometer and a polarization beam splitter divides the two polarization components. A data acquisition unit collects the output of the photodetectors and estimates the fiber signatures. Commercial systems are engineered to collect the amount of light backscattered by SMFs, with power in the order of −110 dBm, and therefore do not require additional components. However, recent systems such as the sensing network proposed by Beisenova *et al* (2019) make use of fibers with enhanced backscattering to increase the amount of light returned to the photodetectors and achieve a multiplexed network.

The OBR system has a spatial resolution of $c/(2n_{\text{eff}}\Delta f)$; for a 80 pm scanning laser, the spatial resolution is equal to 10 μm, which can be achieved over tens of meters of fiber, with frequency $\geqslant$ 3 Hz. Local averaging is typically used to improve accuracy while maintaining a millimeter-level resolution in space and 1 to 10 pm resolution in wavelength shift detection. Since the OBR works at short distances it is an excellent candidate for distributed sensing in medical devices. An OBR connected to a single fiber (Froggatt and Moore 1998) or to a network of fibers (Beisenova *et al* 2019) allows the detection of strain/temperature over a short distance, with narrow spatial resolution and the possibility to configure the fibers for sensing in 2D or 3D geometries. One of the most important applications of OBR in medical devices is the detection of multidimensional strain data and the reconstruction of shapes. The system reported by Issatayeva *et al* (2019) proposes a four-fiber configuration mounted at the cardinal points of an epidural catheter, in order to estimate the shape of an epidural needle as it pierces through the skin through the epidural space. By detecting the strain over 40 sensing points per fiber and using a trigonometric model to convert strain to shape, the authors reconstruct the shape of the needle with bending displacement up to 2 mm over 80 mm needle length. Parent *et al* (2018) reported a shape-sensing system based on a fiber triplet connected to the OBR for use in cardiovascular devices. Compared to the previous method, this arrangement allows the detection of wider curvatures with much narrower bending radii and compares the results with imaging methods to perform intra-arterial diagnostic. Beisenova *et al* (2019) exploited a network of four fibers to map the temperature of radiofrequency ablation thermotherapy on a grid, with 2.5 × 5.0 mm pixel size.

6.7 Biosensors

Fiber optic biosensors are an emerging technology that makes use of an optical fiber structure to detect the presence and concentration level of analytes, performing selective biorecognition (Guo *et al* 2017). The principle of operation of an optical fiber biosensor starts from an RI-sensitive device, as the fiber device is sensing the surrounding environment. Hence fiber optic biosensors operate using grating-based structures such as etched FBGs, tilted FBGs, or long-period gratings (Chiavaioli *et al* 2017), or through plasmonic devices (Loyez *et al* 2018), or interferometer-based structures (Udd and Spillman 2011). Then the fiber is functionalized to a specific detection by means of surface chemistry, which allows selective detection to be performed. Since the fiber is a dielectric, which does not allow direct chemical bonding, the fiber is first coated with a metallic thin film that allows a better surface chemistry. The two main processes that are used are the deposition of thin metallic films through sputtering or electrodeless plating, as described by Sypabekova *et al* (2019), where layers up to 50 nm thick of gold are deposited on the fiber surface, or the silanization process reported by Bekmurzayeva *et al* (2018) which covers the fiber surface with organofunctional alkoxysilane molecules. The surface-modified fiber can then successfully host bioreceptors, which bind to the analyte under test by means of covalent bonding, as shown in figure 6.10(a) (Guo *et al* 2017). The two main scenarios for biosensors involve the antibody–antigen interaction, where the antibody act as a bioreceptor for the antigen and the surface functionalization through aptamers that is particularly common for the detection of proteins. Surface-assembled monolayer (SAM) biosensors have also been reported as biorecognition elements.

The interaction between the fiber surface modification and the sensing mechanism can be divided into two areas. On one side, the metallic film and the surface functionalization simply act as a passive surface, having minimal effects on the photonic structure of the sensor. Since surface modifications have a much smaller size than the wavelength, the optical structure can be approximated as a fiber device surrounded by a liquid with different values of RI. This is the case for Bragg grating biosensors (etched, tilted, and long-period gratings) as well as most interferometers. On the other side, surface plasmon resonance (SPR) methods (Guo *et al* 2017), which can be used as a stand-alone method or in conjunction with a tilted FBG,

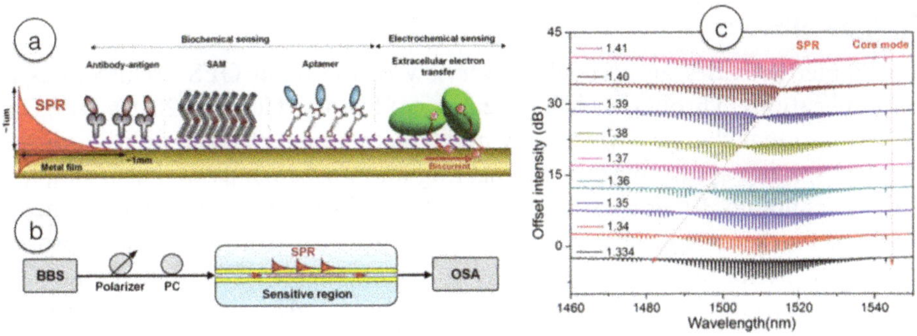

Figure 6.10. Schematic and results of fiber optic biosensors, using the fiber as a sensing method for biological analytes. (a) Schematic of the fiber modifications through a metallic thin-film and a diagram of typical surface functionalization for a plasmonic biosensor. (b) Interrogation scheme for a plasmonic fiber biosensor. (c) Transmission spectra for a tilted FBG-SPR biosensor at different values of surrounding RI. Reproduced from Guo *et al* (2017) with permission of MDPI.

convey the sensing mechanism into the metallic thin film which is part of the sensing device. When light with parallel polarization travels through the biosensing region, the surface plasmon condition is excited at a specific incidence angle between the light and the substrate, which converts into a biocurrent that vanishes in the metallic substrate. Hence the SPR mode appears as a dip in the transmission spectrum, as it corresponds to the evanescent wave that leaks out of the fiber. Using an interrogation scheme as in figure 6.10(b), based on a broadband source, a polarizer that aligns with the p-polarization, and an optical spectrum analyzer (OSA) that detects the transmission spectrum, it is possible to evaluate the SPR effect.

Modern biosensors combine the SPR phenomenon with a tilted FBG. The tilt angle of the grating induces cladding modes which interact with the substrate at a wide range of effective refractive indices and some of the light incident on the substrate excites the resonance plasmonic condition. Figure 6.10(c) shows the spectrum of a tilted FBG-SPR structure in which we observe the cladding modes of the grating modulated by the SPR envelope. This structure is interrogated by detecting the change of the SPR mode.

Biosensors are used to perform the specific and selective detection of analytes, and therefore require specific figures of merit, described by Chiavaioli *et al* (2017). The relationship between wavelength shift measured in the optical spectrum and the concentration of analyte is application-dependent. However, the most important figure is the limit of detection (LoD), which defines the smallest analyte concentration that the system can effectively detect. The LoD depends on the quality of the biological functionalization, on the sensitivity of the sensing device, and on the wavelength shift detection method used for the demodulation. The selectivity of the biosensor is measured by comparing the response to the analyte to specific controls, chosen between proteins with similar chemical composition. This test quantifies the amount of selectivity of the sensing method, with an ideal sensor that responds only to the analyte and not to the control.

6.8 Conclusions

This chapter provides an accessible overview of the main OFS architectures and their applications. Through their half century of history (Culshaw 2019), OFSs have seen generational changes, which now make them a mature technology, addressing advanced sensing applications in engineering and medicine, as well as a simple yet reliable technology for low-cost sensing. The main trends in the development of OFSs rely on the specific characteristics of the fiber for sensing and have provided exceptional performance enhancement in recent years. Distributed and semi-distributed sensing have reached a high maturity. Although the interrogation systems are expensive, the sensors are made of inexpensive fibers and rely on their scattering properties. Recent advances have shown the possibility of tailoring the scattering properties in order to spatially modulate the scattering characteristics of the fiber and implement massive sensing networks. Prototype measurements using OBR and nanoparticle-doped fibers have shown great potential (Beisenova *et al* 2019). The recent possibility to design advanced biosensors using the fiber as a platform to sense its surrounding environment can provide an excellent tool for diagnostics, offering a great opportunity for fiber surface functionalization. Inexpensive smartphone-based and wearable sensors are scaling down the size and cost of the systems to values comparable to handheld devices, enabling applications through portable instruments.

Spatial resolution and advanced signal processing represent key opportunities for future developments. As OFSs are narrowing the spatial resolution down to a few micrometers in the latest sensors, each measurement acquires up to millions of data points that can be reinterpreted and can train sensing networks. In this framework sensors are breaking the boundary between sensing and imaging, as distributed sensors can acquire data on 2D and 3D structures with a narrow spatial resolution in each dimension. Multi-dimensional signal processing then becomes a key solution to render physical data, or even biological data, along spatial dimensions.

References

Agrawal G P 2012 *Fiber-optic Communication Systems* (New York: Wiley)

Aitkulov A and Tosi D 2019 Optical fiber sensor based on plastic optical fiber and smartphone for measurement of the breathing rate *IEEE Sensors J.* **19** 3282–7

Albert J, Shao L Y and Caucheteur C 2013 Tilted fiber Bragg grating sensors *Laser Photonics Rev.* **7** 83–108

Arkwright J W, Blenman N G, Underhill I D, Maunder S A, Szczesniak M M, Dinning P G and Cook I J 2009 *In-vivo* demonstration of a high resolution optical fiber manometry catheter for diagnosis of gastrointestinal motility disorders *Opt. Express* **17** 4500–8

Baldwin C 2018 Fiber optic sensors in the oil and gas industry: current and future applications *Opto-Mechanical Fiber Optic Sensors* (Oxford: Butterworth-Heinemann) pp 211–36

Bao X and Chen L 2012 Recent progress in distributed fiber optic sensors *Sensors* **12** 8601–39

Bekmurzayeva A, Dukenbayev K, Shaimerdenova M, Bekniyazov I, Ayupova T, Sypabekova M, Molardi C and Tosi D 2018 Etched fiber Bragg grating biosensor functionalized with aptamers for detection of thrombin *Sensors* **18** 4298

Beisenova A, Issatayeva A, Iordachita I, Blanc W, Molardi C and Tosi D 2019 Distributed fiber optics 3D shape sensing by means of high scattering NP-doped fibers simultaneous spatial multiplexing *Opt. Express* **27** 22 074–87

Bremer K and Roth B 2015 Fibre optic surface plasmon resonance sensor system designed for smartphones *Opt. Express* **23** 17179–84

Chiavaioli F, Gouveia C A, Jorge P A and Baldini F 2017 Towards a uniform metrological assessment of grating-based optical fiber sensors: from refractometers to biosensors *Biosensors.* **7** 23

Culshaw B 2019 Half a century of fiber sensing *Opt. Eng.* **58** 072 006

Dakin J P and Pratt D J 1985 Distributed optical fibre Raman temperature sensor using a semiconductor light source and detector *Electron. Lett.* **21** 569–70

Froggatt M and Moore J 1998 High-spatial-resolution distributed strain measurement in optical fiber with Rayleigh scatter *Appl. Opt.* **37** 1735–40

Gasulla I, Barrera D, Hervás J and Sales S 2017 Spatial division multiplexed microwave signal processing by selective grating inscription in homogeneous multicore fibers *Sci. Rep.* **7** 1–0

Gowri A and Sai V V 2016 Development of LSPR based U-bent plastic optical fiber sensors *Sensors Actuators* B **230** 536–43

Guo T, Liu F, Guan B O and Albert J 2016 Tilted fiber grating mechanical and biochemical sensors *Opt. Laser Technol.* **78** 19–33

Guo T, González-Vila Á, Loyez M and Caucheteur C 2017 Plasmonic optical fiber-grating immunosensing: a review *Sensors* **17** 2732

Issatayeva A *et al* 2019 Multiplexing of distributed temperature sensing achieved by nanoparticle-doped fibers *Proc. SPIE* **11190** 111900H

Jáuregui-Vázquez D, Estudillo-Ayala J M, Rojas-Laguna R, Vargas-Rodríguez E, Sierra-Hernández J M, Hernández-García J C and Mata-Chávez R I 2013 An all fiber intrinsic Fabry–Perot interferometer based on an air-microcavity *Sensors* **13** 6355–64

Kersey A D, Davis M A, Patrick H J, LeBlanc M, Koo K P, Askins C G, Putnam M A and Friebele E J 1997 Fiber grating sensors *J. Lightwave Technol.* **15** 1442–63

Korganbayev S *et al* 2018 Detection of thermal gradients through fiber-optic chirped fiber Bragg grating (CFBG): medical thermal ablation scenario *Opt. Fiber Technol.* **41** 48–55

Lee B H, Kim Y H, Park K S, Eom J B, Kim M J, Rho B S and Choi H Y 2012 Interferometric fiber optic sensors *Sensors* **12** 2467–86

Leng J and Asundi A 2003 Structural health monitoring of smart composite materials by using EFPI and FBG sensors *Sensors Actuators* A **103** 330–40

Leung A, Shankar P M and Mutharasan R 2007 A review of fiber-optic biosensors *Sensors Actuators* B **125** 688–703

Loyez M, Albert J, Caucheteur C and Wattiez R 2018 Cytokeratins biosensing using tilted fiber gratings *Biosensors.* **8** 74

Lu Y, Zhu T, Chen L and Bao X 2010 Distributed vibration sensor based on coherent detection of phase-OTDR *J. Lightwave Technol.* **28** 3243–9

Massaroni C, Saccomandi P and Schena E 2015 Medical smart textiles based on fiber optic technology: an overview *J. Funct. Biomater* **6** 204–21

Mihailov S J 2012 Fiber Bragg grating sensors for harsh environments *Sensors* **12** 1898–918

Othonos A S and Kalli K 1999 *Fiber Bragg Gratings: Fundamentals and Applications in Telecommunications and Sensing* (Norwood, MA: Artech House)

Parent F, Gérard M, Monet F, Loranger S, Soulez G, Kashyap R and Kadoury S 2018 Intra-arterial image guidance with optical frequency domain reflectometry shape sensing *IEEE Trans. Med. Imaging* **38** 482–92

Poeggel S, Duraibabu D, Lacraz A, Kalli K, Tosi D, Leen G and Lewis E 2015 Femtosecond-laser-based inscription technique for post-fiber-Bragg grating inscription in an extrinsic Fabry–Perot interferometer pressure sensor *IEEE Sensors J.* **16** 3396–402

Rodrigues C, Félix C, Lage A and Figueiras J 2010 Development of a long-term monitoring system based on FBG sensors applied to concrete bridges *Eng. Struct.* **32** 1993–2002

Sai V V, Kundu T and Mukherji S 2009 Novel U-bent fiber optic probe for localized surface plasmon resonance based biosensor *Biosens. Bioelectron.* **24** 2804–9

Schena E, Tosi D, Saccomandi P, Lewis E and Kim T 2016 Fiber optic sensors for temperature monitoring during thermal treatments: an overview *Sensors* **16** 1144

Schenato L 2017 A review of distributed fibre optic sensors for geo-hydrological applications *Appl. Sci.* **7** 896

Sypabekova M, Korganbayev S, González-Vila Á, Caucheteur C, Shaimerdenova M, Ayupova T, Bekmurzayeva A, Vangelista L and Tosi D 2019 Functionalized etched tilted fiber Bragg grating aptasensor for label-free protein detection *Biosens. Bioelectron.* **146** 111 765

Tosi D, Olivero M and Perrone G 2008 Low-cost fiber Bragg grating vibroacoustic sensor for voice and heartbeat detection *Appl. Opt.* **47** 5123–9

Tosi D, Perrone G and Vallan A 2013 Performance analysis of a noncontact plastic fiber optical fiber displacement sensor with compensation of target reflectivity *J. Sensors* 2013

Tosi D, Saccomandi P, Schena E, Duraibabu D B, Poeggel S, Leen G and Lewis E 2016 Intra-tissue pressure measurement in *ex vivo* liver undergoing laser ablation with fiber-optic Fabry–Perot probe *Sensors* **16** 544

Tosi D 2017 Review and analysis of peak tracking techniques for fiber Bragg grating sensors *Sensors* **17** 2368

Tosi D, Schena E, Molardi C and Korganbayev S 2018 Fiber optic sensors for sub-centimeter spatially resolved measurements: review and biomedical applications *Opt. Fiber Technol.* **43** 6–19

Udd E and Spillman W B Jr (ed) 2011 *Fiber Optic Sensors: An Introduction for Engineers and Scientists.* (New York: Wiley)

Vallan A, Casalicchio M L, Olivero M and Perrone G 2012 Assessment of a dual-wavelength compensation technique for displacement sensors using plastic optical fibers *IEEE Trans. Instrum. Meas.* **61** 1377–83

IOP Publishing

Optical Fiber Technology and Applications
Recent advances
Mário F S Ferreira and Mukul Chandra Paul

Chapter 7

Recent advances and perspectives on pulsed fiber lasers

Qian Li, Xuanyi Liu, Renlai Zhou, M S Aruna Gandhi and H Y Fu

Fiber lasers have received a great deal of attention in recent times, in addition to their remarkable applications in the industrial manufacturing, military, medical, and aerospace fields, etc. Compared to solid-stated lasers, the unique advantages of miniaturization, high efficiency, low manufacturing cost, good stability, and excellent heat dissipation have been demonstrated in fiber lasers. In addition to continuous-wave (CW) operation, pulsed fiber lasers have been designed and achieved. Among them, Q-switched fiber lasers are promising sources for obtaining high-energy pulses. By modulating the laser cavity loss, the pulse energy directly produced by Q-switched fiber lasers reaches the order of millijoules. The most notable feature of gain-switched fiber lasers is that they can deliver ultrashort pulses in the mid-infrared wavelength. This is undoubtedly of great significance for applications in the long-wavelength region. Although Q-switched and gain-switched fiber lasers can output high-energy pulses, their applications are limited due to their wide pulse width (at the nanosecond level). By employing mode-locking techniques the performance of fiber lasers has been fully optimized to unprecedented levels. Mode-locked fiber lasers exhibit state-of-the-art performance (femtosecond pulse width, high peak power, and low noise), which has allowed their application range to expand significantly. However, the single-pulse energy directly output by the fiber laser is generally low. For various applications it is necessary to amplify the pulse energy. Amplifying the femtosecond pulse requires many new amplification techniques. In this chapter we focus on the progress of Q-switched, gain-switched, and mode-locked fiber lasers. Recent advances in pulse amplification technology are briefly reviewed. We also discuss our latest results on the nonlinear amplifying loop mirror (NALM) fiber lasers.

doi:10.1088/978-0-7503-3243-9ch7

7.1 *Q*-switched fiber lasers

The *Q*-switched technique was first introduced in fiber lasers by Alcock in 1986 [1] and the *Q*-switched fiber laser has since become a research focus in the field of laser technology. The *Q*-switched fiber laser has the advantages of narrow pulse width, high peak power, high pulse energy, good beam quality, etc. In recent years *Q*-switched fiber lasers have shown their unique application value in medical treatment, environmental remote sensing, laser ranging, industry production, and other fields. Two methods, the actively *Q*-switched technique and passively *Q*-switched technique, have been utilized to achieve *Q*-switched pulses. The actively *Q*-switched technique, that is, the change of the *Q* value in the resonant cavity, needs to be realized by inserting an electro-optic or acousto-optic modulator (EOM or AOM). This structure is relatively complicated and the insertion loss between the fiber and non-fiber *Q* switch is large, which cause difficulty in forming a high-energy pulse output. A passively *Q*-switched fiber laser without an additional modulator has the advantages of compact structure, low cost, a simple system, and high efficiency, and is amongst the most important choices to obtain high repetition frequency, and nanosecond and sub-nanosecond laser pulses.

The *Q*-switched technology aims to change the loss in the cavity, that is, the *Q* value changes to form an inverse accumulation of the number of particles in the upper level within a certain control time. When the energy storage is high enough the *Q* value of the cavity suddenly increases to achieve a giant laser pulse output.

The pump source of a *Q*-switched fiber laser is usually a laser diode (LD). For the working medium, the fundamental difference between fiber lasers and traditional all-solid-state lasers is that the former uses optical fibers with unique waveguide characteristics as the gain media, which directly determine the advantages of fiber lasers in thermal management, mode, and efficiency. Therefore, the development of fiber lasers is inseparable from the advancement of the fiber, considering the development from the original ordinary fibers to the double-clad fiber, the large-mode field area double-clad fiber, and the photonic crystal fiber. Recently, the single-pulse energy of *Q*-switched fiber lasers has reached the level of millijoules, where the gain medium mostly uses a large-mode-field rare-earth ion-doped fiber and photonic crystal fiber. The advantage of the acousto-optic *Q* switch is that the repetition frequency of the pulse can be artificially controlled to obtain a more stable pulse sequence. The limit of the Q-switched pulse width is about the order of L/c (L is the cavity length and c is the speed of light), which is generally on the order of the nanosecond level.

Considering the stability and system integration, the development of *Q*-switched fiber lasers tends to an all-fiber structure and various fiber devices have been fabricated. Different to a solid-state laser, the pump coupling in an all-fiber *Q*-switched laser involves directly fusing the LD pigtail and the gain fiber, or using wavelength division multiplexing (WDM) as a coupler. The resonator is composed of a pair of equivalent function fiber Bragg gratings (FBGs), which can directly write on the fiber. An AOM/EOM has also been spliced through the pigtail fiber. To prevent backscattered laser light and residual pump light from reversely transmitting and damaging the pump source, isolators are often added for unidirectional isolation. The overall structure of this device is more compact and has better stability than previous devices. A diagram of a typical device is shown in figure 7.1.

The key points to consider in Q-switched fiber laser design are the following:

(i) *Pulse control and generation.* The factors that affect the pulse energy include the modulator diffraction loss, core mode field area, upper-level lifetime of doped rare-earth ions, and modulation depth. Among them, the effect of the activation level of the upper level of the activated ion on the repetition rate during the oscillation process needs to be selected according to the pulse energy.

(ii) *Selection of pump coupling.* Side-pumping and end-pumping are typical pump methods. Side-pumping always has greater technical requirements, but the coupling efficiency is high. There are three ways of end-pumping: forward, backward, and bidirectional pumping. The selection of pump coupling affects the output characteristics of the fiber laser and has to be carefully chosen according to different pulse repetition frequencies.

(iii) *Fiber fusion.* The devices are all made with pigtails and gain fibers. The all-fiber type has advantages, but fusion is a problem, in particular for double-clad fibers or photonic crystal fibers of different shapes and sizes. The accuracy of the process needs to be ensured to control its loss, otherwise the pump efficiency is much reduced.

(iv) *Mode control.* For lasers with large-mode field area fibers, improving the laser beam quality requires careful control of the transverse mode. To date, the techniques for mode selection include fiber bending, tapering, gain control, and mode selection by a cavity mirror. Among these, the most directly feasible method is fiber bending. Compared to the fundamental mode, the higher-order mode is more sensitive to bending. By bending the optical fiber within a radius, the loss of the higher-order mode increases as the bending radius decreases and, consequently, higher-order modes are suppressed to optimize beam quality. Therefore choosing an appropriate bending radius is key, but the bending of the fiber may also affect the laser wavelength.

(v) *Suppression of nonlinearity effects.* Nonlinearity effects can affect the quality of the pulsed beam and may also reduce the output power. Over the years, several methods have been proposed to suppress the nonlinearity effects, and the gain fibers have improved in terms of geometry, material composition, manufacturing process, and working conditions.

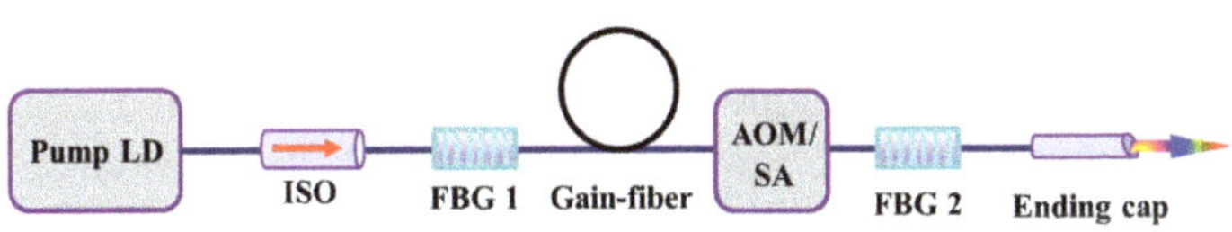

Figure 7.1. A typical configuration of a Q-switched fiber laser.

From the Nd^{3+}/Yb^{3+}-doped gain fibers to the Tm^{3+}/Ho^{3+}-doped gain fibers, Q-switched fiber lasers have realized kilowatt and millijoule levels in peak power and single-pulse energy, respectively. In 2001, Renaud *et al* [2] obtained a 7.7 mJ pulse energy, 500 Hz repetition rate pulse laser with a Yb^{3+}-doped Q-switched fiber laser. In 2004 Fan *et al* [3] achieved a 4.2 ns pulse duration in a tunable wavelength range from 1080.8 to 1142.7 nm, and the maximum peak power and single-pulse energy were 175 kW and 1.57 mJ, respectively. To solve the contradiction between the larger mode field area and the single-mode output, various photonic crystal fibers have been designed and fabricated. In 2007, a PCF with a 70 μm core diameter was used to achieve a high average power output of 100 W at a high repetition frequency of 100 kHz, and its single-pulse energy of 2 mJ and the pulse width of 7.3 ns made peak power up to 275 kW [4].

The Tm^{3+} and Tm^{3+}:Ho^{3+} co-doped Q-switched fiber lasers have recently attracted a lot of attention. In 2011 researchers achieved an average power of 10 W in a nearly 3 μm wavelength band for the first time [5]. In the short-wave direction, the researchers utilized the second harmonic to generate the third harmonic, which extended the laser wavelength to 515 and 343 nm, and the single-pulse energy could reach the millijoule level [6]. In addition, the IPG has reported a 200 W ytterbium-doped fiber laser pulsed with a maximum energy of 10 mJ.

With the development of fiber devices and high-power LDs, fiber laser systems now have compact structure, stable performance, relatively low cost, and ease of thermal management. Therefore they have gradually been industrialized and are widely used in the fields of optical communications, laser processing, laser medical treatment, and national defense and security. Nanosecond fiber lasers are developing towards high repetition frequencies and high average powers.

The Q-switched technique has been applied in solid-state lasers to generate nanosecond pulses with higher pulse energy (tens of millijoules). Despite the development and fabrication of Q-switched fiber devices, further improvements in many aspects, such as coupling efficiency, insertion loss, and thermal management, are required.

7.2 Gain-switched fiber lasers

Unlike the principle of Q-switched laser pulse generation, the gain-switched technique does not require a change in the cavity loss to achieve the accumulation of the upper-level populations and the inversion populations are quickly achieved using a nanosecond pump pulse with sufficient peak power. The loss does not change in the cavity during the process of laser pulse formation, therefore the dynamic of a gain-switched pulse laser is completely different to that of the Q-switched laser. Figure 7.2(a) shows a typical two-level system. The populations in the lower level have inversed to the upper level under the action of pump light due to the stimulated amplifications. Figure 7.2(b) shows the process of gain-switched laser pulse formation. Under the action of a pump pulse with sufficient power, the populations in the upper level accumulate quickly and the maximum value is achieved at $t = t_1$.

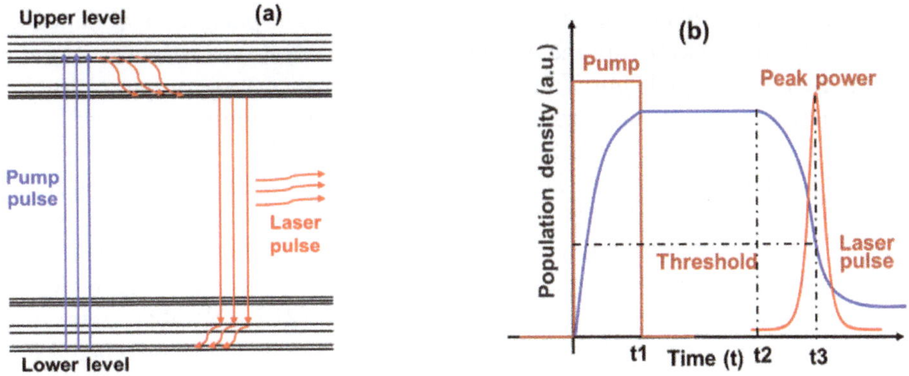

Figure 7.2. The principle of a gain-switched fiber laser: (a) populations transition in a two-level system and (b) building up of the gain-switched laser pulse.

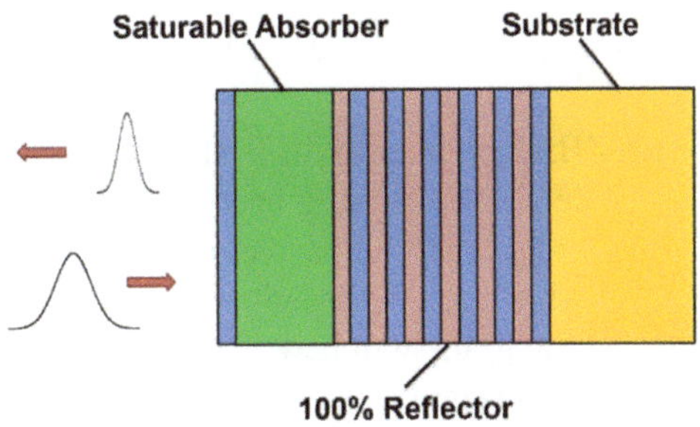

Figure 7.3. The structure of a semiconductor saturable absorber mirror.

Due to the longer upper-level life (generally more than one order of magnitude of the pump pulse), the initial photons in the cavity are fewer, the probability of stimulated emission is lower, and the variation of populations in the upper level is negligible. With the enhancement of stimulated emission, the inversional populations in the upper level undergo an avalanche process at $t = t_2$ and the photons in the cavity increase rapidly. When $t = t_3$ the inversional populations are reduced to the threshold, photons in the cavity reach the maximum value, and the laser pulse is generated in the cavity.

Gain-switched pulse fiber lasers in the 1 μm regime were demonstrated by Zenteno *et al* in 1989 [7] and the peak power was several orders higher than that of continuous-wave lasers under the same pump power. The 2 μm regime pulse lasers have potential to be applied in laser radar, medical research, and 3–5 μm optical parametric oscillators. Jiang *et al* [8] first reported stable short pulse (10 ns) generation from a gain-switched Tm^{3+}-doped fiber laser in the 2 μm spectral region. A modulated erbium-doped fiber amplifier with 100 ns pulse duration was used as the pump source. When the pump energy was four times greater than the threshold energy, the output pulse width was 10 ns, the slope efficiency was 50%, the peak power rate was greater than 1 kW, and the frequency could be up to 500 kHz. Simakov *et al* [9] demonstrated a linearly polarized gain-switched Tm^{3+}-doped fiber laser. Via two-stage fiber amplification, an 18 W 1.55 μm pulse laser obtained at a repetition rate of 600 kHz with the output power of nearly 8 W was achieved, and the laser wavelength was 2.044 μm. Tang *et al* [10] realized an average output of 100 W at a wavelength of 2.02 μm for a gain-switched Tm^{3+}-doped double-clad fiber and the beam quality was close to the diffraction limit.

Recently, high-power pulsed lasers at 3 μm have attracted considerable interest because of their proximity to the fundamental absorption of liquid water, and two atmospheric transmission windows of 3–5 and 8–13 μm [13]. Pulsed lasers near 2.8 μm have already been demonstrated by mode-locked, *Q*-switched, or gain-switched fiber lasers based on Er^{3+}- and Ho^{3+}-doped fluoride (ZBLAN) fiber lasers in the recent decade. Gorjan *et al* [11] demonstrated a high-power gain-switched 2.8 μm Er^{3+}-doped ZBLAN fiber laser, generating a maximum average (peak) power of 2 W (68 W) with a repetition rate of 100 kHz and a pulse width of 300 ns. Wei *et al* [12] reported a tunable pulse laser in the range of 2.7–2.87 μm. Paradis *et al* [13] developed an all-fiber gain-switched Er^{3+}-doped ZBLAN fiber laser at 2.83 μm, yielding a record average power of 11.2 W with a pulse width of 170 ns and a repetition rate of 140 kHz.

Based on Ho^{3+}-doped ZBLAN fiber with a slightly red-shifted emission spectrum, dual-waveband cascade gain-switching at 2.93 and 2.07 μm was realized [14]. For the 3.5 μm emission band of Er^{3+} ions based on the F49/2→I49/2 transition, gain-switched dual-wavelength pumped Er^{3+}-doped ZBLAN fiber lasers have been demonstrated. Luo reported a gain-switched fiber laser giving a maximum peak power of 204 W [15] and an average power of 1.04 W [16].

To date, gain-switching has proven a feasible approach to achieve a nanosecond pulse laser in the 2–5 μm regime. Benefiting from their simple geometry, gain-switched fiber lasers are convenient for integration and commercialization. With the development of high-power LDs, gain-switched fiber lasers are excellent candidates for producing millijoule-energy, nanosecond-duration pulsed lasers.

7.3 Mode-locked fiber lasers

Ultrafast mode-locked fiber lasers have a wide range of applications in both the scientific and industrial fields due to their excellent spatial mode quality, thermal effects, compact structure, and low manufacturing costs. Various types of mode-locking technologies have been developed to obtain pulses with higher energy and shorter duration, as well as to ensure a low mode-locking threshold and long-term stability. The generation of ultrashort pulses includes two processes, namely pulse picking and pulse shaping. Pulse picking converts the fiber laser from continuous-wave (CW) operation into pulsed operation. Pulse shaping is responsible for the formation of the pulse shape and has several stages, such as soliton, dispersion-managed soliton, and self-similar and all-normal dispersion.

7.3.1 Pulse picking in mode-locked fiber lasers

Real saturable absorbers and artificial saturable absorbers have been employed to realize pulse picking. In 1992 Keller *et al* achieved mode-locking with a semiconductor film containing a saturable absorber, which was grown directly on the saturable Bragg reflector (SBR) using epitaxy, as shown in figure 7.3 [17]. It is also called a semiconductor saturable absorber mirror (SESAM). SESAMs are now widely used to mode-lock fiber lasers with a high repetition rate. In 2010 Byun *et al* constructed an SBR-based fiber laser generating 187 fs pulses at a repetition rate of 967 MHz [18]. In 2014 an ultra-stable mode-locked 12 GHz fiber laser was developed using polarization-maintaining (PM) fibers and an SESAM [19]. More recently, a passively mode-locked fiber laser based on SESAM was demonstrated to deliver more energetic pulses [20]. There are two major problems in SESAM-based fiber lasers. One is that *Q*-switched mode-locking is prone to occur instead of stable mode-locked pulse trains. The other is a low optical damage threshold. However, as long as the parameters of the SESAM are suitably designed, stable mode-locked pulse trains with a high destruction threshold can be achieved. Figure 7.4 shows the schematic of a SESAM-based fiber laser.

In recent years, some new saturable absorbers such as single-walled carbon nanotubes (SWCNTs) and graphene have been used in mode-locked fiber lasers. In 2003 SWCNTs were used as saturable absorbers for the first time in an erbium-doped fiber laser to achieve mode-locking [21]. In 2019 Guo *et al* demonstrated experimentally a tunable ytterbium-doped fiber laser mode-locked with SWCNTs [22]. Compared to SESAMs, SWCNTs-based saturable absorbers have the advantages of low cost, short recovery time, tunable wavelength, and a high damage threshold. Zero bandgap graphene has inter-band optical absorption under strong excitation, which makes it a suitable saturable absorber for mode-locking. In 2009, using atomic layer graphene as a saturable absorber in a mode-locked fiber laser, ultrashort soliton pulses (756 fs) were generated in the telecommunications band. The modulation depth was been investigated by varying the graphene thickness [23]. In 2018 a dual-wavelength passively mode-locked erbium-doped fiber laser employing a graphene/polymethyl-methacrylate saturable absorber was fabricated [24].

Artificial saturable absorption mechanisms, including nonlinear polarization rotation (NPR) with a nonlinear amplifying loop mirror (NALM), were used to generate mode-locked pulses. Using the Kerr nonlinearity in the fiber, the NPR mechanism evolves the change of polarization state into amplitude modulation. Figure 7.5 shows the concept of NPR if the incident light is linearly polarized low-intensity CW light, which is parallel to the particular axis of the fiber. Due to the natural birefringence in the fiber, the light exiting the linear birefringence fiber may be elliptically polarized with an angle α from the x-axis of the fiber. If the incident light contains high-intensity pulses, the projection of the pulse on the x- and the y-axis is different, resulting from various nonlinear phase shifts due to the Kerr effect.

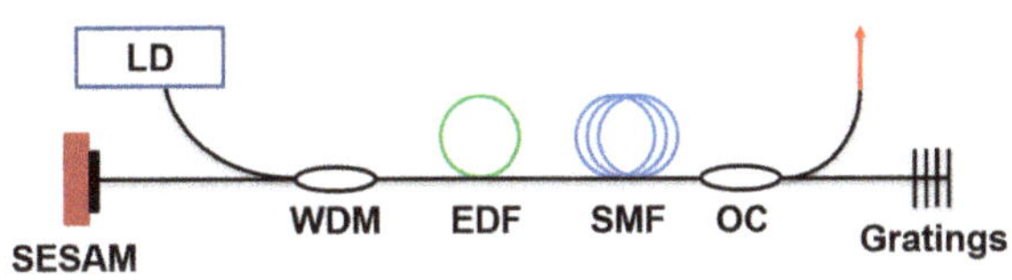

Figure 7.4. Schematic of a SESAM-based fiber laser.

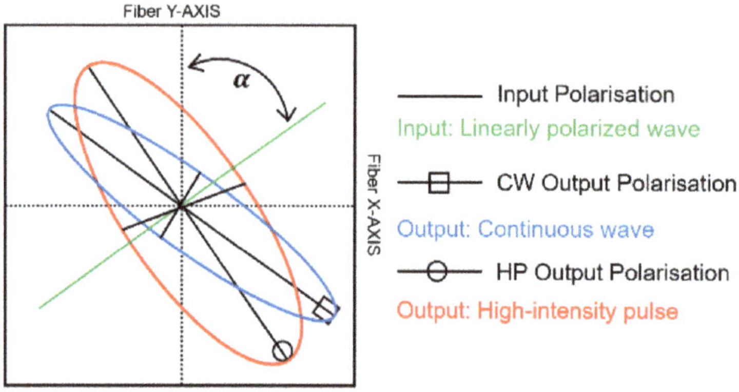

Figure 7.5. The principle of the NPR technique.

The optical axis of the synthesized elliptically polarized light may rotate. Potentially, by aligning the analyzer (output polarizer) its transmission polarization direction is consistent with the polarized optical axis caused by high-intensity pulses. In this way it only allows high-intensity pulses to pass through the polarizer and rejects the CW light. This effect is similar to saturable absorption. NPR-based fiber lasers can output pulses with ultrashort pulse widths and high intra-cavity pulse energy. In 2010, Ma *et al* reported an NPR-based ring-cavity fiber laser, as demonstrated in figure 7.6, generating 37.4 fs pulses at a fundamental repetition frequency of 225 MHz [25]. In 2017, utilizing the NPR technique, a thulium-doped fiber laser was reported to deliver 350 fs pulses with high energy (7.8 nJ) [26]. In 2019 a bidirectional pumped fiber laser mode-locked with NPR can produce 331 nJ rectangular pulses [27]. However, NPR mode-locking is highly dependent on the polarization. When the fiber is subjected to external pressure, vibrations, and temperature fluctuations, the resulting birefringence causes the polarization to change accordingly, which changes the original polarization state and makes the laser lose mode-locking. This brings great uncertainty to the NPR mode-locked fiber laser, restricting its use under less stable conditions (figure 7.6).

Unlike real saturable absorbers and NPR mode-locked fiber lasers, fiber lasers based on NALM can employ PM fibers. The main advantage of this kind of fiber laser is its environmental robustness. The nonlinear optical loop mirror (NOLM) was proposed as an all-optical switch in 1988 [28]. Subsequently, it has been used as a mode-locker in fiber lasers. The NALM extended the concept of NOLM by adding an amplifier at the end of the loop close to the coupler [29, 30]. Figure 7.7 shows the basic principle of the NALM. This mode-locking process is based on the interference between two counter-propagating lights inside the fiber loop connected by a beam splitter. However, these kinds of oscillators lack self-starting, and one method to achieve easier mode-locking is to introduce a non-reciprocal phase bias into the NALM loop. The phase bias allows the laser to have a greater chance to produce high-intensity spikes, which lowers the self-starting threshold. Figure 7.7 also shows two types of NALM-based fiber lasers: figure-8 and figure-9 shaped lasers.

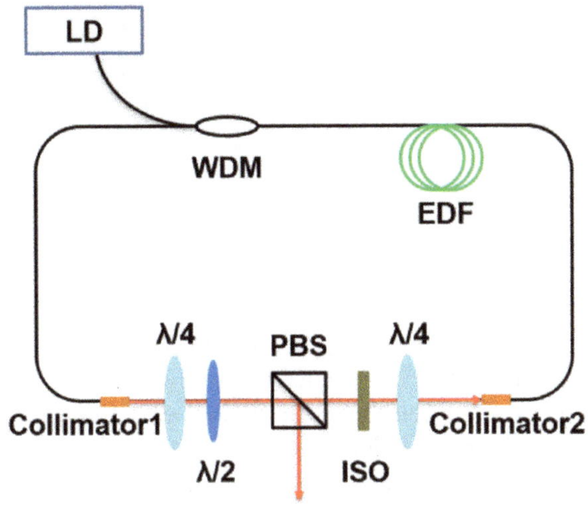

Figure 7.6. An NPR mode-locked fiber laser.

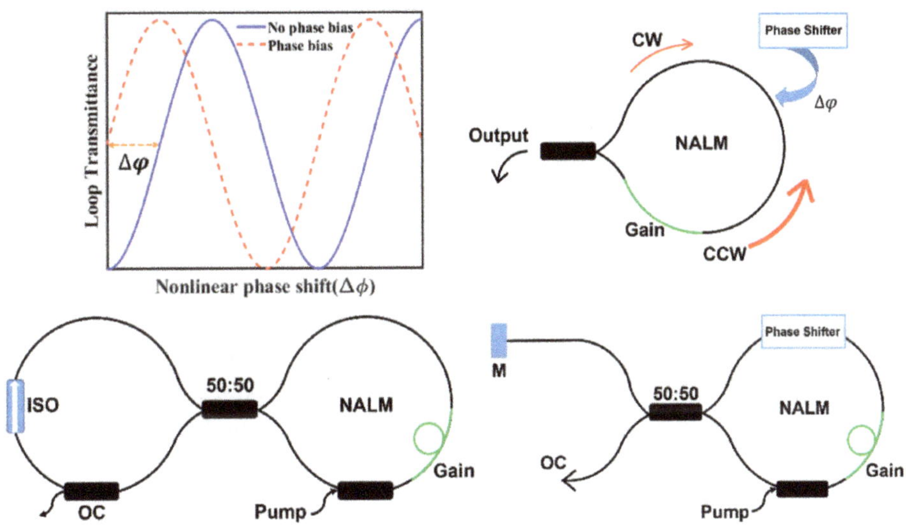

Figure 7.7. The principle of NALM and NALM-based figure-8 and figure-9 shaped fiber lasers [31], reproduced with permission, copright The Optical Society.

Figure-8 mode-locked fiber lasers exhibit excellent optical performance, as well as long-term stability. In 2012 Erkintalo *et al* constructed an all-PM fiber laser based on an NALM. Before entering the NALM, the pulses propagate along a single-mode PM fiber and obtain high pulse energy. The laser produces 16 nJ pulses with a repetition rate of 1.7 MHz, with a pulse width after extra-cavity compression of 400 fs [32]. In 2016 Szczepanek *et al* reported a figure-8 mode-locked fiber laser using a large splitting ratio coupler, where the proposed laser can output 3.5 nJ pulses at the repetition rate of 15 MHz [33]. In 2018 Yu *et al* demonstrated an NALM-based figure-8 fiber laser that can directly generate 10 nJ, 93 fs pulses [34].

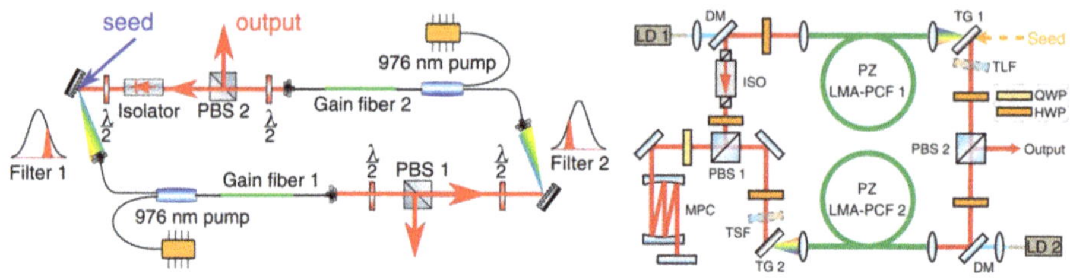

Figure 7.8. Mamyshev fiber oscillators [38, 40], reproduced with permission, copright The Optical Society.

Compared to figure-8 configurations, figure-9 fiber lasers offer many advantages, such as a tunable repetition rate, free-space flexibility, and low noise characteristics. Hänsel *et al* proposed all-PM fiber lasers mode-locked with a reflective phase biased NALM working at different wavelengths (1030, 1565, and 2050 nm) [35]. In 2016 Kuse *et al* used a graphene modulator to stabilize an all-PM figure-9 fiber laser with below 0.2 rad carrier envelope-offset frequency phase noise [36]. In 2018 a 700 MHz mode-locked fiber laser based on a phase biased NALM was reported, with the oscillator delivering 215 fs pulses [37].

Recently, interest in Mamyshev fiber lasers has developed from the demand for increasingly higher single-pulse energies (figure 7.8). The mechanism of the Mamyshev oscillator was first proposed in 1998. The offset filter plays a vital role in this kind of fiber laser. There are always two offset filters which are placed at the front and back ends of the gain fiber. The transmission wavelength of these two filters is not at the center wavelength of the laser pulse, but at the two sides of the center wavelength, and there is generally no overlapping part. The offset filters can filter out small pulses and regenerate distorted pulses, making the single-pulse energy break through the limitation of high nonlinear effects. In 2017 Liu *et al* constructed a Mamyshev oscillator delivering 50 nJ, 40 fs pulses [38]. In 2019 a 100 fs Mamyshev fiber laser at 1550 nm was reported to generate high-energy pulse trains up to 31.3 nJ [39]. By using single-polarization, large-mode-area photonic crystal fibers, a Mamyshev oscillator can generate a 1 μJ pulse at a 8 MHz repetition rate [40]. Its state-of-the-art performance makes the Mamyshev oscillator very attractive for various applications.

7.3.2 Pulse shaping in mode-locked fiber lasers

The combination of linear effects and nonlinear effects in mode-locked fiber lasers leads to four major pulse shaping mechanisms, including soliton, dispersion-managed soliton, and self-similar and all-normal-dispersion.

Soliton mode-locked fiber lasers work in the anomalous dispersion region. The balance of anomalous dispersion and self-phase modulation in optical fibers supports the generation of soliton pulses. In addition, self-amplitude modulation plays an important role and stabilizes the pulses. Kelly sidebands can be seen on both sides of the spectrum, which are caused by periodic disturbances of soliton pulses circulating in the laser cavity.

If anomalous dispersion is balanced by normal dispersion, fiber lasers may work in the dispersion-managed soliton regime. The pulse stretches and compresses twice when it propagates in the fiber laser. Minimum pulse durations can be found at the center of each fiber segment with different dispersions. The output pulse is normally transform-limited. Unlike soliton pulses, there are no Kelly sidebands. The increased average pulse duration reduces the nonlinearity, which enables laser output pulses with higher energy.

The pulse energy of a soliton mode-locked fiber laser is limited to picojoules and fiber lasers working in the dispersion-managed soliton regime that generates nanojoule energy pulses. To further increase the single-pulse energy, it is necessary to prevent pulse breaking caused by nonlinearity accumulation in the laser cavity. One method is to operate the laser in net normal dispersion; this pulse shaping mechanism is called self-similar or all-normal-dispersion. The pulses in the cavity are highly chirped, and the leading and trailing edges of the pulses belong to a different wavelength. To ensure stable circulation of the pulse in the cavity, the short and long-wavelength parts of the pulse are filtered using spectral filtering technology. In the time domain, the short and long-wavelength parts are placed on two wings of the pulse. Filtering out these parts also means that the pulse is compressed in the time domain. Self-similar or all-normal-dispersion pulses can tolerate high nonlinearity, enabling higher pulse energy in fiber lasers.

Different pulse evolutions of the above four pulse shaping mechanisms are shown in figure 7.9. The green solid line is the pulse width. The red dotted line represents the second-order dispersion distribution. The blue dashed line shows the spectrum width. The nonlinear phase shift that can be tolerated is shown at the top of each figure.

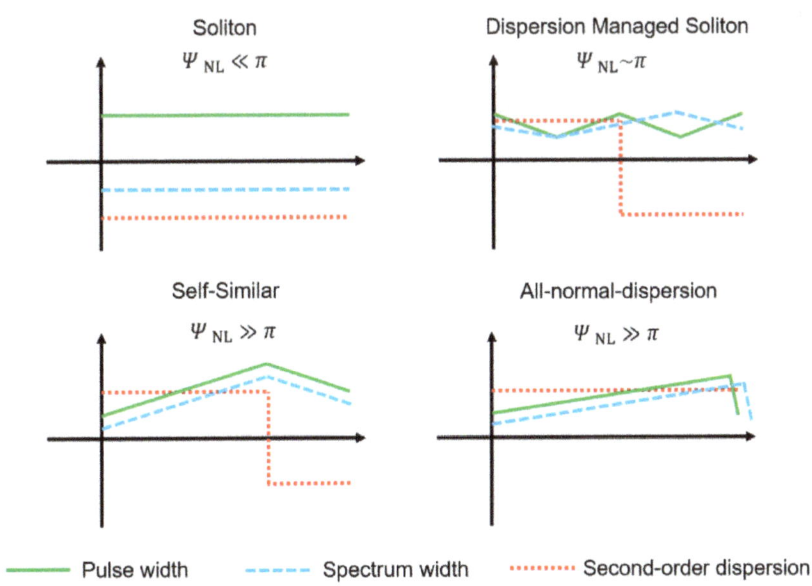

Figure 7.9. Different pulse evolutions of the four pulse shaping mechanisms [41], reproduced with permission, copyright (2008) John Wiley and Sons.

7.4 Chirped pulse fiber laser amplifier

The average output power of a mode-locked fiber laser is generally a few milliwatts to hundreds of milliwatts and the repetition rate is 80–100 MHz. The single-pulse energy is therefore in the order of a few nanojoules. For various applications it is necessary to scale the pulse energy. However, due to the thermal effect of the gain medium, a high repetition rate and high single-pulse energy conflict with each other. Also, as pulses are confined in the fiber core with a diameter of several micrometers, the single-pulse energy of fiber lasers is limited by high nonlinear effects. The chirped pulse amplification (CPA) technique has been developed to reduce the effects of nonlinearity and increase the pulse energy. In 2018 a Nobel prize was awarded for the invention of CPA, which enables high-intensity, ultrashort pulses [42]. The basic principle is shown in figure 7.10. The pulses from a seed oscillator propagate into a grating stretcher and are stretched in the time domain. The AOM is employed to adjust the repetition rate. After that, the pulses are linearly amplified twice in two gain fibers and finally compressed by the grating pair. The width of the output pulse is almost transform-limited. The use of the CPA technique reduces the accumulation of nonlinear phase shift, which avoids damage to the optical devices caused by the high peak power of the pulses. The stretcher and compressor are the most critical devices in CPA systems and their performance has an important influence on the pulse energy and pulse width. In addition, effects such as dispersion mismatch and gain narrowing can also lead to unsatisfactory pulse quality. Several schemes based on CPA have been proposed to improve the optical performance of the pulses.

7.4.1 Nonlinear chirped pulse amplification

Although fiber-based stretchers and compressors may be used simultaneously, the energy processing of fiber-based compressors is limited. Thus bulk grating compressors are used for the generation of higher energy pulses. However, the combination of a fiber stretcher and a bulk grating compressor produces a significant third-order dispersion mismatch. Adding an appropriate amount of self-phase modulation (SPM) to the CPA process can compensate for the third-order dispersion mismatch of fiber stretchers and grating compressors, thereby achieving good pulse compression. The main advantage of nonlinear CPA is that it allows the accumulation of significant

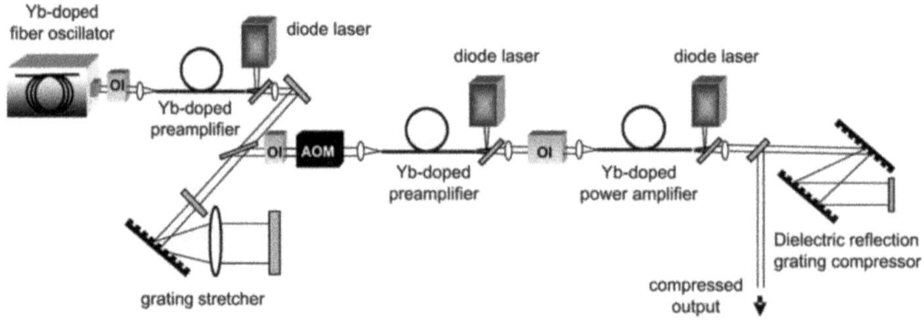

Figure 7.10. The principle of CPA [43], reproduced with permission, copright The Optical Society.

nonlinear phase shift, although the pulse quality may be significantly reduced. In 2005 Shah *et al* proposed a ytterbium-doped fiber CPA system, which can directly generate 200 μJ cubicon pulses. After compression based on the compensation of third-order dispersion mismatch between the stretcher and compressor, 650 fs, 100 μJ pulses have obtained [44]. In 2008 Zaouter *et al* demonstrated an ytterbium-doped fiber chirped pulse amplifier system delivering high quality pulses with 270 fs pulse width and 100 μJ pulse energy at a repetition rate of 300 kHz [45]. In 2010 Kalaycioglu *et al* reported an entirely fiber-integrated amplifier through nonlinear CPA. It produced pulses with up to 4 μJ energy at a 1 MHz repetition rate with a shortest compressed pulse duration of 140 fs [46].

7.4.2 Pre-chirp managed amplification

The pulse evolution in the fiber is determined by self-phase modulation, dispersion, loss, and gain. Before the seed pulse is amplified, by adding a certain amount of negative chirp in advance, the self-phase modulation and dispersion will work together during the amplification of the pulse and the spectrum may be broadened to a greater extent. This amplification technique is called pre-chirp managed amplification (PCMA). The advantage of PCMA is that it can produce pulses with a duration shorter than 100 fs. For a negatively pre-chirped input pulse, its spectrum undergoes initial spectrum compression and then undergoes spectrum broadening. The negative pre-chirp can be generated from a diffraction-grating pair, a chirped fiber Bragg grating, and a hollow-core photonic crystal fiber. In 2013 Lim *et al* reported a frequency comb based on a narrowband ytterbium-doped fiber laser with a repetition rate of 280 MHz. By pre-chirping the pulses at the amplifier input, the carrier envelope-offset frequency is optimized with 0.64 rad (integrated from 1 Hz to 10 MHz) phase noise for the in-loop error signal [47]. In 2015 a pre-chirp managed Yb-doped fiber laser system was demonstrated to generate 75 MHz, 130 W spectrally broadened pulses. After compensation, pulses with a pulse width of 60 fs and an average power of 100 W were obtained [48]. In 2017 Song *et al* proposed a femtosecond polarization-maintaining fiber laser amplification system with a standard double-cladding fiber technique. Using a hybrid amplification scheme this system can deliver 24 fs transform-limited pulses with 1 μJ pulse energy at a 1 MHz repetition rate [49]. In 2018 a PCMA system was built to output pulses with a pulse width of less than 65 fs and an energy of more than 10 nJ, and by filtering the spectrum of the amplification chain the relative intensity noise was reduced to 0.008%, the compressed pulse having a near-perfect transition limit [50].

7.4.3 Self-similar amplification and pre-chirp managed self-similar amplification

In a normal dispersion fiber, the interaction of nonlinear effects and dispersion can cause pulse distortion and pulse breaking, which limits the output energy of the fiber amplifier. In 1993 Anderson *et al* found that parabolic pulses with linear chirp can propagate in a normal dispersion fiber without pulse breaking [51]. Therefore, self-similar amplification has been proposed for obtaining high peak power, short duration pulses. In 2009 Deng *et al* demonstrated a Yb-doped fiber amplifier, which can generate 48 fs pulses with 18 W average power and 226 nJ of pulse energy. The system works in the parabolic pulse regime [52]. However, self-similar amplification is affected by third-order dispersion, limited gain bandwidth, and stimulated Raman scattering, and the pulse width cannot be further compressed. Pre-chirp management self-similar amplification has been developed to further increase the pulse energy. In 2013, Wang *et al* used a nonlinear preshaper to optimize the initial pulses for self-similar evolution. The proposed system can generate transform-limited pulses with ~60 fs [53]. In 2014 a polarization-maintaining large-mode-area photonic crystal fiber amplifier using pre-chirping management was reported. The system can generate ultrashort pulses of 38 fs with an average output power of 80 W [54]. In 2016 Liu *et al* used a pair of prisms as the pre-chirper in a high-power third-order dispersion-managed amplification system, delivering 33 fs pulses of 93.5 W at a repetition rate of 55 MHz [55]. Wang *et al* demonstrated a fiber amplifier based on the spectral and temporal breathing parabolic pulse evolution in the condition of an optimized negative pre-chirp. After compression, the pulse width was 36 fs. This self-similar amplification technology also effectively reduces the relative intensity noise during the amplification process [56].

7.4.4 Parabolic pre-shaping amplification

Optical pulses with a parabolic intensity profile can maintain their characteristics while propagating in a normal dispersion fiber. Pulse breaking will not occur under the influence of nonlinearity. In addition, the combination of normal dispersion and nonlinearity makes the pulses highly linearly chirped in the time domain. In 2007 Finot *et al* found that pulses with a parabolic intensity profile may be formed in a length of the normally dispersive nonlinear fiber and the parabolic profile stabilized in the second highly nonlinear fiber. During amplification the spectral width of pulses becomes wider, resulting in a shorter duration than the seed pulse width after compression [57]. In 2009 Deng *et al* demonstrated a Yb-doped fiber amplifier delivering 48 fs pulses with 18 W average power and 226 nJ of pulse energy. By varying the gain fiber length and pump wavelength, the effects of distributed longitudinal gain on the fiber amplifier performance were discussed [52]. Large pitch fibers (LPFs) are a good choice for obtaining high-energy pulses. In 2013 Pierrot *et al* used Yb-doped rod-type fiber as the gain medium and constructed a novel CPA scheme. This system can convert a 27 ps parabolic pulse into 780 fs pulses with 25 W average output power [58]. More recently, Fu *et al* designed a femtosecond fiber amplifier, which compresses 9 ps seed pulses 33 times and generates a pulse of 4.3 μJ and 275 fs. They found that Raman scattering effects impose a fundamental limitation on pulse energy [59].

7.4.5 Gain-narrowing compensation

The gain medium has a finite fluorescence bandwidth. The spectral width of the pulse is affected by the gain-narrowing effect during energy amplification, which limits the pulse duration. Gain-narrowing compensation is used to overcome the gain-narrowing effect. Compared to the nonlinearity compensation discussed above, the advantage of this method is that the temporal pulse shape of the output pulses has no obvious pedestal, post- and prepulse. In 2006 Takada *et al* used this technique in a Ti:sapphire chirped pulse amplification system. By using multilayer gain-narrowing compensators and an adaptive dispersion compensator, 12 fs pulses were obtained. In 2015 Chiba *et al* applied this technique in a Yb-doped fiber laser system [60]. A gain-narrowing compensator is made of multiple dielectric layers incorporated into the CPA system as a filter. This amplifier can output 65 fs pulses with 100 nJ energy at a 3 MHz repetition rate [61]. In 2017 this system was optimized by using a better fiber oscillator, resulting in a shorter pulse duration (41 fs) [62].

With the invention of CPA technology and the development of all kinds of large-mode-area gain fibers, fiber laser amplifiers are able to output nearly kilowatt average power pulses, and the pulse energy is on the order of millijoules. However, CPA technology is limited by the effect of gain narrowing and a pulse width below 200 fs is difficult to achieve. Using the above-mentioned amplification techniques, such as nonlinear chirped pulse amplification, pre-chirp managed amplification, self-similar amplification, pre-chirp managed self-similar amplification, parabolic pre-shaping amplification, and gain-narrowing compensation, the interaction of non-linearity and dispersion in the amplification process can be controlled. Therefore, a narrower pulse width and higher pulse energy are achieved.

7.5 NALM fiber lasers

Figure-8 and figure-9 mode-locked fiber lasers based on NALM have become effective tools for integrating applications and scientific research. The fiber is used as a Kerr medium in the laser cavity to confine the pulse in the fiber core with a core diameter of several microns. Under high pulse intensity, the phase shift is accumulated within the laser cavity producing high nonlinear effects. Taking figure-8 and figure-9 mode-locked fiber lasers as nonlinear platforms, very distinctive works have been carried out [63–66]. Highly nonlinear fiber has been inserted into the figure-8 and figure-9 mode-locked fiber lasers, and the laser cavities can generate pulses with unique characteristics. The structure diagram of a figure-8 mode-locked fiber laser based on NALM is shown in figure 7.11.

In addition, to gain fiber and ordinary single-mode fiber, the laser contains a long segment of highly nonlinear fiber, which brings high nonlinear effects to the fiber laser. By adjusting the pump power and the polarization states in the cavity, different pulse characteristics and transformation between different pulses can be realized. The output characteristics (spectra, pulse train, autocorrelation trace, and RF spectra) of the stable rectangular-shaped noise-like pulse are shown in figure 7.12. Apart from the rectangular-shaped noise-like pulse, versatile multi-soliton patterns such as single-scale soliton clusters, high-order harmonic pulses, and localized chaotic multiple pulses have also been experimentally investigated in detail. In addition, various low-amplitude sub-pulses have also been observed in the figure-8 fiber laser. The sub-pulses are released by the rectangular noise-like pulse and caused by the interactions among the CW components and the random ultrashort pulses in the noise-like pulse wave packet. The obtained results are of

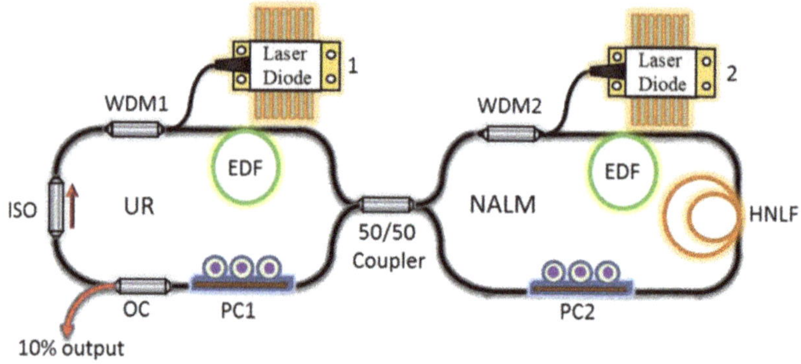

Figure 7.11. The structure diagram of a figure-8 mode-locked fiber laser [63], reproduced with permission, copright The Optical Society.

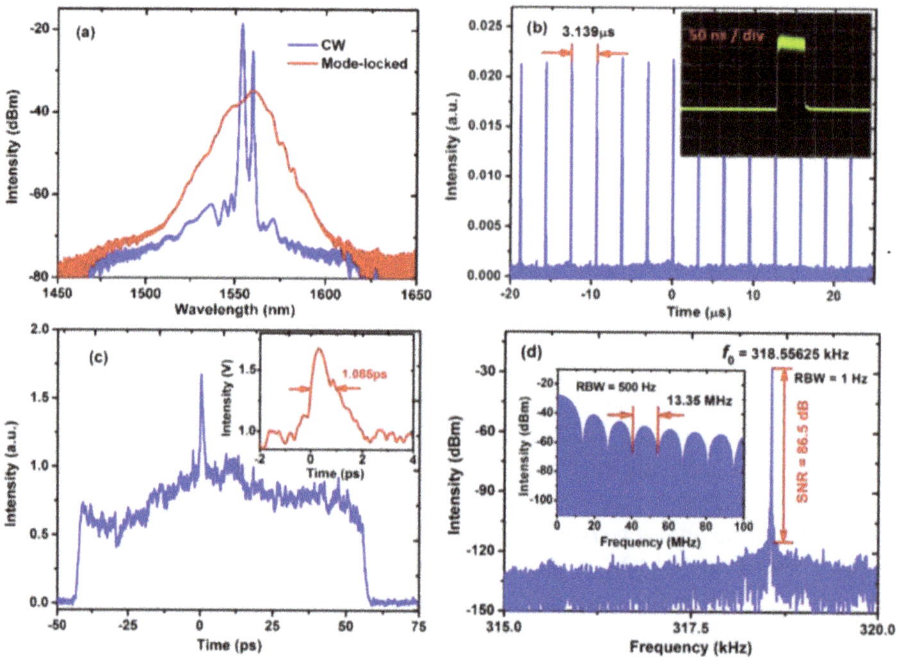

Figure 7.12. The output characteristics of a long cavity fiber laser: (a) spectra, (b) pulse train, (c) autocorrelation trace, and (d) RF spectra [63], reproduced with permission, copright The Optical Society.

great value for studying the physical mechanism of pulse formation and transformation in mode-locked fiber lasers [63, 64].

Figure-9 mode-locked fiber lasers with high nonlinearity have also been constructed, which can generate dark rectangular noise-like pulses with interesting nonlinear characteristics. The experimental configuration is depicted in figure 7.13. When the pump power is 420 mW, this oscillator can output dark rectangular pulses with a pulse energy of 27.2 nJ. The repetition rate of the figure-9 fiber laser is tunable by changing the polarization states of the cavity. More specifically, for these kinds of

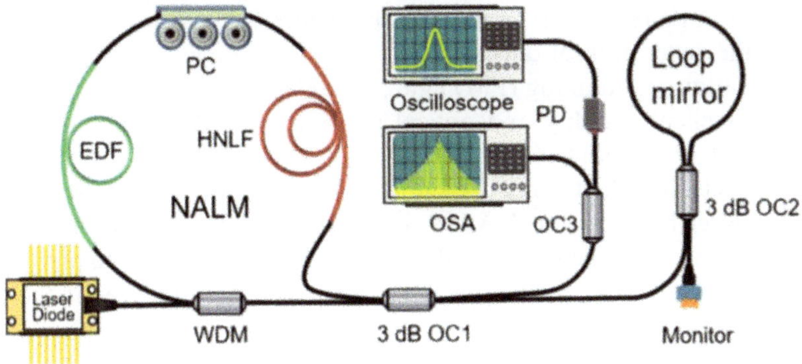

Figure 7.13. The experimental configuration of a figure-9 mode-locked fiber laser [65], reproduced with permission, copright The Optical Society.

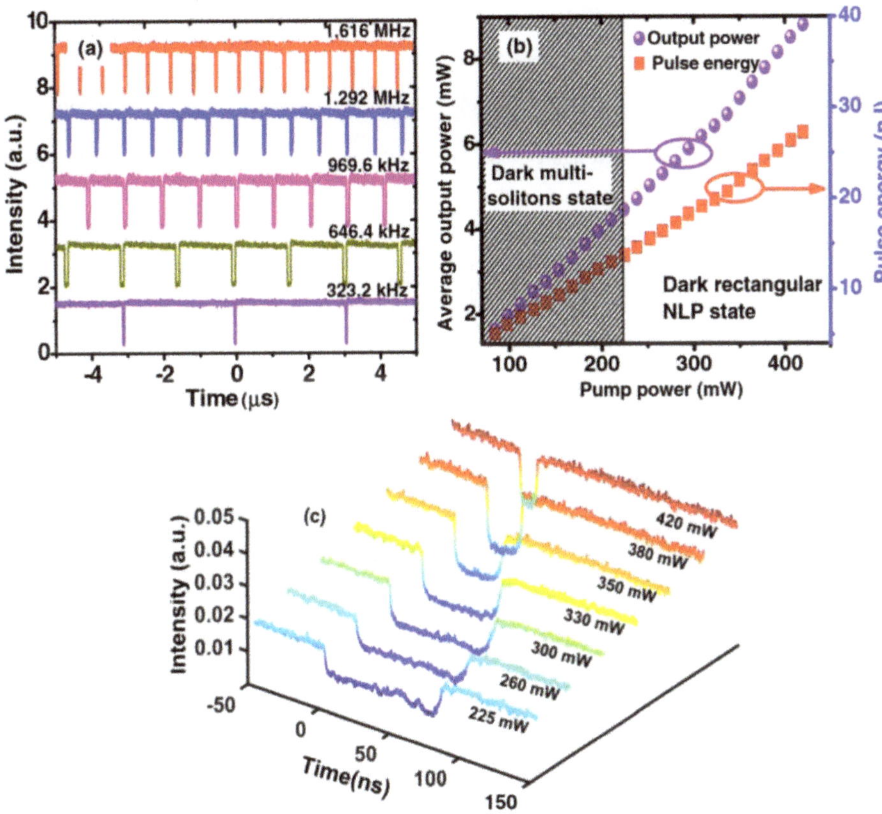

Figure 7.14. The dark rectangular pulse evolution as a function of the pump power [65], reproduced with permission, copright The Optical Society.

pulses, when the pump power increases, the pulse width of the dark rectangular pulses does not increase, but decreases. The experimental results are depicted in figure 7.14. Compared to bright pulses, dark pulses exhibit better stability and provide potential values in an optical communication system. The formation mechanism of dark rectangular pulses in this figure-9 fiber laser promotes understanding of the generation of dark pulses.

All polarization-maintaining (PM) figure-9 mode-locked fiber lasers can generate pulses with excellent stability and optical performance. The non-reciprocal phase bias in figure-9 fiber lasers can enhance self-starting ability. To investigate how phase bias affects the self-starting threshold and noise characterizations, an 88 MHz all-PM fiber laser based on an NALM with different biases has been constructed. As shown in figure 7.15, the laser can deliver 0.42 nJ pulses when the pump power is 450 mW. The noise characteristics of such a laser have been analyzed in the presence

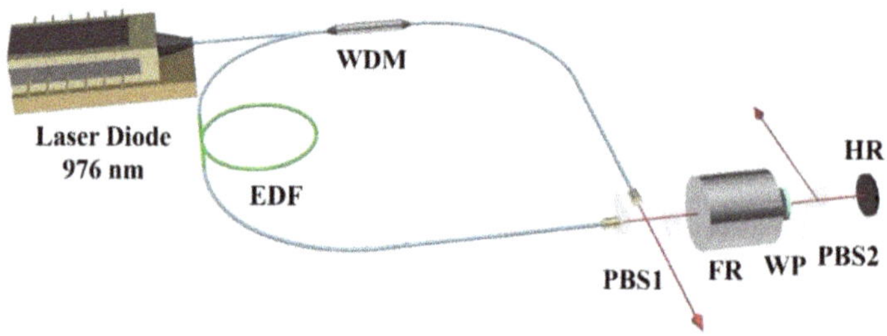

Figure 7.15. The diagram of an all-PM figure-9 fiber laser [66]. Copyright IOP Publishing. All rights reserved.

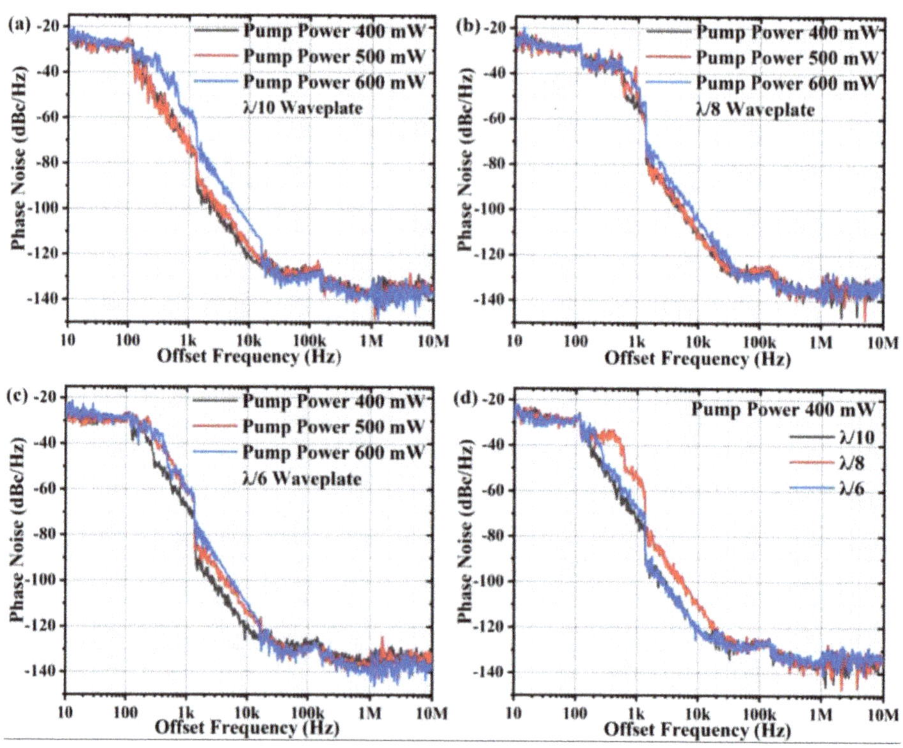

Figure 7.16. Phase noise characteristics of an all-PM NALM-based fiber laser [66]. Copyright IOP Publishing. All rights reserved.

of different phase shifts. The phase noise spectra under different pump powers and phase biases have different levels and are shown in figure 7.16. The difference is caused by external fluctuations or an intra-cavity mechanism. The obtained results may guide the design of NALM fiber lasers.

7.6 Conclusions

In this chapter, sections 7.1–7.3 mainly focus on three different kinds of pulsed fiber lasers, with an emphasis on the basic principles and recent state-of-the-art fiber lasers. Section 7.4 is devoted to five chirped pulse amplification technologies. Section 7.5 provides a brief introduction to our latest works on fiber lasers based on a nonlinear amplifying loop mirror. Section 7.6 gives our conclusion. Due to their excellent characteristics, pulsed fiber lasers have been intensively utilized in fields such as optical communications, high precision micromachining, and medical treatment. Various innovative fiber optic devices and new materials have made fiber lasers more powerful tools with a higher repetition rate, higher average power, and narrower pulse width. In addition, numerous pulse amplification technologies have also accelerated the practical use of pulsed fiber lasers. With continued development, one can speculate that in the future fiber lasers will be used in many applications both in the laboratory and in the field, enabling much greater optical performance.

References

[1] Alcock I P, Tropper A C, Ferguson A I and Hanna D C 1986 *Q*-switched operation of a neodymium-doped monomode fibre laser *Electron. Lett.* **22** 84–5

[2] Renaud C C, Alvarez-Chavez J A, Sahu J K, Nilsson J, Richardson D J and Clarkson W A 2001 7.7 mJ pulses from a large core Yb-doped cladding pumped *Q*-switched fibre laser *Conf. on Lasers and Electro-Optics. Postconference Technical Digest* p 219

[3] Fan Y-X *et al* 2004 Tunable high-peak-power, high-energy hybrid *Q*-switched double-clad fiber laser *Opt. Lett.* **29** 724

[4] Schmidt O *et al* 2007 Millijoule pulse energy *Q*-switched short-length fiber laser *Opt. Lett.* **32** 1551

[5] Tokita S, Murakami M, Shimizu S, Hashida M and Sakabe S 2011 12 W *Q*-switched Er: ZBLAN fiber laser at 2.8 μm *Opt. Lett.* **36** 2812

[6] Laurila M *et al* 2011 *Q*-switching and efficient harmonic generation from a single-mode LMA photonic bandgap rod fiber laser *Opt. Express* **19** 10824

[7] Zenteno L A, Snitzer E, Po H, Tumminelli R and Hakimi F 1989 Gain switching of a Nd^{+3}-doped fiber laser *Opt. Lett.* **14** 671

[8] Jiang M and Tayebati P 2007 Stable 10 ns, kilowatt peak-power pulse generation from a gain-switched Tm-doped fiber laser *Opt. Lett.* **32** 1797

[9] Simakov N, Hemming A, Bennetts S and Haub J 2011 Efficient, polarised, gain-switched operation of a Tm-doped fibre laser *Opt. Express* **19** 14949

[10] Tang Y, Xu L, Yang Y and Xu J 2010 High-power gain-switched Tm^{3+}-doped fiber laser. *Opt. Express* **18** 22964

[11] Gorjan M, Petkovšek R, Marinček M and Čopič M 2011 High-power pulsed diode-pumped Er:ZBLAN fiber laser *Opt. Lett.* **36** 1923

[12] Wei C, Luo H, Shi H, Lyu Y, Zhang H and Liu Y 2017 Widely wavelength tunable gain-switched Er^{3+}-doped ZBLAN fiber laser around 2.8 μm *Opt. Express* **25** 8816
[13] Paradis P, Fortin V, Aydin Y O, Vallée R and Bernier M 2018 10 W-level gain-switched all-fiber laser at 2.8 μm *Opt. Lett.* **43** 3196
[14] Luo H, Li J, Zhu C, Lai X, Hai Y and Liu Y 2017 Cascaded gain-switching in the mid-infrared region *Sci. Rep.* **7** 1–7
[15] Jobin F, Fortin V, Maes F, Bernier M and Vallée R 2018 Gain-switched fiber laser at 3.55 μm *Opt. Lett.* **43** 1770
[16] Luo H *et al* 2019 Watt-level gain-switched fiber laser at 3.46 μm *Opt. Express* **27** 1367
[17] Keller U, Miller D A B, Boyd G D, Chiu T H, Ferguson J F and Asom M T 1992 Solid-state low-loss intracavity saturable absorber for Nd:YLF lasers: an antiresonant semiconductor Fabry–Perot saturable absorber *Opt. Lett.* **17** 505
[18] Byun H *et al* 2010 Compact, stable 1 GHz femtosecond Er-doped fiber lasers *Appl. Opt.* **49** 5577–82
[19] Thapa R, Nguyen D, Zong J and Chavez-Pirson A 2014 All-fiber fundamentally mode-locked 12 GHz laser oscillator based on an Er/Yb-doped phosphate glass fiber *Opt. Lett.* **39** 1418
[20] Wang S *et al* 2020 Obtaining more energetic modelocked pulses from a SESAM-based fiber laser *Opt. Express* **28** 20345
[21] Set S *et al* 2003 A noise suppressing saturable absorber at 1550 nm based on carbon nanotube technology *OFC 2003 Optical Fiber Communications Conf. 2003* **2** 723–5
[22] Guo H *et al* 2019 Tunable ytterbium-doped mode-locked fiber laser based on single-walled carbon nanotubes *J. Lightwave Technol.* **37** 2370–74
[23] Bao Q *et al* 2009 Atomic-layer graphene as a saturable absorber for ultrafast pulsed lasers *Adv. Funct. Mater.* **19** 3077–83
[24] Lau K Y *et al* 2018 Dual-wavelength, mode-locked erbium-doped fiber laser employing a graphene/polymethyl-methacrylate saturable absorber *Opt. Express* **26** 12790
[25] Ma D, Cai Y, Zhou C, Zong W, Chen L and Zhang Z 2010 37.4 fs pulse generation in an Er fiber laser at a 225 MHz repetition rate *Opt. Lett.* **35** 2858
[26] Gao C, Wang Z, Luo H and Zhan L 2017 High energy all-fiber Tm-doped femtosecond soliton laser mode-locked by nonlinear polarization rotation *J. Lightwave Technol.* **35** 2988–93
[27] Gao J *et al* 2019 Generation of high-energy rectangular pulses in a nonlinear polarization rotation mode-locked ring fiber laser *Appl. Opt.* **58** 7897
[28] Doran N J and Wood. D 1988 Nonlinear-optical loop mirror *Opt. Lett.* **13**
[29] Fermann M E, Haberl F, Hofer M and Hochreiter H 1990 Nonlinear amplifying loop mirror *Opt. Lett.* **15** 752
[30] Fermann M E, Turi L, Hofer M, Haberl F and Schmidt A J 1991 Additive-pulse-compression mode locking of a neodymium fiber laser *Opt. Lett.* **16** 244
[31] Kim J and Song Y 2016 Ultralow-noise mode-locked fiber lasers and frequency combs: principles, status, and applications *Adv. Opt. Photonics* **8** 465
[32] Erkintalo M, Aguergaray C, Runge A and Broderick N G R 2012 Environmentally stable all-PM all-fiber giant chirp oscillator *Opt. Express* **20** 22669
[33] Szczepanek J, Kardaś T M, Michalska M, Radzewicz C and Stepanenko Y 2015 Simple all-PM-fiber laser mode-locked with a nonlinear loop mirror *Opt. Lett.* **40** 3500

[34] Yu Y *et al* 2018 Highly-stable mode-locked PM Yb-fiber laser with 10 nJ in 93-fs at 6 MHz using NALM *Opt. Express* **26** 10428
[35] Hänsel W *et al* 2017 All polarization-maintaining fiber laser architecture for robust femtosecond pulse generation *Appl. Phys.* B **123** 41
[36] Kuse N, Jiang J, Lee C-C, Schibli T R and Fermann M E 2016 All polarization-maintaining Er fiber-based optical frequency combs with nonlinear amplifying loop mirror *Opt. Express* **24** 3095
[37] Liu G, Jiang X, Wang A, Chang G, Kaertner F and Zhang Z 2018 Robust 700 MHz mode-locked Yb fiber laser with a biased nonlinear amplifying loop mirror *Opt. Express* **26** 26003
[38] Liu Z, Ziegler Z M, Wright L G and Wise F W 2017 Megawatt peak power from a Mamyshev oscillator *Optica* **4** 649
[39] Olivier M, Boulanger V, Guilbert-Savary F, Sidorenko P, Wise F W and Piché M 2019 Femtosecond fiber Mamyshev oscillator at 1550 nm *Opt. Lett.* **44** 851
[40] Liu W *et al* 2019 Femtosecond Mamyshev oscillator with 10-MW-level peak power *Optica* **6** 194
[41] Wise F W, Chong A and Renninger W H 2008 High-energy femtosecond fiber lasers based on pulse propagation at normal dispersion *Laser Photonics Rev.* **2** 58–73
[42] Strickland D and Mourou G 1985 Compression of amplified chirped optical pulses *Opt. Commun.* **55** 447–49
[43] Röser F *et al* 2007 90 W average power 100 μJ energy femtosecond fiber chirped-pulse amplification system **32** 2230–32
[44] Shah L, Liu Z, Hartl I, Imeshev G, Cho G C and Fermann M E 2005 High energy femtosecond Yb cubicon fiber amplifier *Opt. Express* **13** 4717
[45] Zaouter Y, Boullet J, Mottay E and Cormier E 2008 Transform-limited 100 μJ, 340 MW pulses from a nonlinear-fiber chirped-pulse amplifier using a mismatched grating stretcher–compressor *Opt. Lett.* **33** 1527
[46] Kalaycioglu H, Oktem B, Şenel Ç, Paltani P P and Ilday F Ö 2010 Microjoule-energy, 1 MHz repetition rate pulses from all-fiber-integrated nonlinear chirped-pulse amplifier *Opt. Lett.* **35** 959
[47] Lim J, Chen H-W, Chang G and Kärtner F X 2013 Frequency comb based on a narrowband Yb-fiber oscillator: pre-chirp management for self-referenced carrier envelope offset frequency stabilization *Opt. Express* **21** 4531
[48] Liu W *et al* 2015 Pre-chirp managed nonlinear amplification in fibers delivering 100 W, 60 fs pulses *Opt. Lett.* **40** 151
[49] Song H *et al* 2017 Practical 24-fs, 1-μJ, 1-MHz Yb-fiber laser amplification system *Opt. Express* **25** 7559
[50] Hua Y, Chang G, Kärtner F X and Schimpf D N 2018 Pre-chirp managed, core-pumped nonlinear PM fiber amplifier delivering sub-100-fs and high energy (10 nJ) pulses with low noise *Opt. Express* **26** 6427
[51] Anderson D, Desaix M, Karlsson M, Lisak M and Quiroga-Teixeiro M L 1993 Wave-breaking-free pulses in nonlinear-optical fibers *J. Opt. Soc. Am.* B **10** 1185
[52] Deng Y, Chien C-Y, Fidric B G and Kafka J D 2009 Generation of sub-50 fs pulses from a high-power Yb-doped fiber amplifier *Opt. Lett.* **34** 3469
[53] Wang S *et al* 2013 Self-similar evolution in a short fiber amplifier through nonlinear pulse preshaping *Opt. Lett.* **38** 296

[54] Zhao J, Li W, Wang C, Liu Y and Zeng H 2014 Pre-chirping management of a self-similar Yb-fiber amplifier towards 80 W average power with sub-40 fs pulse generation *Opt. Express* **22** 32214

[55] Liu Y, Li W, Luo D, Bai D, Wang C and Zeng H 2016 Generation of 33 fs 935 W average power pulses from a third-order dispersion managed self-similar fiber amplifier *Opt. Express* **24** 10939

[56] Wang S, Chen W, Qin P, Song Y, Hu M and Liu B 2016 Spectral and temporal breathing self-similar evolution in a fiber amplifier for low-noise transform-limited pulse generation *Opt. Lett.* **41** 5286

[57] Finot C, Provost L, Petropoulos P and Richardson D J 2007 Parabolic pulse generation through passive nonlinear pulse reshaping in a normally dispersive two segment fiber device *Opt. Express* **15** 852

[58] Pierrot S and Salin. F 2013 Amplification and compression of temporally shaped picosecond pulses in Yb-doped rod-type fibers *Opt. Express* **21** 20484

[59] Fu W, Tang Y, McComb T S, Lowder T L and Wise F W 2017 Limits of femtosecond fiber amplification by parabolic pre-shaping *J. Opt. Soc. Am.* B **34** A37

[60] Takada H and Torizuka K 2006 Design and construction of a TW-class 12-fs Ti:sapphire chirped-pulse amplification system *IEEE J. Sel. Top. Quantum Electron.* 12 201–12

[61] Chiba Y, Takada H, Torizuka K and Misawa K 2015 65-fs Yb-doped fiber laser system with gain-narrowing compensation *Opt. Express* **23** 6809

[62] Takada H, Chiba Y, Yoshitomi D, Torizuka K and Misawa K 2017 41-fs, 35-nJ, green pulse generation from a Yb-doped fiber laser system *Opt. Express* **25** 2115

[63] Zhou R, Liu X, Yu D, Qian L and Fu H Y 2020 Versatile multi-soliton patterns of noise-like pulses in a stable passively mode-locked fiber laser *Opt. Express* **28** 912–23

[64] Zhou R, Qian L and Fu H Y 2020 Sub-pulses releasing from rectangular-shaped noise-like pulses in a passively mode-locked fiber laser *IEEE Photonics Technol. Lett.* **32** 925–28

[65] Zhou R, Yu D, Liu X, Qian L and Fu H Y 2019 Dark rectangular noise-like pulses in figure-nine fiber laser based on nonlinear amplifying loop mirror *Opt. Lett.* **44** 3717–20

[66] Liu X, Liu G, Zhou R, Yu D, Wu J, Fu H Y, Zhang Z and Li Q 2020 An all-PM fiber laser mode locked by nonlinear amplifying loop mirror with different biases *Laser Phys.* **30** 085104

IOP Publishing

Optical Fiber Technology and Applications
Recent advances
Mário F S Ferreira and Mukul Chandra Paul

Chapter 8

Nonlinear fiber optics

Mário F S Ferreira

Nonlinear effects in optical fibers impose different limitations on communications systems, and an understanding of such effects is almost a prerequisite for actual lightwave-system designers. On the other hand, they offer a variety of possibilities to perform diverse all-optical signal processing functions. The nonlinear effects are enhanced dramatically and new phenomena are observed in the so-called highly nonlinear fibers, made from silica or from other glasses with a greater intrinsic material nonlinearity. This chapter provides an overview of the main nonlinear effects both in conventional and in highly nonlinear fibers.

8.1 Introduction

Silica-based optical fibers are available at very low cost and have been acknowledged as indispensable transmission media for the global telecommunications network due to their attenuation, as low as 0.148 dB km^{-1} [1]. They have a broad low-loss optical window, which expands to about 60 THz and ranges from 1260 to 1675 μm [2] and much optical equipment is designed to operate in this spectral region. Silica fibers are available in a wide range of different forms with the development during the last few years of several types of microstructured fibers.

A number of third-order nonlinear processes can occur when high-power light is launched into the optical fibers [3]. These can grow to appreciable magnitudes over the long lengths available in fibers, even though the nonlinear index of silica glass is very small ($n_2 = 2.35 \times 10^{-20}$ m^2 W^{-1} at 1550 nm [4]).

The third-order nonlinear processes occurring in optical fibers can be classified into two main categories [3]. The first category is related to the dependence of the silica refractive index on the signal intensity (Kerr effect). This determines three main nonlinear effects: self-phase modulation (SPM) [5], where an optical signal alters its own phase; cross-phase modulation (XPM) [6], where one signal affects the phases of all other optical signals and vice versa; and four-wave mixing (FWM) [7], where signals with different frequencies interact to produce mixing sidebands.

doi:10.1088/978-0-7503-3243-9ch8

The second category of third-order nonlinear effects corresponds to stimulated scattering processes, namely the stimulated Brillouin scattering (SBS) [8] and stimulated Raman scattering (SRS) [9] effects.

The nonlinear effects generated in the fibers affect severely the performance of optical communications systems, since they impose limits on the launched power of the signals, channel bit rate, channel spacing, transmission bandwidth, and hence the entire information capacity of such systems [10–12]. However, the same nonlinear effects have been positively utilized in various types of fiber lasers, such as Raman fiber lasers [13], Brillouin fiber lasers [14], parametric oscillators [15], and soliton-SRS wavelength tunable lasers [16]. In addition, due to their femtosecond-order response time [17] and broad operation bandwidth [3], the fiber nonlinearities offer a variety of novel possibilities for optical amplification and all-optical processing, which are very promising for near-future high-capacity networks.

One of the most impressive nonlinear phenomena that can be observed in highly nonlinear fibers is supercontinuum generation (SCG). It corresponds to the extremely wide spectrum achieved by an optical pulse while propagating in a nonlinear medium, and results generally from the synergy between several fundamental nonlinear processes, such as self-phase modulation, cross-phase modulation, stimulated Raman scattering, and four-wave mixing [3, 18]. Soliton related effects also play a significant role whenever light with sufficient power propagates in the anomalous dispersion regime [18]. A supercontinuum source is very useful for several applications, namely in the area of spectroscopy, optical coherence tomography [19], metrology [20], and multichannel optical communications [21].

The fiber nonlinearity is commonly characterized by the nonlinear parameter γ, which in standard single-mode silica fibers has a typical value of $\gamma \approx 1.3\ \mathrm{W^{-1}\,km^{-1}}$ [22]. Such a value is too small for some applications requiring highly efficient nonlinear processes, namely for optical signal processing. As a consequence, a fiber length of several kilometers would generally be necessary in such applications. The desire to reduce such fiber lengths has led to the development of some types of specialized nonlinear fibers based on conventional fiber technology. Such fibers have an increased refractive index contrast between the core and cladding, up to around 3%, and well controlled dispersion characteristics. These fibers are called highly nonlinear fibers (HNLFs), since they present a smaller effective mode area and hence a larger nonlinear coefficient ($\gamma \approx 10 - 20\ \mathrm{W^{-1}\,km^{-1}}$) [23].

Microstructured optical fibers (MOFs), also called photonic crystal fibers (PCFs), represent a new class of optical fibers that are characterized by the fact that the silica cladding presents an array of embedded air holes. A reduced core size and a large air-filling fraction in the cladding determine an increase of the mode confinement in such fibers, which provides an enhancement of the fiber nonlinearity [24, 25]. Moreover, the waveguide contribution to the group-velocity dispersion (GVD) becomes particularly important in these fibers, which allow the convenient adjustment of their dispersive characteristics [26]. In particular, it is possible to achieve anomalous GVD and hence conditions for soliton formation at shorter wavelengths than is possible with standard fibers [27]. Silica nanowires can be considered as a special case of highly nonlinear MOFs [24, 25], in which the air-filling fraction of the cladding has reached unity.

Much higher values of γ can be achieved by combining tight mode confinement with the use of glasses with a greater intrinsic material nonlinearity coefficient than that of silica. Examples of suitable glasses that have been used include bismuth oxide ($\gamma \approx 1100\ \mathrm{W^{-1}\,km^{-1}}$) [28], lead-silicate ($\gamma \approx 1860\ \mathrm{W^{-1}\,km^{-1}}$), [29] and chalcogenide ($\gamma \approx 2450\ \mathrm{W^{-1}\,km^{-1}}$) [30] glasses.

8.2 Fiber dispersion

The mode-propagation constant $\beta(\omega)$ is related to the refractive index $n(\omega)$ in the form

$$\beta(\omega) = \frac{\omega n(\omega)}{c}. \tag{8.1}$$

Mathematically, the effects of fiber dispersion are accounted for by expanding $\beta(\omega)$ in a Taylor series about the carrier frequency ω_0 at which the pulse spectrum is centered:

$$\beta(\omega) = \beta_0 + \beta_1(\omega - \omega_0) + \frac{1}{2}\beta_2(\omega - \omega_0)^2 + \dots , \tag{8.2}$$

where

$$\beta_j = \left(\frac{d^j\beta}{d\omega^j}\right)_{\omega=\omega_0} \quad (j = 0, 1, 2, \dots). \tag{8.3}$$

The GVD is characterized by the parameter β_2

$$\beta_2 = \frac{d^2\beta}{d\omega^2} = -\frac{dv_g/d\omega}{v_g^2}, \tag{8.4}$$

where $v_g = c/n_g$ and $n_g = n + \omega(dn/d\omega)$ represent the group velocity and the group refractive index, respectively. In practice, the GVD is often characterized by another parameter, D, given by

$$D = -\frac{2\pi c\beta_2}{\lambda^2} = -\frac{\lambda}{c}\frac{d^2n}{d\lambda^2}. \tag{8.5}$$

For a typical single-mode fiber, we have $D = 0$ (16 ps (nm-km)$^{-1}$) at $\lambda = 1.31\ \mu$m (1.55 μm). This indicates that two wavelength components of a pulse separated by 1 nm arrive with a time delay on the order of a few picoseconds when they propagate over a distance of 1 km in a typical fiber.

The fiber dispersion is given by the sum of the material dispersion, D_M, and waveguide dispersion, D_W. The material dispersion D_M is related to the rate of variation of the group index with wavelength, $D_M = c^{-1}(dn_g/d\lambda)$. In the case of fused

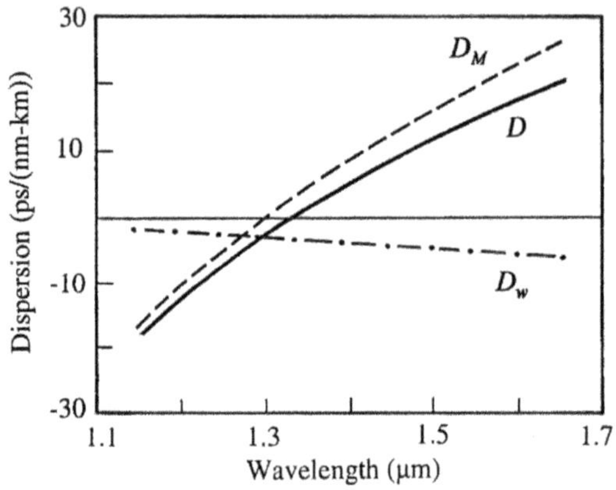

Figure 8.1. Relative contributions of material dispersion D_M and waveguide dispersion D_w, as well as total dispersion $D = D_M + D_w$ for a typical single-mode fiber.

silica $dn_g/d\lambda = 0$ and consequently $D_M = 0$ at $\lambda = \lambda_{ZD} = 1.276\ \mu m$, which is referred to as the material zero-dispersion wavelength.

The waveguide dispersion D_w depends on the fiber parameters, such as the core radius and the index difference between the core and the cladding, and is negative in the range 0–1.6 μm. Figure 8.1 shows D_M, D_w, and their sum $D = D_M + D_w$ for a typical single-mode fiber. The main effect of waveguide dispersion is to shift λ_{ZD} so that the total dispersion is zero near 1.31 μm.

It is possible to design the fiber such that the zero-dispersion wavelength λ_{ZD} is shifted to the vicinity of 1.55 μm, where the fiber attenuation is minimum. Such fibers are referred to as dispersion-shifted fibers (DSFs). It is also possible to design the fiber such that the total dispersion D is relatively small over a wide wavelength range extending from 1.3 to 1.6 μm. Such fibers are called dispersion-flattened fibers.

8.3 The Kerr effect

Nonlinear effects are attributed to the dependence of the susceptibility on the electric field, which becomes important at high field strengths. As a result, the total polarization vector **P** can be written in the frequency domain as a power series expansion in the electric field vector [31]:

$$\mathbf{P}(r, \omega) = \varepsilon_0[\chi^{(1)}.\ \mathbf{E} + \chi^{(2)}\mathbf{:}\ \mathbf{EE} + \chi^{(3)}\ \vdots\ \mathbf{EEE} + \ldots\ ] = \mathbf{P}_L(r, \omega) + \mathbf{P}_{NL}(r, \omega), \quad (8.6)$$

where $\chi^{(j)}$ $(j = 1, 2, \ldots)$ is the jth order susceptibility. To account for the light polarization effects $\chi^{(j)}$ is a tensor of rank $j + 1$. The linear susceptibility $\chi^{(1)}$ determines the linear part of the polarization $\mathbf{P}_L$. On the other hand, terms of second and higher order in equation (8.6) determine the nonlinear polarization $\mathbf{P}_{NL}$. Since Sio_2 is a symmetric molecule, the second-order susceptibility $\chi^{(2)}$ vanishes for silica glasses. As a consequence, virtually all nonlinear effects in optical fibers are determined by the third-order susceptibility $\chi^{(3)}$. In the time domain, the form of the expansion is identical to (8.6) if the nonlinear response is assumed to be instantaneous.

The presence of $\chi^{(3)}$ implies that the refractive index depends on the field intensity, I, in the form [3]

$$n = \sqrt{1 + \chi^{(1)} + \frac{3}{2}\frac{\chi^{(3)}}{c\varepsilon_0 n_0} I} = n_0 + n_2 I, \tag{8.7}$$

where $n_0 = \sqrt{1 + \chi^{(1)}}$ is the linear refractive index and $n_2 = 3\chi^{(3)}/(4c\varepsilon_0 n_0^2)$ is the refractive index nonlinear coefficient, also known as the Kerr coefficient.

8.4 The nonlinear Schrödinger equation

The electric field associated with a bit stream propagating in a single-mode fiber along the z-direction can be written in the form

$$E = \frac{1}{2}(F(r, \phi)U(z, t)e^{i(\beta_0 z - \omega_0 t)} + \text{c. c.}), \tag{8.8}$$

where $U(z,t)$ describes the complex amplitude of the field envelope at a distance z inside the fiber, $F(r, \phi)$ gives the transverse distribution of the fundamental fiber mode, and β_0 is the mode-propagation constant at the carrier frequency. The vector nature of the waves can be neglected if they are linearly polarized in the same direction.

The only quantity that changes during propagation is the amplitude $U(z,t)$. Since each frequency component of the optical field propagates with a slightly different propagation constant, it is useful to work in the spectral domain. The evolution of a specific spectral component $\tilde{U}(z, \omega)$ is given by

$$\tilde{U}(z, \omega) = \tilde{U}(0, \omega) \exp[i\beta(\omega)z - i\beta_0 z], \tag{8.9}$$

where $\tilde{U}(0, \omega)$ is the Fourier transform of the input signal $U(z = 0, t)$ and $\beta(\omega)$ is the propagation constant, which can be written as the sum of a linear part, β_L, and a nonlinear part, β_{NL}:

$$\beta(\omega) = \beta_L(\omega) + \beta_{NL}(\omega_0) \tag{8.10}$$

with

$$\beta_L(\omega) = \frac{\tilde{n}(\omega)\omega}{c}, \quad \beta_{NL}(\omega_0) = n_2 I \frac{\omega_0}{c}, \tag{8.11}$$

$\tilde{n}$ being the effective mode index. In equation (8.11) β_{NL} has been written as a frequency-independent quantity, considering that the spectrum of optical pulses used as bits in communication systems is usually much narrower than the carrier frequency ω_0.

Using the expansion given by equation (8.2) for the linear part of the propagation constant and retaining only terms up to second order in $(\omega - \omega_0)$, substituting equations (8.10) and (8.11) in equation (8.9), calculating the derivative $\partial\tilde{U}/\partial z$, and converting the resultant equation into the time domain by using the inverse Fourier transform and the relation

$$\Delta\omega \leftrightarrow i\frac{\partial}{\partial t}, \tag{8.12}$$

we obtain the equation [3]

$$i\left(\frac{\partial}{\partial z} + \frac{1}{v_g}\frac{\partial}{\partial t}\right)U - \frac{1}{2}\beta\frac{\partial^2}{\partial t^2}U + \beta_{\mathrm{NL}}U = 0. \tag{8.13}$$

The amplitude U in equation (8.13) can be normalized such that the power is $P = |U|^2$. Making such normalization, we have $\beta_{\mathrm{NL}} = \gamma\,|U|^2$, where

$$\gamma = \frac{\omega_0 n_2}{cA_{\mathrm{eff}}} \tag{8.14}$$

is known as the fiber nonlinear parameter and A_{eff} is the effective mode area.

Considering a moving frame propagating with the group velocity and using the new time variable

$$\tau = t - \frac{z}{v_g}, \tag{8.15}$$

equation (8.13) can be written in the form

$$i\frac{\partial U}{\partial z} - \frac{1}{2}\beta_2\frac{\partial^2 U}{\partial \tau^2} + \gamma|U|^2 U = 0. \tag{8.16}$$

Equation (8.16) is usually called the nonlinear Schrödinger equation (NLSE) due to its similarity to the Schrödinger equation of quantum mechanics. The NLSE describes the propagation of pulses in optical fibers taking into account both the group-velocity dispersion and the fiber nonlinearity.

8.5 Nonlinear effects in optical fibers

The third-order nonlinear processes occurring in optical fibers can be classified into two main categories [3]. The first category is related to the dependence of the silica refractive index on the signal intensity (Kerr effect), whereas the second category corresponds to stimulated scattering processes. In the following we review the nonlinear processes belonging to both categories.

8.5.1 Self-phase modulation

Assuming that the pulse temporal width is sufficiently large, we can neglect the dispersive term in the NLSE in equation (8.16), which reduces to

$$\frac{\partial U}{\partial z} = i\gamma PU, \tag{8.17}$$

where $P = |U|^2$ is the pulse power. Neglecting the fiber loss, this corresponds to the input power of the pulse. Equation (8.17) has the following general solution for the field amplitude at the output of a fiber of length L:

$$U(L, \tau) = U(0, \tau)\exp[i\phi_{NL}(L, \tau)], \tag{8.18}$$

where $U(0, \tau)$ is the input pulse envelope and

$$\phi_{NL}(L, \tau) = \gamma P(\tau)L \tag{8.19}$$

is the nonlinearity-induced phase change. The form of this general solution shows clearly that the fiber nonlinearity modifies the phase shift across the pulse, but not the intensity envelope.

Using (8.19) the electric field transmitted in the fiber can be written in the form

$$E(L, t) = \frac{1}{2}[\hat{E}(0, t)\exp\{i[(\beta_0 + \gamma P)L - \omega_0 t]\} + \text{c. c.}], \tag{8.20}$$

where ω_0 is the pulse central frequency. The phase of the wave described by equation (8.20) is

$$\phi = (\beta_0 + \gamma P)L - \omega_0 t. \tag{8.21}$$

According to equation (8.21) the phase alteration due to the nonlinearity is proportional to the power P of the wave. If the incident wave is a pulse with a given power temporal profile, the power variation within the pulse leads to its own phase modulation. Hence, this phenomenon is appropriately called self-phase modulation (SPM).

The instantaneous frequency within the pulse described by equation (8.19) is given by

$$\omega(t) = -\frac{\partial\phi}{\partial t} = \omega_0 - \gamma L\frac{\partial P}{\partial t}. \tag{8.22}$$

According to equation (8.22) in the leading edge of the pulse, where $dP/d\tau > 0$, the instantaneous frequency is down-shifted from ω_0, whereas in the tailing edge, where $dP/d\tau < 0$, the instantaneous frequency is up-shifted from ω_0. The frequency at the center of the pulse remains unchanged from ω_0.

The chirping due to nonlinearity without any increase in pulse width leads to increased spectral broadening. Figure 8.2 illustrates the SPM-induced spectral broadening experienced by Gaussian pulse, at a distance $z = 11L_{NL}$, where $L_{NL} = 1/(\gamma P_0)$ is the nonlinear length. The spectrum develops multiple peaks such that the two outermost peaks are the most intense.

The spectral broadening determined by the self-phase modulation effect produces different results depending on the dispersion regime of the propagating pulse. In the normal dispersion regime the chirpings due to GVD and SPM act in the same direction, leading to a stronger temporal broadening of the pulse and thus to a more significant reduction of the system capacity. In the anomalous dispersion regime the chirpings due to GVD and SPM act in opposite directions, resulting in a

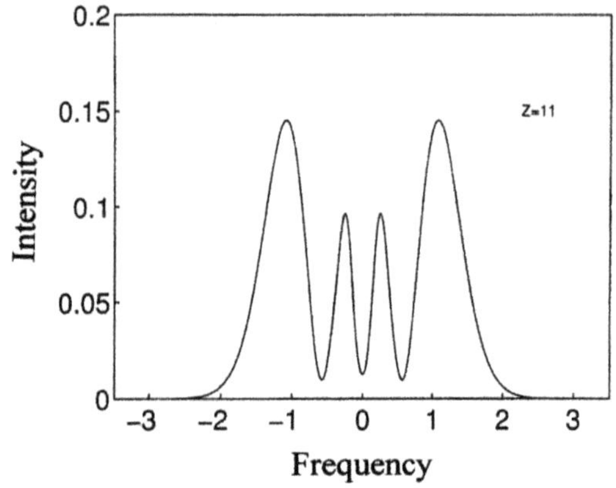

Figure 8.2. SPM-induced spectral broadening of an initially unchirped Gaussian pulse at a distance $z = 11L_{NL}$.

compression of the pulse. In particular, we can have a total cancellation of the two types of chirping and the pulse neither broadens in time nor in its spectrum. Such pulse is called a fundamental soliton.

8.5.2 Cross-phase modulation

Waves with different wavelengths propagating in the same fiber can interact with each other since the refractive index that a wave experiences depends on the intensities of all other waves. Hence, a pulse at one wavelength has an influence on the phase of a pulse at another wavelength. This nonlinear phenomenon is known as cross-phase modulation (XPM), and it can limit significantly the performance of wavelength-division multiplexing (WDM) systems [44–52].

Let us consider two channels propagating at the same time in the fiber. Using an approach similar to that used in section 8.4 to derive the NLSE, assuming that the phase-matching conditions between the involved waves are not satisfied and neglecting the dispersion effects, it is found that the slowly varying amplitudes U_1 and U_2 of the pulses in the two channels satisfy the following coupled differential equations [3]:

$$\frac{\partial U_1}{\partial z} = i\gamma_1(P_1 + 2P_2)U_1 \tag{8.23}$$

$$\frac{\partial U_2}{\partial z} = i\gamma_2(P_2 + 2P_1)U_2, \tag{8.24}$$

where $P_j = |U_j|^2$ is the power of the wave and γ_j is the nonlinear parameter at frequency ω_j, $j = 1, 2$.

If, for simplicity, the powers are assumed as constant, the solution of equation (8.23) is given by

$$U_1(L) = U_1(0)\exp\{i\gamma_1(P_1 + 2P_2)L\}. \tag{8.25}$$

A similar solution can be obtained from equation (8.24) for U_2. From equation (8.25) it is apparent that the phase of the signal at frequency ω_1 is modified not only due to its own power—which corresponds to the SPM effect—but also due to the power of the signal at frequency ω_2. This phenomenon is referred to as XPM and equation (8.25) shows that it is twice as effective as SPM.

The nonlinear phase shift of the signal at frequency ω_1 resulting from the combination of SPM and XPM at the output of a fiber with length L is given by

$$\phi_{1\mathrm{NL}} = \gamma_1(P_1 + 2P_2)L. \tag{8.26}$$

The XPM-induced frequency shift $\Delta\omega_{1\mathrm{XPM}}$ is given by

$$\Delta\omega_{1\mathrm{XPM}} = -2\gamma_1 L\frac{\partial P_2}{\partial t}. \tag{8.27}$$

Equation (8.27) shows that the part of the signal at ω_1 that is affected by the leading edge of the signal at ω_2 will be down-shifted in frequency, whereas the part overlapping with the trailing edge will be up-shifted in frequency. This determines a spectral broadening of the signal at ω_1 that is twice the spectral broadening caused by SPM.

The effect of XPM is different in amplitude- and in phase-modulated systems. In the last case, since the power in each channel is the same for all bits, the main limitation results from arbitrary phase fluctuations, which lead directly to a deterioration of the signal-to-noise ratio. Such phase fluctuations can be induced via the XPM by intensity variations, as happens if semiconductor lasers are directly phase-modulated.

In the case of amplitude-modulated direct detection systems, the XPM has no effect on the system performance if the dispersion is neglected. However, since the phase alteration due to XPM is associated with a frequency alteration, the dispersion determines an additional temporal broadening or compression of the spectral broadened pulses, which affects the system performance

XPM can be useful to realize several optical signal processing functions, such as all-optical switching [53–57], wavelength conversion [58–62], and full pulse regeneration [63–65].

8.5.3 Four-wave mixing

Four-wave mixing (FWM) is a parametric process in which four waves or photons interact with each other due to the third-order nonlinearity of the material. As a result, when several channels with frequencies $\omega_1, \dots, \omega_n$ are transmitted simultaneously over the same fiber, the intensity dependence of the refractive index leads not only to phase shifts within a channel, as discussed in previous sections, but also gives rise to signals at new frequencies.

The efficiency of the four-wave mixing process depends on the relative phase among the interacting optical waves. In the quantum mechanical description, FWM occurs when photons from one or more waves are annihilated and new photons are created at different frequencies. In this process, the rules of conservation of energy and momentum must be fulfilled. The conservation of momentum leads to the phase-matching condition.

Considering the case in which two photons at frequencies ω_1 and ω_2 are annihilated with simultaneous creation of two photons at frequencies ω_3 and ω_4, the conservation of energy imposes the condition

$$\omega_1 + \omega_2 = \omega_3 + \omega_4. \tag{8.28}$$

On the other hand, the phase matching is given by the condition $\Delta k = 0$, where

$$\begin{aligned}\Delta k &= \beta_3 + \beta_4 - \beta_1 - \beta_2 \\ &= (n_3\omega_3 + n_4\omega_4 - n_1\omega_1 - n_2\omega_2)/c.\end{aligned} \tag{8.29}$$

In practice, it is relatively easy to satisfy the phase-matching condition in the degenerate case $\omega_1 = \omega_2$. In this situation, a strong pump at $\omega_1 = \omega_2 \equiv \omega_\mathrm{p}$ generates a low-frequency side band at ω_3 and a high-frequency side band at ω_4, when we assume $\omega_4 > \omega_3$. These sidebands are referred to as the signal and idler bands. The frequency shift of the two sidebands is given by

$$\Omega_S = \omega_\mathrm{p} + \omega_3 = \omega_4 + \omega_\mathrm{p}. \tag{8.30}$$

The phase mismatch given by equation (8.29) can be expressed in terms of the frequency shift Ω_S if we use an expansion of Δk about the pump frequency ω_p. By retaining terms up to the quadratic in Ω_S in this expansion, one obtains

$$\Delta k = \beta_2 \Omega_\mathrm{S}^2, \tag{8.31}$$

where $\beta_2 = d^2\beta/d\omega^2$ is the group-velocity dispersion (GVD) coefficient at the pump frequency.

When $\beta_2 = 0$ we have a perfect phase matching and thus an efficient FWM. This situation is undesirable in WDM transmission systems, because FWM causes a transfer of power from each channel to its neighbors. Such a power transfer not only results in the power loss for the channel but also induces interchannel crosstalk that degrades the system performance severely [3]. This is particularly true for multi-channel systems with equal channel spacing, since in this case most new frequencies coincide with the original channel frequencies. The interference between the original and the new generated waves depends on the bit pattern and leads to significant fluctuations in the detected signal at the receiver, thus increasing the bit error rate (BER) in the system. This problem can be minimized using the technique of dispersion management. In this case, fibers with normal and anomalous dispersion are combined to form a periodic dispersion map, such that the local GVD is high but its average value is kept low. Due to its simplicity of implementation, the dispersion-management technique has become quite common since 1996 to control the FWM-induced limitations in WDM systems [66].

Taking into account the fiber nonlinearity, it can be shown that the effective phase mismatch is given by [3]

$$\kappa = \Delta k + 2\gamma P_\mathrm{p}, \tag{8.32}$$

where P_p is the pump power and and γ is an averaged nonlinear parameter, which ignores the small variation due to the slightly different frequencies of the involved waves.

For a pump wavelength in the anomalous dispersion region ($\beta_2 < 0$), the negative value of Δk can be compensated by the fiber nonlinearity. Using equation (8.31) in equation (8.32), for phase matching one has

$$\kappa \approx \beta_2 \Omega_S^2 + 2\gamma P_p = 0. \tag{8.33}$$

From this, we obtain that the frequency shift is given by

$$\Omega_S = \sqrt{2\gamma P_p I |\beta_2|}\,. \tag{8.34}$$

When $|\beta_2|$ is small, we must take into account the fourth-order term in the expansion of Δk, which becomes

$$\Delta k \approx \beta_2 \Omega_S^2 + \frac{\beta_4}{12}\Omega_S^4, \tag{8.35}$$

where $\beta_4 = d^4\beta / d\omega^4$. In this case the phase-matching condition is given by

$$\kappa \approx \beta_2 \Omega_S^2 + \frac{\beta_4}{12}\Omega_S^4 + 2\gamma P_0 = 0. \tag{8.36}$$

We observe from equation (8.36) that phase matching can be achieved even for a pump wavelength in the normal-GVD regime ($\beta_2 > 0$) if $\beta_4 < 0$.

An efficient FWM process is very useful for applications such as all-optical switching [67–71], wavelength conversion [72–76], pulse regeneration [77–80], optical phase conjugation (OPC) [81], and optical amplification [82–85].

8.5.4 Stimulated Raman scattering

In processes of light scattering in different media, we can observe the existence of a frequency-shifted field in addition to the incident field. The frequency shift of the scattered light is determined by the vibrational oscillations that occur between constituent atoms within the molecules of the material. A slight excitation of the molecular resonances due to the presence of the input field results in spontaneous Raman scattering. Stimulated Raman scattering (SRS) occurs as a consequence of the coherent interaction between the optical fields of the incident wave (also called the pump wave) and of the new frequency-shifted wave (also called the Stokes wave). This interaction originates a driving force that excites the molecular resonances. In a quantum mechanical description, one has simultaneously the absorption of a photon from the pump beam at frequency ω_p and the emission of a photon at the Stokes frequency ω_S. The difference in energy is taken up by a high energy phonon (molecular vibration) at frequency ω_V.

The pump wave intensity (I_p) and the Stokes wave intensity (I_S) propagating along an optical fiber satisfy the following equations [3]:

$$\frac{dI_S}{dz} = g_R I_S I_p - \alpha I_S \tag{8.37}$$

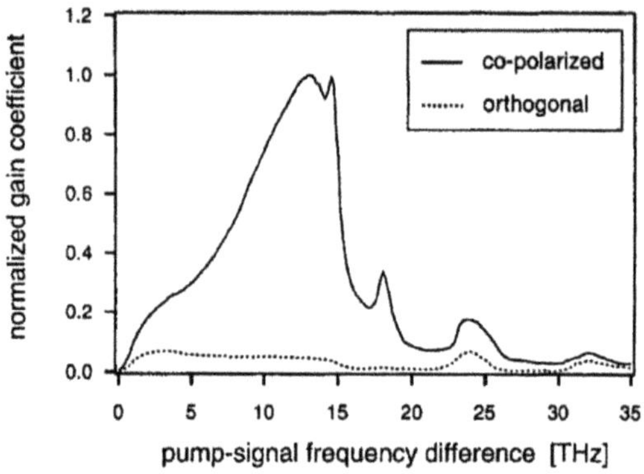

Figure 8.3. Normalized Raman gain coefficient for copolarized (full curve) and orthogonally polarized (dotted curve) pump and signal beams (after [86]).

$$\frac{dI_{\mathrm{S}}}{dz} = -\frac{\omega_{\mathrm{p}}}{\omega_{\mathrm{S}}} g_{\mathrm{R}} I_{\mathrm{S}} I_{\mathrm{p}} - \alpha I_{\mathrm{p}}, \tag{8.38}$$

where α takes into account the fiber losses and g_{R} is the Raman gain coefficient. Figure 8.3 shows the normalized Raman gain spectra for bulk fused silica, measured when the pump and signal light were either copolarized (full curve) or orthogonally polarized (dotted curve) [86]. The most significant feature of the Raman gain in silica fibers is that g_{p} extends over a large frequency range (up to 40 THz). Optical signals whose bandwidths are of this order or less can be amplified using the Raman effect if a pump wave with the right wavelength is available. The Raman shift, corresponding to the location of the main peak in figure 8.3, is close to 13 THz for silica fibers. The Raman gain depends on the relative state of polarization of the pump and signal fields. The copolarized gain is almost an order of magnitude larger than the orthogonally polarized gain near the peak of the gain curve. The peak value of the Raman gain decreases with increasing pump wavelength, and it is about 6×10^{-14} m W^{-1} in the wavelength region around 1.5 μm.

When the input Stokes wave intensity is weak, such that $I_{\mathrm{S0}} \ll I_{\mathrm{p0}}$, the evolution of the Stokes wave intensity is given approximately by

$$I_{\mathrm{S}}(z) \approx I_{\mathrm{S0}} \exp\left\{\frac{g_{\mathrm{R}} I_{\mathrm{p0}}[1 - \exp(-\alpha z]}{\alpha} - \alpha z\right\}. \tag{8.39}$$

If an input signal I_{S0} is not present, the Stokes wave is generated from spontaneous Raman scattering. It has been shown that this process is equivalent to injecting one fictitious photon per mode at the input end of the fiber [87]. The threshold for stimulated Raman scattering is defined as the input pump power at which the output powers for pump and Stokes wave become equal. It is given approximately by [87]

$$P_{\mathrm{p0}}^{\mathrm{th}} = 16 \frac{A_{\mathrm{eff}}}{g_{\mathrm{R}} L_{\mathrm{eff}}}, \tag{8.40}$$

where

$$L_{\text{eff}} = \frac{1 - \exp(-\alpha L)}{\alpha} \tag{8.41}$$

is the effective fiber length. Considering a sufficiently long fiber, such that $L_{\text{eff}} \approx 22$ km, and an effective core area $A_{\text{eff}} = 50\ \mu\text{m}^2$, we find a threshold power $P_{\text{p}0}^{\text{th}} \approx 600$ mW at $\lambda_{\text{p}} = 1.55\ \mu$m.

Because SRS has a relatively high threshold, it is not of concern for single-channel systems. However, in WDM systems SRS can cause crosstalk between channels signals whose wavelength separation falls within the Raman gain curve. The Raman-induced power transfer between two channels depends on the bit pattern, which leads to power fluctuations and determines additional receiver noise.

Instead of being generated from spontaneous Raman scattering, light at the Stokes wavelength can be input (along with the pump) and is consequently amplified through the stimulated conversion process. This is the basis of a fiber Raman amplifier, which allows amplification for a signal at any wavelength, provided a pump wave with an appropriate wavelength is used. The large bandwidth of fiber Raman amplifiers makes them attractive for fiber optic communication applications. Raman amplifiers can be used both in lumped and in distributed configurations. Moreover, their gain profile can be made very large and uniform using appropriate pumping sources [88–94].

8.5.5 Stimulated Brillouin scattering

Stimulated Brillouin scattering (SBS) is similar to stimulated Raman scattering, in that a Stokes wave that is downshift in frequency from that of a strong pump wave is also generated in this case. However, instead of the material resonances, which play a fundamental role in the Raman scattering, the interaction between the pump and Stokes waves in the SBS process occurs by way of an acoustic wave. In the classical picture, the pump wave creates a pressure wave in the medium through electrostriction. A material density wave is established that propagates at the velocity of sound in the medium in the direction of the pump. The periodic changes in material density appear as a moving refractive index grating. Thus the incident wave pumps the acoustic wave which scatters it and the scattering creates a Stokes wave.

The SBS process involves three waves: the incident (pump) wave of frequency ω_{p}, the Stokes wave of frequency ω_{S}, and an acoustic wave of frequency ω_{a}. The pump creates a pressure wave in the medium owing to electrostriction, which determines a periodic modulation of the refractive index. In this process the energy conservation requires that $\omega_{\text{p}} - \omega_{\text{S}} = 2\pi f_{\text{a}}$, where f_{a} is the linear frequency of the acoustic wave. On the other hand, the momentum conservation requires that the wave vectors of the three waves satisfy $\mathbf{K}_{\text{a}} = \mathbf{K}_{\text{p}} - \mathbf{K}_{\text{S}}$. Since the acoustic wave velocity $V_{\text{a}} \approx 5.96$ km s^{-1} is far smaller than the light velocity, we have $|\mathbf{K}_{\text{a}}| = 2\pi f_{\text{a}}/v_{\text{a}}|\mathbf{K}_{\text{p}}| \approx |\mathbf{K}_{\text{S}}|$. As a consequence, we conclude that the Stokes and the pump waves must propagate in opposite directions.

The pump wave intensity (I_p) and the Stokes wave intensity (I_S) propagating along an optical fiber satisfy the following equations [3]:

$$\frac{dI_S}{dz} = -g_R I_p I_S - \alpha I_S \tag{8.42}$$

$$\frac{dI_p}{dz} = -g_B I_S I_p - \alpha I_p, \tag{8.43}$$

where (g_B) is the Brillouin gain coefficient, which for typical fibers is estimated to be about 2.5×10^{-11} m.W^{-1}. This value is between two and three orders of magnitude larger than the Raman gain coefficient at $\lambda_p = 1.55\ \mu$m.

The spectrum of the Brillouin gain is Lorentzian with an FWHM $\Delta_{\nu B}$ is determined by the acoustic phonon lifetime. For bulk silica the width is expected to be about 17 MHz at 1.55 μm. However, several experiments have shown much larger bandwidths for silica-based fibers, which in some cases can exceed 100 MHz. Such larger bandwidths are generally attributed both to the guided nature of acoustic modes and to inhomogeneities in the fiber-core cross-section along the fiber length.

If an input signal is not present, the Stokes wave is generated from spontaneous Brillouin scattering. It has been shown that this process is equivalent to injecting one fictitious photon per mode at a distance where the gain is equal to the fiber loss [87]. The threshold pump power is defined as the input pump power that is equal to the output power of the Stokes wave, which gives the result [87]

$$P_{p0}^{th} = 21\frac{A_{eff}}{g_B L_{eff}}. \tag{8.44}$$

Considering an effective area $A_{eff} = 50\ \mu m^2$, an effective length $L_{eff} \approx 22$ km, and a gain coefficient $g_B = 2.5 \times 10^{-11}$ mW^{-1}, we obtain a threshold of $P_{p0}^{th} \approx 2$ mW. This value is about three orders of magnitude smaller than the threshold required for Raman scattering, which makes stimulated Brillouin scattering the dominant nonlinear effect in some circumstances.

The value reported above for the Brillouin gain coefficient is valid only when the spectral width of the pump beam ($\Delta\nu_p$) is much narrower that the Brillouin linewidth ($\Delta\nu_B$). In general, however, to take into account the finite spectral width of the pump beam, the convolution of the power spectrum of the pump field with the intrinsic Brillouin frequency spectrum must be calculated. Assuming that the pump spectrum has a Lorentzian profile, such a convolution is a Lorentzian of width $\Delta\nu_B + \Delta\nu_p$ and thus the Brillouin gain coefficient is reduced and given by

$$\tilde{g}_B = \frac{\Delta\nu_B}{\Delta\nu_B + \Delta\nu_p} g_B. \tag{8.45}$$

SBS can be detrimental to such systems in a number of ways: by introducing a severe signal attenuation, by causing multiple frequency shifts, and by introducing interchannel crosstalk in bidirectional transmission systems. However, Brillouin gain can also find some useful applications, namely for optical amplification and lasing [95–97], channel selection [98], optical phase conjugation [99], temperature and strain sensing [100, 101], slow-light control [102, 103], optical storage [104], etc.

8.6 Optical solitons

The formation of solitons in optical fibers is the result of a balance between the negative (anomalous) group-velocity dispersion (GVD) of the glass fiber, which occurs for wavelengths longer than 1.3 μm in a standard fiber, and the Kerr nonlinearity. The existence of fiber solitons was suggested for the first time by Hasegawa and Tappert [105] in 1973, while their first experimental observation was made by Mollenauer *et al* [106] in 1980. Optical solitons offer the possibility of transmitting optical pulses over extremely large distances without distortion [107–109].

Considering the NLSE given by equation (8.16), let us introduce a normalized amplitude Q given by

$$U(z, t) = \sqrt{P_0 Q(z, \tau)}, \tag{8.46}$$

where P_0 is the peak power of the incident pulse. $Q(z, \tau)$ is found to satisfy the equation

$$i\frac{\partial Q}{\partial z} - \frac{1}{2}\beta_2\frac{\partial^2 Q}{\partial \tau^2} + \gamma P_0|Q|^2 Q = 0. \tag{8.47}$$

Let us define a nonlinear distance, L_{NL}, as

$$L_{\mathrm{NL}} = \frac{1}{\gamma P_0}. \tag{8.48}$$

L_{NL} is the propagation distance required to produce a nonlinear phase change rotation of one radian at a power P_0. A dispersion distance, L_{D}, can also be defined as

$$L_{\mathrm{D}} = \frac{t_0^2}{|\beta_2|}, \tag{8.49}$$

where t_0 is the input pulse width. These two characteristic distances provide the length scales over which nonlinear or dispersive effects become important for pulse evolution.

Using a distance Z normalized by the dispersion distance L_{D} and a time scale T normalized by t_0, equation (8.47) becomes

$$i\frac{\partial Q}{\partial z} \pm \frac{1}{2}\frac{\partial^2 Q}{\partial \tau^2} + N^2|Q|^2 Q = 0, \tag{8.50}$$

where

$$N^2 = \frac{L_{\mathrm{D}}}{L_{\mathrm{NL}}} = \frac{\gamma P_0 t_0^2}{|\beta_2|}. \tag{8.51}$$

In the second term of equation (8.50) the plus signal corresponds to the case of anomalous GVD (sgn $(\beta_2) = -1$), where the minus signal corresponds to normal GVD (sgn $(\beta_2) = +1$). The parameter N can be eliminated from equation (8.50) by introducing a new normalized amplitude $q = NQ$, with which equation (8.50) takes the standard form of the NLSE:

$$i\frac{\partial q}{\partial z} + \frac{1}{2}\frac{\partial^2 q}{\partial T^2} + |q|^2\, q = 0. \tag{8.52}$$

The case of anomalous GVD ($\beta_2 < 0$) was assumed in writing equation (8.52).

Figure 8.4(a) shows the numerical solution of the NLSE obtained for an initial condition $q(0, T) = \text{sech}(T)$, which corresponds to the fundamental soliton. This figure illustrates the most important property of the fundamental soliton—as long as the fiber loss is negligible, it propagates undistorted without any change in shape for arbitrarily long distances. It is this feature of fundamental solitons that makes them attractive for optical communications systems. Figure 8.4(b) illustrates the evolution of an initially rectangular pulse towards the fundamental soliton, demonstrating the truly stable nature of this solution. In the first stage of propagation some energy of the initial pulse is radiated until the pulses achieve the fundamental soliton profile.

The nonlinear Schrödinger equation given by equation (8.52) is a completely integrable equation that can be solved exactly using the inverse scattering method [110, 111], as shown by Zakharov and Shabat [112]. In the case $N = 1$ we have the fundamental soliton solution, which can be written in the form

$$q(T, Z) = \eta\ \text{sech}\,[\eta(T + \kappa Z - T_0)]\exp\left(-i\kappa T + \frac{i}{2}(\eta^2 - \kappa^2)Z + i\sigma\right). \tag{8.53}$$

This solution is characterized by four parameters: the amplitude η (also the pulse width), the frequency κ (also the pulse speed), the time position T_0, and the phase σ.

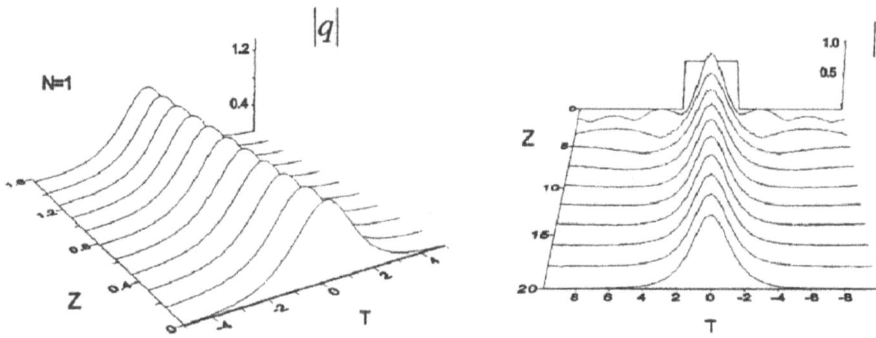

(a) (b)

Figure 8.4. (a) Numerical solution of the NLSE for an initial condition $q(0, t) = \text{sech}(T)$. (b) Numerical solution of the NLSE for an initial rectangular pulse, of height 0.7626 and width 4.

The N-soliton solutions are asymptotically given in the form of N separated solitons:

$$q(Z, T) = \sum_{j=1}^{N} \eta_j \operatorname{sech}\left[\eta_j\left(T + \kappa_j Z - T_{0j}\right)\right] \exp\left[-i\kappa_j T + i\frac{1}{2}\left(\eta_j^2 - \kappa_j^2\right)Z - i\sigma_j\right]. \quad (8.54)$$

If the input pulse shape is symmetrical, all the output solitons propagate at exactly the same speed. This type of solution is called the *bound-soliton solution* and its shape evolves periodically during the propagation due to the phase interference among the constituting solitons. For example, considering an input pulse shape given by

$$q(0, T) = N \operatorname{sech}(T) \quad (8.55)$$

the period of oscillation of the higher-order solitons is

$$Z_0 = \frac{\pi}{2}. \quad (8.56)$$

According to equation (8.51) the peak power of a soliton in an optical fiber is

$$P_0 = \frac{N^2 |\beta_2|}{\gamma t_0^2}. \quad (8.57)$$

Considering a temporal width $t_0 = 6$ ps and using typical parameter values $\beta_2 = -1\ \text{ps}^2\ \text{km}^{-1}$ and $\gamma = 3$ W–1 km^{-1} for dispersion-shifted fibers we obtain $P_0 \sim 10$ mW for fundamental solitons ($N = 1$) in the C-band ($\lambda \approx 1550$ nm). The required power to launch the Nth-order soliton is N^2 times that for the fundamental one. Moreover, higher-order solitons compress periodically, resulting in soliton chirping and spectral broadening. In contrast, fundamental solitons preserve their shape during propagation. This fact, together with the lower power required for their generation, makes fundamental solitons the preferred option in signal processing and soliton communication systems.

In the case of ultrashort solitons ($t_0 < 1$ ps), their spectral width is large enough that it becomes necessary to take into account the Raman effect and the frequency dependence of the nonlinear parameter γ. Moreover, it is also useful to include the higher-order terms in the expansion of the propagation constant, given by equation (8.2). This gives rise to three main higher-order effects which affect the propagation of ultrashort solitons, namely the third-order dispersion (TOD), self-steepening (SST), and intrapulse Raman scattering (IRS) effects.

In the presence of the above higher-order effects, the propagation of ultrashort solitons can be described by the following generalized NLSE [3]:

$$i\frac{\partial q}{\partial Z} + \frac{1}{2}\frac{\partial^2 q}{\partial T^2} + |q|^2 q = i\delta_3\frac{\partial^3 q}{\partial T^3} - is\frac{\partial}{\partial T}\left(|q|^2 q\right) + \tau_R q\frac{\partial |q|^2}{T}, \quad (8.58)$$

where the parameters δ_3, s, and τ_R, govern, respectively, the effects of third-order dispersion, self-steepening, and intrapulse Raman scattering, The three parameters are inversely proportional to the initial pulse width t_0. Assuming a value $t_0 = 30$ fs for pulses propagating at 1550 nm in a standard silica fiber, we have $\delta_3 \approx 0.03$, $s \approx 0.03$, and $\tau_R \approx 0.1$. In fact, the IRS effect is often the most relevant among the higher-order effects.

Considering only the IRS effect, it is found that the soliton amplitude remains constant, whereas its frequency varies with distance as [3]

$$\kappa(Z) = -\frac{8}{15}\tau_R \eta^4 Z. \tag{8.59}$$

The IRS effect leads to a continuous downshift of the carrier frequency, an effect known as the *soliton self-frequency shift* (SSFS) [113]. This effect was observed for the first time by Mitschke and Mollenauer in 1986 using 0.5 ps pulses obtained from a passively mode-locked color-center laser [114]. The SSFS effect occurs as a consequence of the Raman effect affecting the different spectral components of ultrashort solitons. In fact, the high-frequency components of the spectrum lose energy to the low-frequency components, which determines a red shift of the soliton spectrum during propagation. Since the amplitude and width are inversely related, the frequency shift scales as t_0^4, indicating that it can become quite large for ultrashort short pulses.

8.7 Highly nonlinear fibers

Using conventional optical fibers for nonlinear optical signal processing, a length of several kilometers is usually required due to its relatively small nonlinear parameter. Such long fibers pose some practical limitations, concerned namely with the size and stability of the system. In order to minimize these problems, several highly nonlinear silica fibers have been developed during the last few years.

8.7.1 Highly nonlinear silica fibers

The nonlinear-index coefficient of fused silica is relatively small. A value of $n_2 = 2.35 \times 10^{-20}$ m^2 W^{-1} has been measured at 1550 nm [4]. Despite this small value, the effects of the nonlinear refractive index become significant due to the very long interaction lengths provided by the low-loss optical fibers.

The fiber nonlinearity is commonly characterized by the nonlinear parameter γ, given by equation (8.14). Such a parameter can be increased, reducing the effective mode area A_{eff} or increasing the nonlinear refractive index n_2. In fact, it has been shown that n_2 increases by doping the bulk glass with GeO_2[114]. On the other hand, n_2 decreases in the case of a F-doped bulk glass.

In order to enhance the nonlinear parameter γ, the best option will be provided by a fiber heavily doped with GeO_2 in the core and having a large refractive index difference between the core and the cladding, because such fibers exhibit a large n_2 as well as a small effective area A_{eff} [116]. Optical fibers designed in this way are known as highly nonlinear fibers (HNLFs). In typical HNLFs, the relative refractive index difference of the core to the outer cladding is around 3%, while the core diameter is around 4 μm. The values of the nonlinear parameter γ for most of these fibers are in the range 10–20 W^{-1} km^{-1} [117]. Increasing γ much above these values is not possible in these types of fibers, since the optical mode confinement is lost when the core diameter is further reduced.

A main parameter characterizing an optical fiber is the normalized frequency V, given by

$$V = k_0 a n_1 \sqrt{2\Delta}, \tag{8.60}$$

where k_0 is the vacuum wavenumber, a is the core radius, n_1 is the core refractive index, and

$$\Delta = (n_1 - n_c)/n_1 \tag{8.61}$$

is the relative index difference between the core and the cladding of the fiber, n_c being the cladding refractive index. A single-mode fiber is achieved when $V < 2.405$. If Δ is kept constant and the core radius a is reduced, the normalized frequency V decreases and the mode confinement is lost. Equation (8.60) shows that, in order to maintain a given value of V when decreasing a, Δ must increase such that $a^2\Delta$ remains constant.

The maximum value of Δ is achieved when the cladding material is replaced by air ($n_c \approx 1$), in which case the fiber becomes a so-called jacketed air-suspended rod. These kinds of fibers have been produced by tapering standard optical fibers, with an original cladding diameter of 125 μm [118–120]. Tapering a fiber involves heating and stretching it to form a narrow waist connected to an untapered fiber by taper transitions. If the transitions are gradual, light propagating along the fiber suffers very little loss.

In a tapered fiber the mode remains confined to the core even if the core diameter is close to 1 μm. However, such confinement is reduced for smaller values of the core diameter. Figure 8.5 shows the fundamental mode of a silica fiber surrounded by air for two different fiber diameters: (a) $d = 2$ μm and (b) $d = 0.5$ μm. In the case of figure 8.5(a), the field is completely confined by the fiber. However, in the case of figure 8.5(b) the fiber cannot completely confine the propagating wave, leading to a relatively large evanescent field.

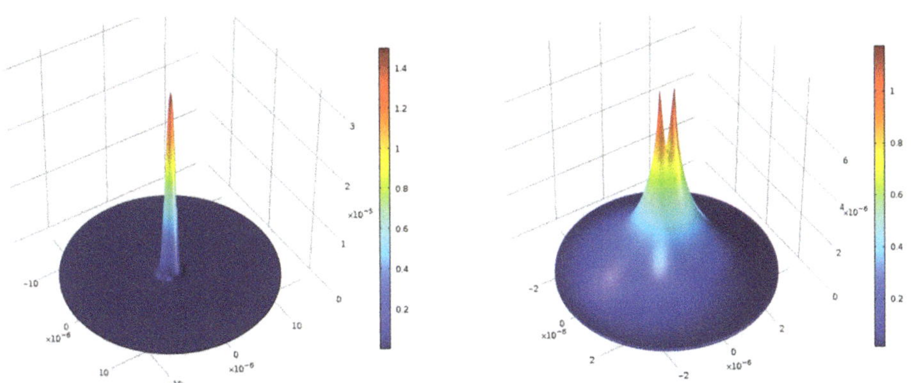

Figure 8.5. The fundamental mode of a tapered silica fiber surrounded by air with a diameter: (a) $d = 2$ μm and (b) $d = 0.5$ μm.

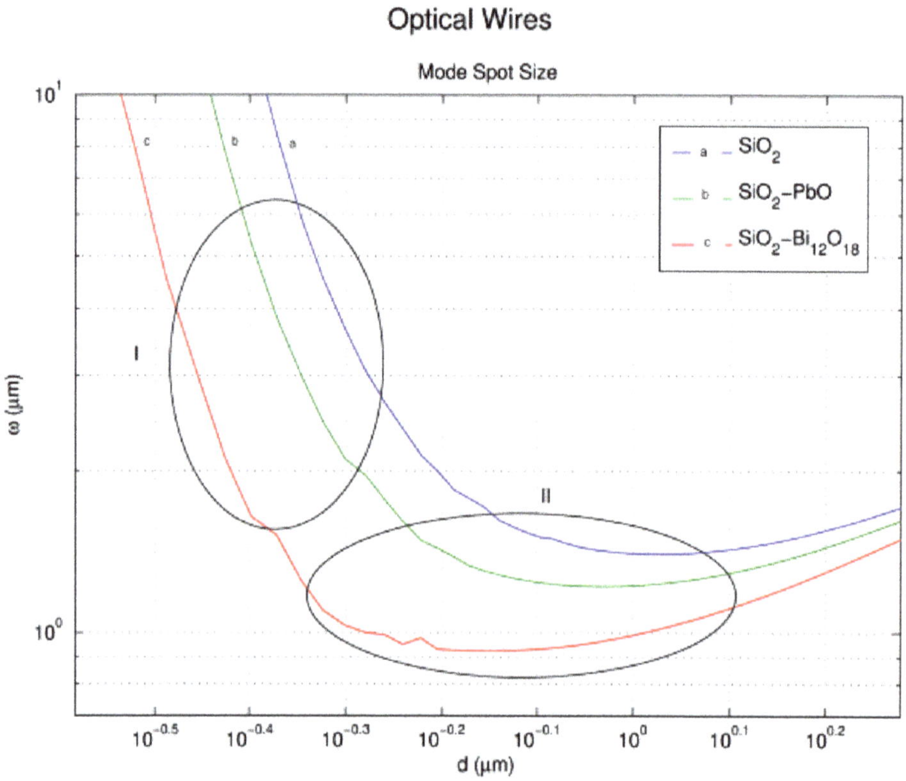

Figure 8.6. Variation of the spot size of a tapered fiber with the fiber diameter.

Figure 8.6 shows the mode spot size as a function of the fiber diameter, considering different fiber materials: silica (SiO_2), a lead-silicate glass (SiO_2-PbO), and a bismuth silicate glass ($SiO_2-Bi_{12}O_{18}$). We observe the existence of a minimum beam waist for a given value of the fiber diameter, which depends on the fiber material. Two regions are identified in figure 8.6. Region I corresponds to the large evanescent-field region, which may be useful for sensing applications. Region II corresponds to the high-confinement region, which is more appropriate for non-linear applications.

The dispersive characteristics of a tapered fiber are very sensitive and can be adjusted by changing the fiber diameter [121, 122]. This is illustrated in figure 8.7, which shows the calculated dispersion spectra for several values of the taper diameter. We observe that, by reducing fiber diameter, the zero-dispersion wavelength (ZDW) can be shifted to the visible range and, in some cases, a second ZDW appears at longer wavelengths, defining a wavelength window between the two ZDWs, which shows anomalous GVD.

For efficient supercontinuum generation in an optical fiber it is generally desirable that the dispersion of the fiber guided mode be zero or anomalous at the pump wavelength. Er^{3+}-based sources at 1550 nm are suitable for conventional step-index silica fibers. However, pump sources based around 1060 and 800 nm can also be

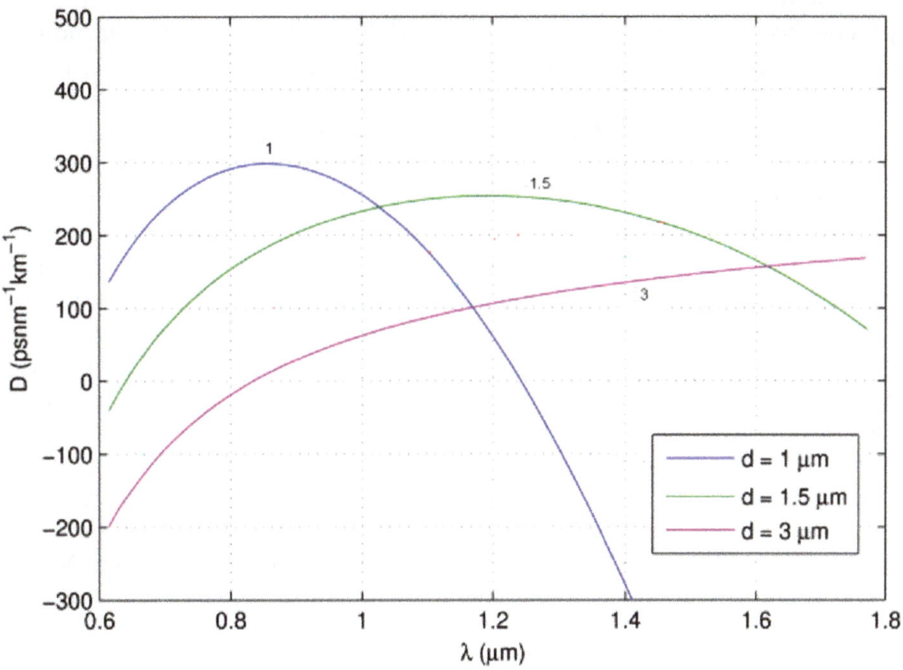

Figure 8.7. Dispersion properties of a tapered fiber for several values of the diameter.

used in the case of a fiber with a high index difference between the core and cladding, by adjusting their core diameter appropriately.

8.7.2 Microstructured optical fibers

A new class of optical fibers is represented by the so-called photonic crystal fibers (PCFs), or microstructured optical fibers (MOFs). These fibers have a microstructured geometry in the cross-section that is repeated over the entire fiber's length. PCFs were first realized in 1996 [123] and their designation is due to the photonic crystals, realized one decade before, which consist of a periodically structured transparent material. This structure can exhibit photonic band gaps, such that light with a wavelength within those bands cannot propagate in the crystal. In fact, there are many kinds of photonic crystal fibers, even if some of them do not use the photonic bandgap effect. Therefore, the designation of microstructured optical fibers is more generally applicable.

There are two categories of microstructured optical fibers: the solid-core MOFs (SC-MOFs) and the hollow-core MOFs (HC-MOFs) [3]. Hollow-core MOFs were realized for the first time in 1999 [124] and present a central region containing air. The photonic crystal that constitutes the cladding only allows propagation of light in certain frequency bands. Light guidance in the fiber core occurs at frequencies within the photonic bandgap of the cladding periodic structure. Extremely low losses are expected from these types of fibers, since the light travels predominantly in empty space. On the other hand, the nonlinear effects are strongly reduced and the dispersion becomes negligible in air-core MOFs [125]. However, the nonlinear effects in hollow-core MOFs can be greatly enhanced if air is replaced with a suitable gas or liquid [126].

The second category of MOFs has a solid core, in which the guidance of the light is generally due to the modified total internal reflection effect, similar to the guidance mechanism of conventional optical fibers. In fact, the periodic nature of air holes in the cladding is not crucial in SC-MOFs, as long as they provide an effective reduction of its refractive index below that of the silica core [127]. These types of MOFs can exhibit a very high nonlinear parameter, particularly if the core size is adequately reduced and the mode confinement is increased using a large air-filling fraction in the cladding. Moreover, when an asymmetric structure is used, such a fiber also exhibits a high birefringence, which provides for the polarization maintenance of the propagated field. Figure 8.8 illustrates the cross-section of a solid-core MOF with a hexagonal pattern of holes. There are two main structural parameters in an MOF: the hair-hole diameter d and the hole-to-hole spacing Λ.

The dispersion characteristics of a microstructured optical fiber can be adjusted by controlling its structural parameters appropriately [128]. In a solid-core MOF, as the holes become larger the core becomes more and more isolated until it resembles an isolated strand of silica glass. A semi-infinite anomalous dispersion region above the zero-dispersion wavelength (ZDW) is observed in MOFs with larger cores. The ZDW moves to shorter wavelengths and can reach the visible region if the core size is reduced. In such case, a second ZDW can appear in the longer wavelength side, such that the GVD is anomalous in the spectral window between the two ZDWs and normal outside it. A conventional tapering process can be used to fabricate MOFs with core diameters below one micron [129].

Figure 8.9 shows the dispersion characteristics of an MOF with a hexagonal pattern of holes spaced by (i) $\Lambda = 1$ μm and (ii) $\Lambda = 2.5$ μm, considering a

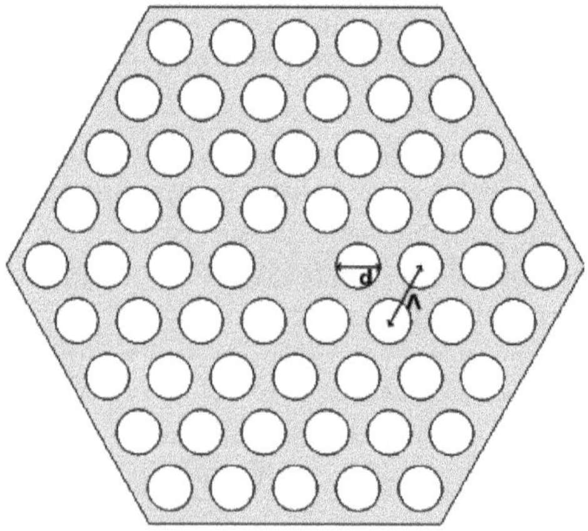

Figure 8.8. Schematic of a solid-core microstructured fiber with a hexagonal geometry. The two main parameters are the hair-hole diameter d and the hole-to-hole spacing Λ.

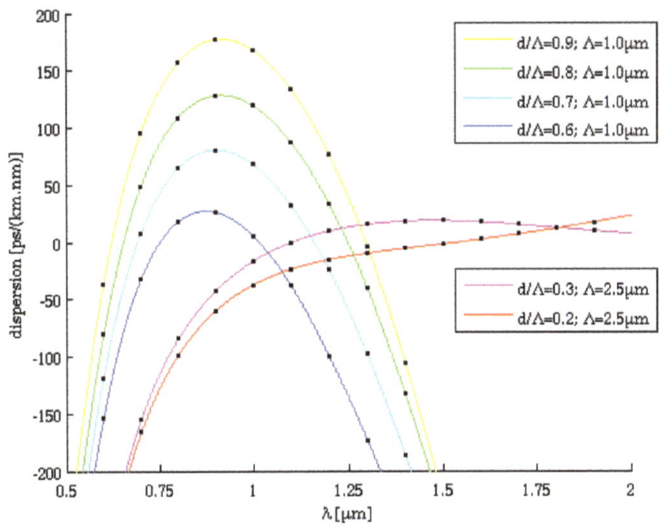

Figure 8.9. Dispersion properties of a microstructured fiber with a hexagonal pattern of holes spaced by (i) $\Lambda = 1\ \mu$m and (ii) $\Lambda = 2.5\ \mu$m, considering a wavelength $\lambda = 1.55\ \mu$m, for different hole–pitch ratios.

wavelength $\lambda = 1.55\ \mu$m, for different hole–pitch ratios. In the case (i), two ZDWs are observed, defining a wavelength window with anomalous GVD. The separation between the two ZDWs increases and the anomalous GVD becomes larger when the hole–pitch ratio increases. For $d/\Lambda = 0.9$ this is essentially the same as a tapered fiber with a diameter $\sim 2\Lambda - d$. On the other hand, for $\Lambda = 2.5\ \mu$m and $d/\Lambda = 0.2$ the waveguide dispersion is relatively weak and the dispersion approaches that of the bulk silica. In the case $d/\Lambda = 0.3$ a nearly zero-dispersion and flattened behavior is achieved. This is particularly important for supercontinuum generation and other applications making use of the four-wave mixing effect.

Figure 8.10 shows the effective mode area for the same cases as figure 8.9. We observe that the effective area is significantly smaller in the case $\Lambda = 1\ \mu$m, compared with the case $\Lambda = 2.5\ \mu$m. In both cases, increasing the hole–pitch ratio provides a reduced effective area. Consequently, the nonlinear parameter is increased, according to equation (8.14).

8.7.3 Non-silica fibers

The value of the nonlinear parameter γ can be enhanced if the optical fiber is made using glasses with higher nonlinearities than silica, such as lead-silicate, tellurite, bismuth oxide, and chalcogenide glasses. Considering several optical glasses, it has been observed that the nonlinear-index coefficient n_2 increases with the linear refractive index, n [130, 131]. Introducing heavy atoms or ions with a large ionic radius (i.e. using chalcogen elements S, Se, or Te to replace oxygen) acts to increase the polarizability of the components in the glass matrix and also increase the nonlinear index n_2. In fact, the nonlinear index can be increased by a factor of almost 1000 compared to its value for silica fibers using some types of highly nonlinear

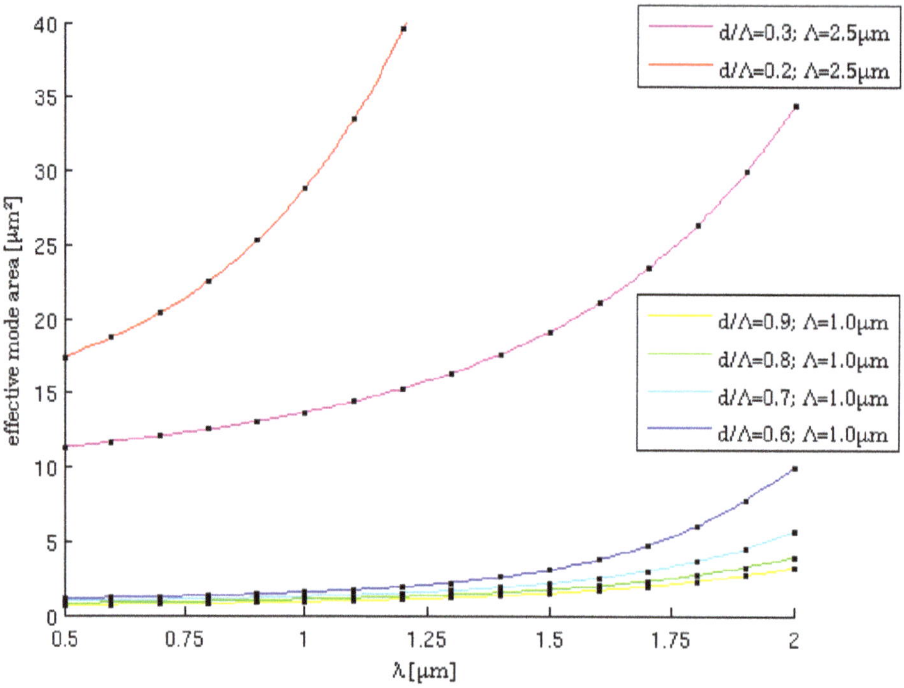

Figure 8.10. Effective mode area for the same cases as in figure 8.9.

glasses. The ZDW of a glass also shifts to longer wavelengths with increasing linear refractive index. Heavy metal oxide glasses (lead-silicate, bismuth oxide, tellurite) have linear indices in the range 1.8–2.0, nonlinear indices ~10 times higher than silica, and material ZDWs of 2–3 μm. Chalcogenide glasses (GLS, As_2S_3) have linear indices of 2.2–2.4, nonlinear indices significantly higher than those of the oxide glasses, and material ZDWs larger than 4 μm.

Lead-silicate and tellurite glasses can offer values of n_2 which are about 10–20 times larger than that of silica. For example, a value $n_2 \approx 2.5 \times 10^{-19}\ \mathrm{m^2\ W^{-1}}$ has been measured for tellurite glasses [132], while a value $n_2 = 4.1 \times 10^{-19}\ \mathrm{m^2\ W^{-1}}$ at 1.06 μm was measured for Schott SF57 glass [133]. Both types of materials have been used for making microstructured fibers. The extrusion technique for fiber preform manufacture is particularly suitable when using compound glasses, since they exhibit low soft temperatures of ~500 °C, as opposed to ~200 °C for silica. In addition to being simpler than the more common stacking technique developed for silica MOFs, extrusion allows the realization of cladding structures consisting of mostly air, as required to increase the fiber nonlinearity.

A nonlinear parameter $\gamma = 1860\ \mathrm{W^{-1}\,km^{-1}}$ at 1.55 μm was measured in a 144 cm long MOF made of lead-silicate glass (SF57) and presenting an effective mode area $A_{\mathrm{eff}} = 1.1\ \mu\mathrm{m}^2$[133]. Higher values of the nonlinear parameter are expected at shorter wavelengths due to a combination of the $1/\lambda$ dependence of γ and the fact that shorter wavelengths can be confined to the core down to smaller core dimensions.

Among the different types of nonlinear fibers, those based on bismuth oxide (Bi_2O_3) have a relatively mature technology and the fibers can be spliced to standard single-mode fibers. They also possess good chemical, mechanical, and thermal stability and have a low photosensitivity compared to chalcogenide fibers. Depending on the concentration of bismuth in the fiber, the nonlinear refractive index n_2 can be varied from 30×10^{-20} m² W^{-1} to 1.1×10^{-18} m² W^{-1}, representing a 10–40 enhancement compared to that of silica [134–137]. By 2008 a bismuth oxide fiber with a mode field diameter of 1.97 μm exhibited a nonlinear parameter $\gamma \approx 1100$ W^{-1} km^{-1} at 1550 nm. Only a 35 cm length of such a highly nonlinear fiber was needed to demonstrate several applications of nonlinear processing of optical signals [17].

A nonlinear parameter $\gamma = 2450$W^{-1} km^{-1} was measured in a 2004 experiment for a 85 cm long chalcogenide fiber with a core diameter of 7 μm [138]. This corresponds to a nonlinear refractive index $n_2 = 2.4 \times 10^{-17}$ m^2 W^{-1}, which is larger by a factor of about 1000 relative to the silica case.

In recent years MOFs have been fabricated with various types of chalcogenide glasses for several applications. Employment of photonic crystal geometry in the cladding of As_2Se_3 fibers can give rise to a strong overlap between the pump and signal, providing higher Raman gain efficiency than conventional fibers, as demonstrated for silica-based MOF structures [139, 140]. It was shown that the Raman gain efficiency in As_2Se_3 MOFs can be improved by a factor of more than four compared to conventional As_2Se_3 fibers [141].

8.8 Supercontinuum generation

The supercontinuum generation (SCG) corresponds to an extremely wide spectrum achieved by an optical pulse while propagating in a nonlinear medium and results generally from the synergy between several fundamental nonlinear processes, such as self-phase modulation, cross-phase modulation, stimulated Raman scattering, and four-wave mixing [3, 18]. The spectral location and power of the pump, as well as the nonlinear and dispersive characteristics of the medium, determine the relative importance and the interaction between these nonlinear processes.

In conventional dispersion-shifted fibers the pulses are usually launched in the 1550 nm window and the resulting supercontinuum spectrum extends at most from 1300 to 1700 nm. However, microstructures optical fibers (MOFs) can be designed with a ZDW of anywhere from 1550 down to 565 nm [142]. Using such fibers it becomes possible to generate a supercontinuum extending from the visible to the near infrared region.

A supercontinuum source can find applications in the area of bio-medical optics, where it allows the improvement of longitudinal resolution in optical coherence tomography by more than an order of magnitude [143, 144], in optical frequency metrology [145, 146], in all kinds of spectroscopy, and as a multiwavelength source in the telecommunications area [147].

Supercontinuum generation can be broadly divided in two categories, depending on the duration of the pump pulses. The first category corresponds to long (picosecond, nanosecond, and continuous wave) pump pulses, while the second category is obtained with short (femtosecond) pump pulses. The dispersion regime (normal or anomalous) at the pump wavelength is also an important factor, at least in the initial propagation phase. Soliton related effects play a significant role whenever light with sufficient power propagates in the anomalous dispersion regime [3, 18].

Pumping with long duration and high peak power pulses in the anomalous dispersion regime near the ZDW, the spontaneous modulation instability (MI) becomes the dominant process in the initial propagation phase. In the frequency domain the modulation instability causes the spread of energy from the central wave frequency to sidebands with a frequency separation given by [3]

$$\Omega_S = \sqrt{\frac{2\gamma P_p}{|\beta_2|}}, \tag{8.62}$$

where β_2 is the group-velocity dispersion, γ is the nonlinear parameter, and P_p is the pump peak power. The MI period can be obtained from the above frequency separation as

$$T_S = \frac{2\pi}{\Omega_S} = \sqrt{\frac{2\pi^2 |\beta_2|}{\gamma P_p}}. \tag{8.63}$$

The modulated picosecond pulse can then break up into a train of solitons with durations given approximately by T_S. The subsequent evolution of these solitons then leads to additional spectral broadening and supercontinuum formation through a variety of mechanisms. If such solitons are sufficiently short, they are red-shifted due to intrapulse Raman scattering effect [113, 114]. As seen in section 8.6, the rate of frequency shift is inversely proportional to the fourth power of the soliton width. If the solitons resulting from the initial pulse break up are not sufficiently short, they can undergo soliton collisions, which determine the transfer of energy among them due to Raman scattering. In this process some solitons achieve a higher peak power and then undergo a larger red shift.

The solitonic effects are absent, at least in the initial propagation phase, if pumping is realized far into the normal dispersion regime. In this case the spectral broadening occurs initially due to stimulated Raman scattering. The contribution of the SRS process for the supercontinuum generation is particularly effective on the long-wavelength side of the spectrum, which tends to make the overall spectrum asymmetric.

The four-wave mixing becomes progressively more important if the pumping is closer to the ZDW, since the parametric gain is higher than the Raman gain [148]. Using pump pulses of subkilowatt peak power and a duration of some tens of picoseconds, a spatially single-mode SC more than 600 nm wide can be generated by the interplay of SRS and FWM [149, 150].

The dispersive properties of the fiber become especially important in the case of the FWM process. If the required phase-matching conditions are satisfied, FWM generates sidebands on the short- and long-wavelength sides of the pulse spectrum. This process can produce a wideband supercontinuum even in the normal-GVD region of an optical fiber. In fact, the phase-matching condition for the FWM process requires that dispersive effects compensate nonlinear effects through equation (8.36). This phase-matching condition can be satisfied even for a pump wavelength in the normal-GVD regime ($\beta_2 > 0$) if the fourth-order dispersion coefficient (β_4) is negative. This can be easily realized in tapered and microstructured fibers [151, 152].

The general features observed in the case of long duration pulses can also be extended to the continuous wave (CW) pumping in the anomalous dispersion regime. In this case, the supercontinuum generation also starts with the modulation instability of the continuous pump wave in the anomalous dispersion region [153–156]. This instability evolves from background noise and is often stimulated by the intensity fluctuations which are typical of a continuous wave pump. The modulation instability leads to the splitting of the continuous wave into fundamental solitons.

Pumping in the normal dispersion regime using femtosecond pulses, where soliton formation is not allowed, leads only to spectral broadening through SPM [157, 158]. Soliton dynamical effects play a prominent role when pumping in the anomalous dispersion regime. If the power of the pump pulses is high enough they can evolve as higher-order solitons. These pulses begin experiencing a spectral broadening and temporal compression which are typical of higher-order solitons [3]. However, because of perturbations such as higher-order dispersion or intrapulse Raman scattering, the dynamics of such pulses departs from the behavior expected of ideal high-order solitons and the pulses break up. This process is known as soliton fission and determines the separation of each higher-order soliton pulse into N fundamental solitons, whose widths and peak powers are given by [159]

$$t_k = \frac{t_0}{2N + 1 - 2k} \tag{8.64}$$

$$P_k = \frac{(2N + 1 - 2k)^2}{N^2} P_0, \tag{8.65}$$

where $k = 1$ to $\tilde{N}$, where $\tilde{N}$ is the integer closest to N when N is not an integer. Soliton fission occurs generally after a propagation distance $L_{\text{fiss}} \sim L_D/N$, where $L_D = t_0^2/|\beta_2|$ is the dispersion distance at which the injected higher-order soliton attains its maximum compression and larger bandwidth.

In addition to higher-order dispersion, another main effect affecting the dynamics of a higher-order soliton fission is the intrapulse Raman scattering (IRS). As seen in section 8.6, this phenomenon leads to a continuous downshift of the soliton carrier frequency, an effect known as the soliton self-frequency shift (SSFS) [113]. The rate of frequency shift per propagation length is given by [3]

$$\frac{df}{dz} = -\frac{4t_R|\beta_2|}{15\pi t_0^4} = -\frac{4t_R(\gamma P_0)^2}{15\pi|\beta_2|}, \tag{8.66}$$

where $t_R \approx 5fs$ is the Raman parameter and P_0 is the soliton peak power. Since the SSFS effect is proportional to $(\gamma P_0)^2$, it will be enhanced if short pulses with high peak power are propagated in highly nonlinear fibers.

As a consequence of the SSFS induced by intrapulse Raman scattering, the solitons arising from the fission process separate from each other. Since the SSFS is the largest for the shortest soliton, its spectrum shifts the most toward the red side. In this process new spectral peaks appear at $z = L_{\text{fiss}}$, corresponding to the so-called nonsoliton radiation (NSR) [160] or Cherenkov radiation [161] which is emitted by the solitons resulting from the fission process in the presence of higher-order dispersion. Such radiation was observed for the first time in a 2001 experiment using 110 fs pulses from a laser operating at 1556 nm, which were launched into a polarization-maintaining dispersion-shifted fiber with a GVD $\beta_2 = -0.1\ \text{ps}^2\ \text{km}^{-1}$ at the laser wavelength [162].

The NSR is emitted at a frequency such that its phase velocity matches that of the soliton. The two phases are equal when the following phase-matching condition is satisfied:

$$\beta(\omega) = \beta(\omega_S) + \beta_1(\omega - \omega_S) + \frac{1}{2}\gamma P_S. \tag{8.67}$$

In equation (8.67) P_S and ω_S are the peak power and the frequency, respectively, of the fundamental soliton formed after the fission process. Expanding $\beta(\omega)$ around ω_s this phase-matching condition reduces to

$$\sum_{n\geqslant 2}\frac{\Omega^n}{n!}\beta_n(\omega_S) = \frac{\gamma P_S}{2}, \tag{8.68}$$

where $\Omega = \omega - \omega_S$ is the frequency shift between the dispersive wave and the soliton and β_n $(n = 2, 3)$ are the dispersion parameters, calculated at the soliton central frequency ω_S.

Neglecting the fourth- and higher-order dispersion terms, the frequency shift between the frequency of dispersive waves and that of the soliton is given approximately by [161]

$$\Omega_d \approx -\frac{3\beta_2}{\beta_3} + \frac{\gamma P_0 \beta_3}{3\beta_2^2}. \tag{8.69}$$

For solitons propagating in the anomalous-GVD regime we have $\beta_2 < 0$. In these circumstances equation (4.13) shows that $\Omega > 0$ when $\beta_3 > 0$, in which case the NSR is emitted at wavelengths shorter than that of the soliton. In contrast, if $\beta_3 < 0$ the NSR is emitted at a longer wavelength than that of the soliton and it can result in the suppression of the SSFS effect [163, 164].

Modeling the ultrashort pulse propagation and supercontinuum generation in optical fibers can be realized considering a generalized nonlinear Schrödinger equation (NLSE) that includes higher-order dispersive and nonlinear effects [165–169]. Such an equation can be written as

$$\frac{\partial U}{\partial z} - i\sum_{k\geqslant 2}\frac{i^k\beta_k}{k!}\frac{\partial^k U}{\partial \tau^k} + \frac{\alpha(\omega)}{2}U = i\gamma\left(1 + \frac{i}{\omega_0}\frac{\partial}{\partial \tau}\right) \left(U(z, \tau)\int_{-\infty}^{+\tau} R(t')|\, U(z, \tau - t')|^2\, dt'\right), \tag{8.70}$$

where $U(z, t)$ is the electric field envelope, ω_0 is the center frequency, β_k are the dispersion coefficients at the center frequency, $\alpha(\omega)$ is the frequency-dependent fiber loss, and γ is the nonlinear parameter of the fiber.

The nonlinear response function $R(t)$ in equation (8.70) can be written as

$$R(t) = \left(1 - f_R\right)\delta(t) + f_R h(t), \tag{8.71}$$

where the δ-function represents the instantaneous electron response (responsible for the Kerr effect), $h(t)$ represents the delayed ionic response (responsible for the Raman scattering), and f_R is the fractional contribution of the delayed Raman response to the nonlinear polarization, in which a value $f_R = 0.18$ is often assumed [170]. It is common to approximate $h(t)$ in the form [148, 171]

$$h(t) = \frac{\tau_1^2 + \tau_2^2}{\tau_1\tau_2^2}\exp(-t/\tau_2)\sin\left(t/\tau_1\right), \tag{8.72}$$

where $\tau_1 = 12.2$ fs and $\tau_2 = 32$ fs. More accurate forms of the response function $h(t)$ have also been investigated recently [172]. The Raman parameter, t_R, which characterizes the intrapulse Raman scattering effect is related to $h(t)$ in the form $t_R \equiv f_R \int_{-\infty}^{+\infty} th(t)dt$. Typically, it has a value $t_R \approx 5fs$. Equation (8.70) can be used to describe the propagation of femtosecond pulses in optical fibers, in both the normal and anomalous dispersion regimes. When such pulses have enough power their spectra undergo extreme broadening. In the anomalous dispersion regime this process is mainly influenced by the phenomenon of soliton fission, which occurs whenever a higher-order soliton is affected by third- or higher-order dispersion. The soliton order is given by $N = \sqrt{L_D/L_{NL}}$, where $L_D = t_0^2/|\beta_2|$ is the dispersion distance and $L_{NL} = 1/\gamma P_0$ is the nonlinear length.

Figure 8.11 shows the temporal and spectral evolution of an optical pulse propagating along a silica microstructured optical fiber with a hexagonal pattern of holes of diameter $d = 1.4$ μm spaced by $\Lambda = 1.6$ μm. Figure 8.12 shows the initial and final pulse profiles, both in the frequency and time domains. These results were obtained taking into account the full dispersion curve of such a fiber, which presents a single ZDW at 735 nm. The pump is in the anomalous dispersion regime at 790 nm, where the nonlinear parameter is $\gamma = 117$ W^{-1} km^{-1}. An input pulse $U(0, \tau) = \sqrt{P_0}\,\mathrm{sech}(\tau/t_0)$ is assumed, where $P_0 = 5$ kW, and an intensity full width at half maximum (FWHM) of 50 fs. For the assumed pulse and fiber parameters, the soliton order is $N = 6.02$.

For the pulse and fiber parameters assumed, we have a fission distance $L_{\rm fiss} \sim 1$ cm, which can be confirmed in figure 8.11. A clear signature of soliton fission is the

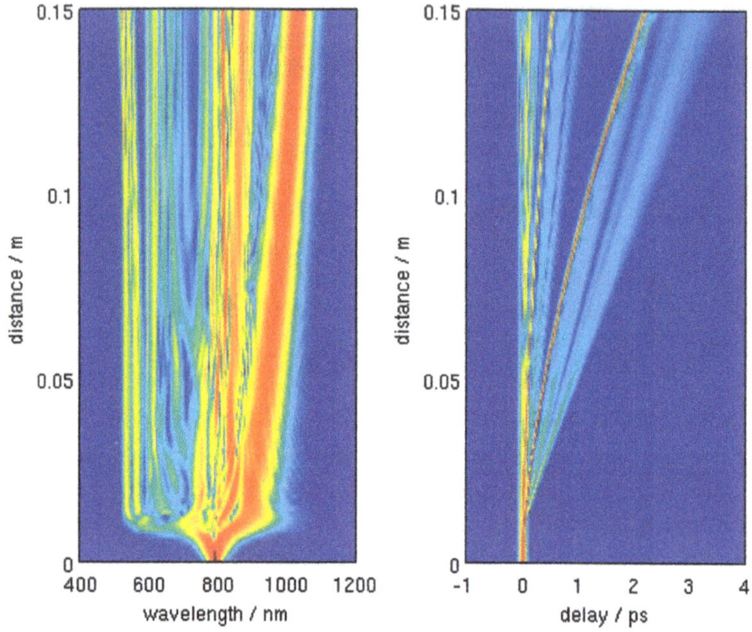

Figure 8.11. (a) Spectral and (b) temporal evolution of an optical pulse along an MOF with $d = 1.4$ μm and $\Lambda = 1.6$ μm. The input pulse has a peak power of 5 kW and an FWHM of 50 fs.

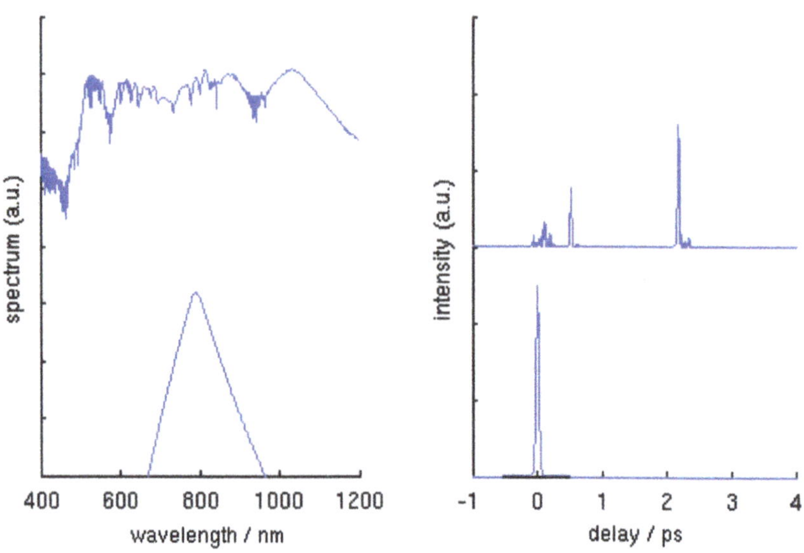

Figure 8.12. Initial and final pulse profiles in the (a) frequency and (b) time domains for the case of figure 8.11.

appearance of a new spectral peak in figure 8.11(a) at $z = L_{fiss}$, which corresponds to the nonsoliton radiation.

As a consequence of the SSFS induced by intrapulse Raman scattering, the solitons arising from the fission process separate from each other. Since the SSFS is the largest for the shortest soliton, its spectrum shifts the most toward the red side in figure 8.11(a). The change of the soliton's frequency determines a reduction in the soliton's speed because of dispersion. This deceleration appears as a bending of the soliton trajectory in the time domain, as observed in figure 8.11(b).

In most SCG studies silica has been used as the MOFs' background material. However, the material loss in silica increases drastically at wavelengths beyond ~2 μm, effectively preventing the spectral evolution of SCG into the mid-infrared (mid-IR) region [173].

Research into other highly nonlinear materials that can expose the importance of the mid-IR region has been increased in recent years. Price *et al* [165] have generated theoretically a mid-IR supercontinuum from 2 to 5 μm using a bismuth-glass MOF. Domachuk *et al* [174] have generated experimentally a mid-IR supercontinuum with a spectral range of 0.8 to 4.9 μm using a tellurite MOF with the same structure. Fluoride, sapphire, As_2Se_3, and $As_{40}Se_{60}$ chalcogenide MOFs have also been used to achieve mid-infrared SCG [175, 176]. A simulated supercontinuum extending from 500 to 3900 nm, achieved after a propagation distance of only 0.3 mm, has been reported recently [177].

Acknowledgements

The author is grateful for funding from the Fundação para a Ciência e a Tecnologia (FCT) (Project UID/CTM/50025/2020).

References

[1] Chigusa Y, Yamamoto Y, Yokokawa T, Sasaki T, Taru T, Hirano M, Kakui M, Onishi M and Sasaoka E 2005 *J. Lightw. Technol.* **23** 3541
[2] ITU-T 2006 *Optical System Design and Engineering Consideration* Recommendation series G, supplement 39
[3] Ferreira M F 2011 *Nonlinear Effects in Optical Fibers* (Hoboken, NJ: Wiley)
[4] Broderick N G, Monroe T M, Bennett P J and Richardson D J 1999 *Opt. Lett.* **24** 1395
[5] Stolen R H and Lin C 1978 *Phys. Rev.* A **17** 1448
[6] Chraplyvy A R and Stone J 1984 *Electron. Lett.* **20** 996
[7] Stolen R H, Bjorkhholm J E and Ashkin A 1974 *Appl. Phys. Lett.* **24** 308
[8] Ippen E P and Stolen R H 1972 *Appl. Phys. Lett.* **21** 539
[9] Stolen R H, Ippen E P and Tynes A R 1972 *Appl. Phys. Lett.* **20** 62
[10] Chraplyvy A 1990 *J. Lightw. Technol.* **8** 1548
[11] Mitra P and Stark J 2001 *Nature* **411** 1027
[12] Wu M and Way W 2004 *J. Lightw. Technol.* **22** 1483
[13] Dianov E M and Prokhorov A M 2000 *IEEE J. Sel. Topics Quantum Electron* **6** 1022
[14] Smith S P, Zarinetchi F and Ezekiel S 1991 *Opt. Lett.* **16** 393
[15] Marhic M E, Wong K K, Kazovsky L G and Tsai T E 2002 *Opt. Lett.* **27** 1439

[16] Nishizawa N and Goto T 2001 *J. Sel. Topics Quantum Electron* **7** 518
[17] Vogel E M, Weber M J and Krol D M 1991 *Phys. Chem. Glasses* **32** 231
[18] Dudley J and Taylor J (ed) 2010 *Supercontinuum Generation in Optical Fibers* (Cambridge: Cambridge University Press)
[19] Humbert G *et al* 2006 *Opt. Express* **14** 1596
[20] Bellini M and Hänsch T 2000 *Opt. Lett.* **25** 1049
[21] Takara H, Ohara T, Yamamoto T, Masuda H, Abe M, Takahashi H and Morioka T 2005 *Electron. Lett.* **41** 270
[22] Kato T, Suetsugu Y, Takagi M, Sasaoka E and Nishimura E 1995 *Opt. Lett.* **20** 988
[23] Hirano M, Nakanishi T, Okuno T and Onishi M 2009 *J. Sel. Topics Quantum Electron* **15** 103
[24] Knight J C, Birks T A, Russell P and Atkin D M 1996 *Opt. Lett.* **21** 1547
[25] Russel P 2006 *J. Lightw. Technol.* **24** 4729
[26] Mogilevtsev D, Birks T A and Russell P 1998 *Opt. Lett.* **23** 1662
[27] Gander M J, McBride R, Jones J, Mogilevtsev D, Birks T A, Knight J C and Russell P 1999 *Electron. Lett.* **35** 63
[28] Fok M and Shu C 2008 *IEEE J. Selected Top. Quantum Electron* **14** 587
[29] Leong J *et al* 2006 *J. Lightw. Technol.* **24** 183
[30] Slusher R E, Lenz G, Hodelin J, Sanghera J, Shaw L B and Aggarwal I D 2004 *J. Opt. Soc. Am.* B **21** 1146
[31] Butcher P N and Cotter D N 1990 *The Elements of Nonlinear Optics* (Cambridge: Cambridge University Press)
[32] Mamyshev P V 1998 *Eur. Conf. Opt. Commun. (ECOC98)* (Madrid, Spain) pp 475–77
[33] Matsumoto M 2004 *J. Lightw. Technol* **23** 1472
[34] Johannisson P and Karlsson M 2005 *IEEE Photon. Technol. Lett.* **17** 2667
[35] Striegler A G and Schmauss B 2006 *J. Lightw. Technol.* **24** 2835
[36] Rochette M, Libin F, Ta'eed V, Moss D J and Eggleton B J 2006 *IEEE J. Select. Topics Quantum Electron* **12** 736
[37] Igarashi K and Kikuchi K 2008 *IEEE J. Sel. Topics Quantum Electron* **14** 551
[38] Fok M P and Shu C 2008 *IEEE J. Sel. Topics Quantum Electron* **14** 587
[39] Liu Z, Ziegler Z M, Wright L G and Wise F W 2017 *Optica* **4** 649
[40] Liu Z, Li C, Zhang Z, Kärtner F X and Chang G 2016 *Opt. Express* **24** 15328
[41] Liu W, Chia S-H, Chung H-Y, Greinert R, Kärtner F X and Chang G 2017 *Opt. Express* **25** 6822
[42] Chung H-Y, Liu W, Cao Q, Kärtner F X and Chang G 2017 *Opt. Express* **25** 15760
[43] Chung H, Liu W, Cao Q, Song L, Kärtner F X and Chang G 2018 *Opt. Express* **26** 3684
[44] Chraplyvy A R and Stone J 1984 *Electron. Lett.* **20** 996
[45] Wang J and Petermann K 1992 *J. Lightw. Technol.* **10** 96
[46] Marcuse D, Chraplyvy A R and Tkach R W 1994 *J. Lightw. Technol.* **12** 885
[47] Chiang T K, Kagi N, Marhic M E and Kazovsky L G 1996 *J. Lightw. Technol.* **14** 249
[48] Belloti G, Varani M, Francia C and Bononi A 1998 *IEEE Photon Technol. Lett.* **10** 1745
[49] Hui R, Demarest K R and Allen C T 1999 *J. Lightw. Technol.* **17** 1018
[50] Nelson L E, Jopson R M, Gnauck A H and Chraplyvy A R 1999 *IEEE Photon. Technol. Lett.* **11** 907
[51] Betti S and Giaconi M 2001 *IEEE Photon. Technol. Lett.* **13** 1304
[52] Thiele H J, Killey R I and Bayvel P 2002 *Opt. Fiber Technol.* **8** 71

[53] Blow K J, Doran N J, Nayar B K and Nelson B P 1990 *Opt. Lett.* **15** 248
[54] Jino M and Matsumoto T 1991 *Electron. Lett.* **27** 75
[55] Sharping J E, Fiorentino M, Kumar P and Windeler R S 2002 *IEEE Photon. Technol. Lett.* **14** 77
[56] Lee J H, Tanemura T, Nagashima T, Hasegawa T, Ohara S, Sugimoto N and Kikuchi K 2005 *Opt. Lett.* **30** 1267
[57] Salem R, Lenihan A S, Carter G M and Murphy T E 2006 *IEEE Photon. Technol. Lett.* **18** 2254
[58] McKinstrie C, Radic S and Raymer M 2004 *Opt. Express* **12** 5037
[59] Fu L B *et al* 2005 *Opt. Express* **13** 7637
[60] Galili M, Oxenlowe L K, Hansen H C H, Clausen A T and Jeppesen P 2008 *IEEE J. Sel. Top. Quantum Electron.* **14** 573
[61] Fernández-Ruiz M, Lei L, Rochette M and Azaña J 2015 *Opt. Express* **23** 22847
[62] Watanabe S, Kato T, Tanimura T, Schmidt-Langhorst C, Elschner R, Sackey I, Schubert C and Hoshida T 2019 *Opt. Express* **27** 16767
[63] Rochette M, Blows J L and Eggleton B J 2006 *Opt. Express* **14** 6414
[64] Suzuki J *et al* 2005 *IEEE Photon. Technol. Lett.* **17** 423
[65] Daikoku M, Yoshikane N, Otani T and Tanaka H 2006 *J. Lightw. Technol.* **24** 1142
[66] Agrawal G P 2002 *Fiber-Optic Communication Systems* 3th edn (New York: Wiley)
[67] Morioka T, Takara H, Kawanishi S, Kitoh K and Saruwatari M 1996 *Electron. Lett.* **32** 833
[68] Hansryd J and Andrekson P A 2001 *IEEE Photon. Technol. Lett.* **13** 732
[69] Sakamoto T, Seo K, Taira K, moon N S and Kikuchi K 2004 *IEEE Photon. Technol. Lett.* **16** 563
[70] Tang R, Lasri J, Devgan P S, Grigoryan V, Kumar P and Vasilyev M 2005 *Opt. Express* **13** 10483
[71] Cetina J P, Kumpera A, Karlsson M and Andrekson P A 2015 *Opt. Express* **23** 33426
[72] Hansryd J, Andrekson P A, Westlund M, Li J and Hedekvist P O 2002 *IEEE J. Sel. Top. Quantum Electron.* **8** 506
[73] Chavez Boggio J M, Windmiller J R, Knutzen M, Jiang R, Brès C, Alic N, Stossel B, Rottwitt K and Radic S 2008 *Opt. Express* **16** 5435
[74] Hirano M, Nakanishi T, Okuno T and Onishi M 2009 *IEEE J. Sel. Top. Quantum Electron.* **15** 103
[75] Nodop D, Jauregui C, Schimpf D, Limpert J and Tünnermann A 2009 *Opt. Lett.* **34** 3499
[76] Víctor J F, Rancaño F, Parmigiani P, Petropoulos P and Richardson D J 2014 *J. Lightw. Technol.* **32** 3027
[77] Cappellini G and Trillo S 1991 *J. Opt. Soc. Am.* B **8** 824
[78] Su Y, Wang L, Agrawal A and Kumar P 2000 *Electron. Lett.* **36** 1103
[79] Matsumoto M 2005 *IEEE Photon. Technol. Lett.* **17** 1055
[80] Matsumoto M 2012 *IEEE J. Select. Topics Quantum Electron* **18** 738
[81] Kuo B P P, Myslivets E, Wiberg A O J, Zlatanovic S, Bres C S, Moro S, Gholami F, Peric A, Alic N and Radic S 2011 *J. Lightw. Technol.* **29** 516
[82] Karlsson M 1998 *J. Opt. Soc. Am.* B **15** 2269
[83] McKinstrie C J, Radic S and Chraplyvy A R 2002 *IEEE J. Sel. Top. Quantum Electron.* **8** 538
[84] Radic S and McKinstrie C J 2003 *Opt. Fib. Technol.* **9** 7

[85] Kalogerakis G, Marhic M E, Wong K K and Kazovsky L G 2005 *J. Lightw. Technol.* **23** 2945
[86] Bromage J 2004 *J. Lightw. Technol.* **22** 79
[87] Smith R G 1972 *Appl. Opt.* **11** 2489
[88] Ikeda M 1981 *Opt. Commun.* **39** 148
[89] Chraplyvy A R, Stone J and Burrus C A 1983 *Opt. Lett.* **8** 415
[90] Nakazawa M 1985 *Appl. Phys. Lett.* **46** 628
[91] Nakazawa M, Nakashima T and Seikai S 1985 *J. Opt. Soc. Am.* B **2** 215
[92] Dakss M L and Melman P 1985 *J. Lightw. Technol.* **3** 806
[93] Olsson N A and Hegarty J 1986 *J. Lightw. Technol.* **4** 391
[94] Aoki Y, Kishida S and Washio K 1986 *Appl. Opt.* **25** 1056
[95] Tkach R W and Chraplyvy A R 1989 *Opt. Quantum Electron.* **21** S105
[96] Ferreira M F, Rocha J F and Pinto J L 1994 *Opt. Quantum Electron.* **26** 35
[97] Stepanov D Y and Cowle G J 1997 *IEEE J. Sel. Top. Quantum Electron.* **3** 1049
[98] Zadok A, Eyal A and Tur M 2007 *J. Lightw. Technol.* **25** 2168
[99] Massey S M and Russel T H 2009 *IEEE J. Sel. Top. Quantum Electron.* **15** 399
[100] Nikles M, Thevenaz L and Robert P A 1996 *Opt. Lett.* **21** 738
[101] Hotate K and Tanaka M 2002 *IEEE Photon. Technol. Lett.* **14** 179
[102] Song K Y, Herraez M G and Thevenaz L 2005 *Opt. Express* **13** 82
[103] Zhu Z, Dawes A M, Gauthier D J, Zhang L and Willner A E 2007 *J. Lightw. Technol.* **25** 201
[104] Zhu Z, Gauthier D J and Boyd R W 2007 *Science* **318** 1748
[105] Hasegawa A and Tappert F D 1973 *Appl. Phys. Lett.* **23** 142
[106] Mollenauer L F, Stolen R H and Gordon J P 1980 *Phys. Rev. Lett.* **45** 1095
[107] Hasegawa A and Kodama Y 1995 *Solitons in Optical Communications* (Oxford: Oxford University Press)
[108] Hasegawa A (ed) 1998 *New Trends in Optical Soliton Transmission Systems* (Dordrecht: Kluwer)
[109] Mollenauer L and Gordon J 2006 *Solitons in Optical Fibers—Fundamentals and Applications* (San Diego, CA: Elsevier Academic)
[110] Gardner C S, Greene J M, Kruskal M D and Miura R M 1967 *Phys. Rev. Lett.* **19** 1095
[111] Ablowitz M J and Clarkson P A 1991 *Solitons, Nonlinear Evolution Equations, and Inverse Scattering* (New York: Cambridge University Press)
[112] Zakharov V E and Shabat A 1972 *Sov. Phys. JETP* **34** 62
[113] Gordon J P 1986 *Opt. Lett.* **11** 662
[114] Mitschke F M and Mollenauer L F 1986 *Opt. Lett.* **11** 659
[115] Nakajima K and Ohashi M 2002 *IEEE Photon. Technol. Lett.* **14** 492
[116] Okuno T, Onishi M, Kashiwada T, Ishikawa S and Nishimura M 1995 *IEEE J. Sel. Topics Quantum Electron* **5** 1385
[117] Hirano M, Nakanishi T, Okuno T and Onishi M 2009 *J. Sel. Topics Quantum Electron* **15** 103
[118] Harbold J, Ilday F, Wise F, Birks T, Wadsworth W and Chen Z 2002 *Opt. Lett.* **27** 1558
[119] Lu F and Knox W 2004 *Opt. Express* **12** 347
[120] Foster M and Gaeta A 2004 *Opt. Express* **12** 3137
[121] Foster M A, Moll K D and Gaeta A L 2004 *Opt. Express* **12** 2880
[122] Cordeiro C M, Wadsworth W J, Birks T A and Russell P St J 2005 *Opt. Lett.* **30** 1980

[123] Knight J C, Birks T A, Russel P S J and Atkin D M 1996 *Opt. Lett.* **21** 1547
[124] Cregan R F, Mangan B J and Knight J C 1999 *Science* **285** 1537
[125] Roberts P J *et al* 2005 *Opt. Express* **13** 236
[126] Russell P 2003 *Science* **299** 358
[127] Eggleton B J, Kerbage C, Westbrook P S, Windeler R S and Hale A 2001 *Opt. Express* **9** 698
[128] Saito K and Koshiba M 2005 *J. Lightw. Technol.* **23** 3580
[129] Leon-Saval S G, Birks T A, Wadsworth W J and Russel P St J 2004 *Opt. Express* **12** 2864
[130] Feng X, Mairaj A K, Hewak D W and Monro T M 2005 *J. Lightw. Technol.* **23** 2046
[131] Price J H V *et al* 2007 *IEEE J. Sel. Top. Quantum Electron.* **23** 738
[132] Wang J S, Vogel E M and Snitzer E 1994 *Opt. Mat.* **3** 187
[133] Leong J *et al* 2006 *J. Lightw. Technol.* **24** 183
[134] Fok M and Shu C 2008 *IEEE J. Selected Top. Quantum Electron* **14** 587
[135] Lee J H, Kikuchi K, Nagashime T, Hasegawa T, Ohara S and Sugimoto N 2005 *Opt. Express* **13** 3144
[136] Lee J H, Nagashima T, Hasegawa T, Ohara S, Sugimoto N and Kikuchi K 2006 *J. Lightw. Technol.* **24** 22
[137] Jauregui C, Ono H, Petropoulos P and Richardson D J 2006 *Proc. Optical Fiber Communication Conf.* paper PDP2
[138] Slusher R E, Lenz G, Hodelin J, Sanghera J, Shaw L B and Aggarwal I D 2004 *J. Opt. Soc. Am.* B **21** 1146
[139] Fuochi M, Poli F, Cucinotta A and Vincetti L 2003 *J. Lightw. Technol.* **21** 2247
[140] Varshney S K, Fujisawa T, Saito K and Koshiba M 2005 *Opt. Express* **13** 9516
[141] Varshney S K, Saito K, Iizawa K, Tsuchida Y, Koshiba M and Sinha R K 2008 *Opt. Lett.* **33** 2431
[142] Reeves W H, Skryabin D V, Biancalana F, Knight J C, Russel P S J, Omenetto F G, Efomov A and Taylor A J 2003 *Nature* **424** 511
[143] Fercher A F, Drexler W, Hitzenberger C K and Lasser T 2003 *Rep. Progr. Phys.* **66** 239
[144] Humbert G *et al* 2006 *Opt. Express* **14** 1596
[145] Bellini M and Hänsch T W 2000 *Opt. Lett.* **25** 1049
[146] Hundertmark H, Wandt D, Fallnich C, Haverkamp N and Telle H 2004 *Opt. Express* **12** 770
[147] Takara H, Ohara T, Yamamoto T, Masuda H, Abe M, Takahashi H and Morioka T 2005 *Electron. Lett.* **41** 270
[148] Agrawal G P 2007 *Nonlinear Fiber Optics* 4th edn (Burlington, MA: Academic)
[149] Coen S, Chau A H L, Leonhardt R, Harvey J D, Knight J C, Wadsworth W J and Russell P S J 2002 *J. Opt. Soc. Am.* B **19** 753
[150] Dudley J M, Provino L, Grossard N, Maillotte H, Windeler R S, Eggleton B J and Coen S 2002 *J. Opt. Soc. Am.* B **19** 765
[151] Wadsworth W J, Joly N, Knight J C, Birks T A, Biancalana F and Russell P J L 2004 *Opt. Express* **12** 299
[152] Wong G K, Chen A Y, Murdoch S G, Leonhardt R, Harvey J D, Joly N Y, Knight J C, Wadsworth W J and Russel P S J 2005 *J. Opt. Soc. Am.* B **22** 2505
[153] Hasegawa A and Brinkman W F 1980 *IEEE J. Quantum Electron.* **16** 694
[154] Itoh H, Davis G M and Sudo S 1989 *Opt. Lett.* **14** 1368

[155] Mussot A, Lantz E, Maillotte H, Sylvestre T, Finot C and Pitois S 2004 *Opt. Express* **12** 2838
[156] Tai K, Hasegawa A and Tomita A 1986 *Phys. Rev. Lett.* **56** 135
[157] Ortigosa-Blanch A, Knight J C and Russel P S J 2002 *J. Opt. Soc. Am.* B **19** 2567
[158] Geny G, Lehtonen M, Ludvigsen H, Broeng J and Kaivola M 2002 *Opt. Express* **10** 1083
[159] Kodama Y and Hasegawa A 1987 *IEEE J. Quantum Electron.* **23** 510
[160] Husakou A and Herrmann J 2001 *Phys. Rev. Lett.* **87** 203901
[161] Akhmediev N and Karlsson M 1995 *Phys. Rev.* A **51** 2602
[162] Nishizawa N and Goto T 2001 *Opt. Express* **8** 328
[163] Skryabin D V, Luan F, Knight J C and Russell P S 2003 *Science* **301** 1705
[164] Biancalana D V, Skryabin D V and Yulin A V 2004 *Phys. Rev.* E **70** 016615
[165] Price J H V *et al* 2007 *IEEE J. Sel. Top. Quantum Electron.* **23** 738
[166] Dudley J M and Coen S 2002 *J. Sel. Top. Quantum Electron.* **8** 651
[167] Gaeta A L 2002 *Opt. Lett.* **27** 924
[168] Husakou A V and Herrmann J 2002 *J. Opt. Soc. Am.* B **19** 2171
[169] Hult J 2007 *J. Lightw. Technol.* **25** 3770
[170] Stolen R H, Gordon J P, Tomlinson W J and Haus H A 1989 *J. Opt. Soc. Am.* B **6** 1159
[171] Blow K J and Wood D 1989 *IEEE J. Quantum Electron.* **25** 2665
[172] Lin Q and Agrawal G P 2006 *Opt. Lett.* **31** 3086
[173] Moller U *et al* 2015 *Opt. Express* **23** 3282
[174] Domachuk P *et al* 2008 *Opt. Express* **16** 7161
[175] Xia C *et al* 2006 *Opt. Lett.* **31** 2553
[176] Saghaei H, Kazem M, Farshi M, Heidari M and Moghadasi M N 2016 *IEEE J. Sel. Topics Quantum Electron.* **22** 4900508
[177] Rodrigues S, Facão M and Ferreira M 2017 *J. Nonlinear Opt. Phys. Mater.* **26** 1750049

IOP Publishing

Optical Fiber Technology and Applications
Recent advances
Mário F S Ferreira and Mukul Chandra Paul

Chapter 9

Pulsating dissipative solitons in optical fiber systems

Sofia C V Latas and Mário F S Ferreira

We introduce the concept of dissipative solitons, which emerge as a result of a double balance: between nonlinearity and dispersion and also between gain and loss. Such dissipative solitons have many unique properties which differ from those of their conservative counterparts and which make them similar to living things. We focus our discussion on dissipative solitons in optical fiber systems, which can be described by the cubic-quintic complex Ginzburg–Landau equation (CGLE). The conditions to have stable solutions of the CGLE are discussed using the perturbation theory. We also use the variational approach to find approximate pulse solutions for the CGLE. We find that the evolution equations for the soliton parameters are similar to those derived using the method of moments. The existence of both stationary and pulsating soliton solutions is indicated by both approaches and confirmed through some numerical examples. In particular, we describe the dynamics of plain pulsating, creeping, and erupting soliton solutions. Moreover, we discuss the effect that some higher-order effects, namely the third-order dispersion, intrapulse Raman scattering, and self-steepening, can have on these pulses. We show that, for some range of the parameter values, these higher-order effects can transform the pulsating solitons into fixed-shape solitons.

9.1 Introduction

The combination of ultrafast lasers and optical fiber technology has given birth to one of the most important classes of optical sources, and today mode-locked fiber lasers play a central role in fields such as medicine, ultrasensitive laser spectroscopy, micromachining, ultrahigh bit rate communication systems, metrology, and others [1, 2]. Ultrashort light pulses produced by mode-locked fiber lasers can help us directly observe some of the fastest processes in nature, along with studies of matter under extreme conditions. Mode-locked fiber lasers are also important test beds to research the nonlinear dynamics of complex systems far from equilibrium, as well as to study a wide variety of dissipative structures and self-organization effects [3, 4].

The concept of temporal and spatial dissipative solitons can be applied to solid-state lasers and fiber lasers as well as to optical pulses propagating in fiber-optic communication systems in which real losses are compensated with fiber-optic amplifiers [3, 5–17]. In fact, dissipative optical solitons have many unique properties which differ from those of their conservative counterparts. For example, except for very few cases [11], they form zero-parameter families and their properties are completely determined by the external parameters of the optical system. They can exist indefinitely in time, as long as these parameters stay constant. These features of dissipative solitons are highly desirable for several applications, such as in-line regeneration of optical data streams and the generation of stable trains of laser pulses by mode-locked fiber lasers

The basic mathematical model for the description of all nonlinear phenomena occurring in mode-locked fiber lasers and long-haul soliton transmission lines is the cubic-quintic complex Ginzburg–Landau equation (CGLE) [11–13]. However, the CGLE has also been used to describe a number of phenomena including second-order phase transitions, superconductivity, superfluidity, Bose–Einstein condensation, liquid crystals, and string theory [18]. In fact, the areas of application and the number of solutions to this equation are so large that the sphere of this knowledge by itself has been called 'the world of the Ginzburg–Landau equation' [19].

Various techniques have been applied to investigate the soliton solutions of the CGLE. Exact solutions were found [11, 20–24], but they can be presented explicitly only for certain relations between the parameters of the equation. Furthermore, so far only stationary solutions of the CGLE are known in analytical form. The soliton perturbation theory can be used for small values of the parameters [24]. Moreover, approximate expressions for some localized solutions can be derived for arbitrary values of the CGLE parameters by reducing this equation to finite-dimensional dynamical models. The reduced models can be obtained by applying the method of moments [25–27] or Lagrangian techniques [28–32].

To fully explore the CGLE, huge numerical simulations must be carried out. Different types of soliton solutions have been obtained in this way, which can be divided in two main classes: localized fixed-shape solutions and localized pulsating solutions [24, 31–33]. Examples of localized fixed-shape solutions are the plain stationary pulses, flat-top pulses, composite pulses, and moving pulses [11]. Among the localized pulsating solutions we may refer to the plain pulsating and the creeping solitons, as well as the erupting solitons, which belong to the class of chaotic solutions [34, 35]. Erupting solitons have been observed experimentally in a solid-state Kerr-lens mode-locked Ti:sapphire laser [36] and in mode-locked fiber lasers [37].

To describe the behavior of ultrashort optical pulses it is necessary to consider several higher-order dispersive and nonlinear effects, namely the third order of dispersion (TOD), self-steepening (SS), and intrapulse Raman scattering (IRS) [24, 38, 39]. The effect of higher-order effects, both isolated or in combination, on localized pulsating solutions has been studied by several authors [40–56]. It has been shown that, under the influence of such higher-order effects, the localized pulsating solutions of the CGLE can be transformed into fixed-shape solutions for a certain range of parameter values [51–54].

In this chapter we will discuss different types of pulsating CGLE soliton solutions, namely plain pulsating, creeping, and erupting solutions. We begin in section 9.2 by presenting the cubic-quintic CGLE, after which we describe in section 9.3 the variational approach to find approximate solutions to this equation. In section 9.4 we present a generalized version of the cubic-quintic CGLE, taking into account several higher-order effects, namely the IRS, self-steepening, and third-order dispersion. The impact of these higher-order effects on pulsating solitons is discussed in the following sections. Section 9.5 is dedicated to the plain pulsating solitons, while section 9.6 describes the creeping solitons. Finally, the erupting solitons are discussed in section 9.7.

9.2 The cubic-quintic complex Ginzburg–Landau equation

The cubic-quintic complex Ginzburg–Landau equation can be written in the form [9, 11, 24]

$$i\frac{\partial q}{\partial Z} + \frac{1}{2}\frac{\partial^2 q}{\partial T^2} + |q|^2 q = i\delta q + i\beta\frac{\partial^2 q}{\partial T^2} + i\varepsilon\,|q|^2 q + i\mu\,|q|^4 q - \nu\,|q|^4 q, \tag{9.1}$$

where Z is the normalized propagation distance, T is the retarded time, q is the normalized envelope of the electric field, β stands for spectral filtering ($\beta > 0$), δ is the linear gain or loss coefficient, ε accounts for nonlinear gain-absorption processes (for example two-photon absorption), μ represents a higher-order correction to the nonlinear gain-absorption, and ν is a higher-order correction term to the nonlinear refractive index.

Equation (9.1) has been used to model the propagation of optical pulses in soliton transmission systems [9, 11, 24]. It has also been used as a master equation to describe the behavior of solid-state lasers [6] and mode-locked fiber lasers [57–59]. The relations between the physical parameters describing a ring fiber laser mode locking through nonlinear polarization rotation and the coefficients of CGLE have been derived in [60].

Equation (9.1) becomes the standard nonlinear Schrödinger equation (NLSE) when the right-hand side is set to zero. In the general case, it is non-integrable and only particular exact solutions can be obtained. Despite having no known conserved quantities, there are two rate equations which can be written for the CGLE: one for the energy,

$$Q = \int_{-\infty}^{\infty} |q|^2 dT, \tag{9.2}$$

and the other for the momentum,

$$M = \mathrm{Im}\left(\int_{-\infty}^{\infty} q\frac{dq^*}{dT}dT\right). \tag{9.3}$$

For a dissipative system, Q is a function of Z rather than a conserved quantity. The rate of change of the energy is [12]

$$\frac{d}{dZ}Q = F[q], \tag{9.4}$$

where the real functional $F[q]$ is given by

$$F[q] = 2\int_{-\infty}^{\infty}\left[\delta\,|q|^2 + \varepsilon\,|q|^4 + \mu\,|q|^6 - \beta\left|\frac{\partial q}{\partial T}\right|^2\right]dT. \tag{9.5}$$

For localized solutions Q is finite and it takes a constant value when the solution is stationary. The rate of change of the momentum is given by [12]

$$\frac{d}{dZ}M = J[q], \tag{9.6}$$

where the real functional $J[q]$ is given by

$$J[q] = 2\mathrm{Im}\int_{-\infty}^{\infty}\left[(\delta + \varepsilon\,|q|^2 + \mu\,|q|^4)q + \beta\frac{\partial^2 q}{\partial T^2}\right]\frac{\partial q^*}{\partial T}dT. \tag{9.7}$$

The rate equations (9.4) and (9.6) can be used for studying different problems related to CGLE solitons. For example, assuming that the coefficients on the right-hand side of equation (9.1) are all small, they can be used to derive the evolution equations for the amplitude and the frequency of the perturbed soliton solution of the NLSE:

$$q(T, Z) = \eta\,\mathrm{sech}\,[\eta(T + \kappa Z)]\exp[-i\kappa T + i(\eta^2 - \kappa^2)Z/2]. \tag{9.8}$$

The equation for the evolution of $\eta(Z)$ can be derived from the equation for the energy (9.4), whereas the equation for $\kappa(Z)$ can be obtained from the equation for the momentum (9.6). It turns out that both results are given by [24]

$$\frac{\partial\eta}{\partial Z} = 2\delta\eta - 2\beta\eta\left(\frac{1}{3}\eta^2 + \kappa^2\right) + \frac{4}{3}\varepsilon\eta^3 + \frac{16}{15}\mu\eta^5 \tag{9.9}$$

$$\frac{d\kappa}{dZ} = -\frac{4}{3}\beta\eta^2\kappa. \tag{9.10}$$

As can be seen from equation (9.10), the soliton frequency approaches asymptotically to $\kappa = 0$ (a stable fixed point) if $\eta \neq 0$. It has been found that in order to have a stable pulse propagating in a stable background the following conditions must be satisfied [24]:

$$\delta < 0,\ \mu < 0,\ \varepsilon > \beta/2,\ 15\delta > 8\mu\eta_s^4, \tag{9.11}$$

where

$$\eta_s^2 = \frac{-5(\varepsilon - \varepsilon_s) - 5\sqrt{(\varepsilon - \varepsilon_s)^2 - 24\delta\mu/5}}{8\mu} \tag{9.12}$$

is the stationary value for the soliton amplitude and $\varepsilon_s = \beta/2$ for small values of β.

Figure 9.1 shows (a) the evolution of the peak amplitude and (b) the final pulse profile obtained numerically from equation (9.1), assuming an input pulse with a sech profile and considering the following parameter values: $\delta = -0.01$, $\beta = 0.15$, $\nu = 0$, $\varepsilon = 0.2$, $\mu = -0.1375$ (dashed curves) or $\varepsilon = 0.4$, $\mu = -0.3875$ (solid curves). According to equation (9.9), these values provide a stationary amplitude $\eta_s = 1$. This prediction and the stability of the stationary solution are confirmed by the numerical results in figure 9.1.

For small values of the parameters on the right-hand side of equation (9.1) the stable soliton solutions of the CGLE have a sech profile, similar to the soliton solutions of the NLSE, and correspond to the so-called plain pulses (PPs). However, quite different pulse profiles can be obtained for non-small values of those parameters. As an example, figure 9.2 illustrates the formation and stable propagation of a flat-top soliton, starting from an initial pulse with a sech profile. A flat-top soliton can be considered as two mirror-imaged interacting fronts.

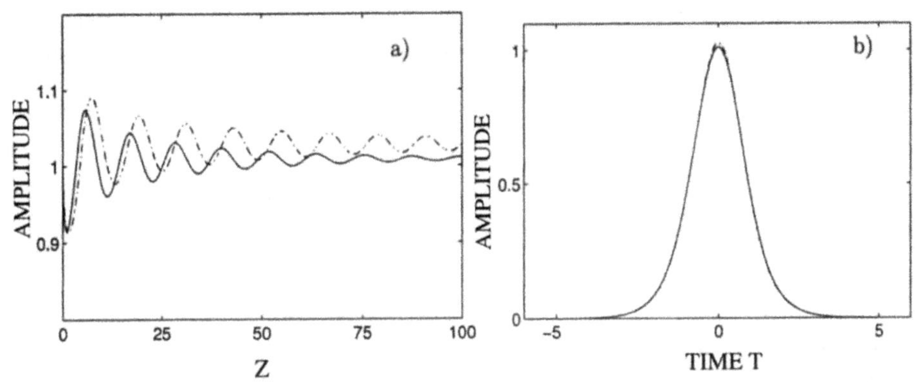

Figure 9.1. (a) Evolution of the peak amplitude and (b) the final pulse profile when $\delta = -0.01$, $\beta = 0.15$, $\varepsilon = 0.2$, $\nu = 0$, $\mu = -0.1375$ (dashed curves) or $\varepsilon = 0.4$, $\mu = -0.3875$ (solid curves), considering an input pulse $q(0, T) = \text{sech}(T)$.

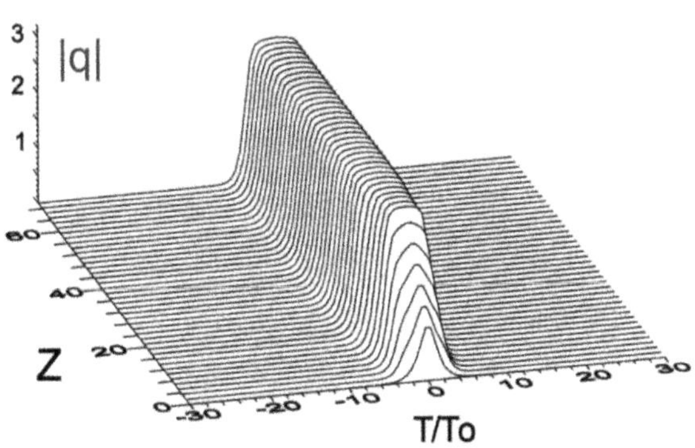

Figure 9.2. Formation and evolution of a flat-top soliton, considering an input pulse $q(0, T) = \text{sech}(T)$, for $\delta = -0.1$, $\beta = 0.5$, $\varepsilon = 0.66$, and $\mu = \nu = -0.01$.

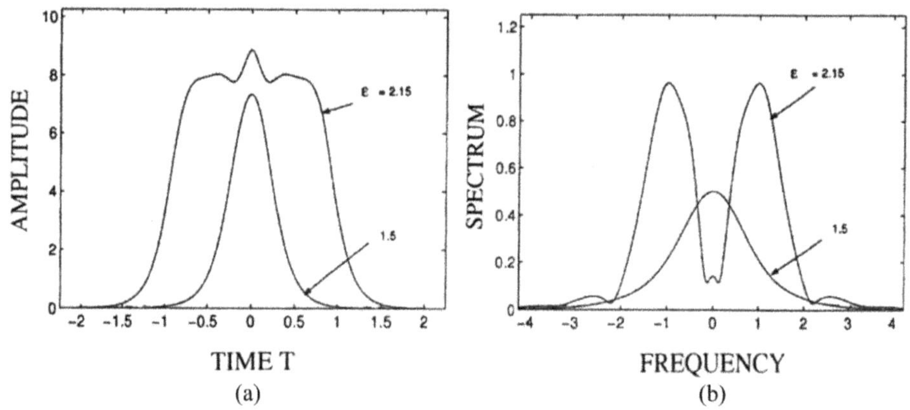

Figure 9.3. (a) Amplitude profiles and (b) spectra of a plain pulse and of a composite pulse for $\delta = -0.0.1$, $\beta = 0.5$, $\mu = -0.03$, $\nu = -0$, $\varepsilon = 1.5$, (plain pulse), and $\varepsilon = 2.5$ (composite pulse).

Increasing the nonlinear gain parameter ε transforms a bell-shaped plain pulse into a composite pulse (CP), as illustrated in figure 9.3(a). A CP can be considered as a combination of a plain pulse and two fronts attached to it from both sides. The 'hill' between the two fronts should be counted as a source, because it follows from the phase profile that energy flows from the center to the CP wings. While the plain pulse has a single peak spectrum, the CP exhibits a dual-frequency but symmetric spectrum (figure 9.3(b)).

9.3 The variational approach to CGLE solitons

The variational technique is particularly useful for describing the propagation of chirped pulses in nonlinear media. Its use in describing pulse propagation in nonlinear dispersive media, such as optical fibers, was first realized by Anderson [61]. This approach can be quite accurate even when the effect of nonlinearity is large [62]. The accuracy depends essentially on the ansatz chosen to describe the evolution of the field experiencing nonlinearity.

The standard variational approach can be modified to allow for dissipative terms in the CGLE. Using such an approach, equation (9.1) can be obtained from the Euler–Lagrange equation [32]

$$\frac{\partial}{\partial \tau}\left(\frac{\partial L_{\mathrm{d}}}{\partial q_T^*}\right) + \frac{\partial}{\partial z}\left(\frac{\partial L_{\mathrm{d}}}{\partial q_Z^*}\right) - \frac{\partial L_{\mathrm{d}}}{\partial q^*} = S, \tag{9.13}$$

where L_{d} is the Lagrangian density, given by

$$L_{\mathrm{d}} = \frac{i}{2}\left(q\frac{\partial q^*}{\partial Z} - q^*\frac{\partial q}{\partial Z}\right) + \frac{1}{2}\left|\frac{\partial q}{\partial T}\right|^2 - \frac{1}{2}|q|^4 - \frac{\nu}{3}|q|^6, \tag{9.14}$$

and S represents the dissipative terms:

$$S = i\delta q + i\beta\frac{\partial^2 q}{\partial T^2} + i\varepsilon\,|q|^2 q + i\mu\,|q|^4 q. \tag{9.15}$$

Considering the Lagrangian

$$L = \int_{-\infty}^{\infty} L_d dT \tag{9.16}$$

and a trial solution containing parameters ν_j, $j = 1, 2, \ldots$, the modified Euler–Lagrange equations are given by [32]

$$\frac{d}{dz}\left(\frac{\partial L}{\partial \nu_z}\right) - \frac{\partial L}{\partial \nu} = 2\,\mathrm{Re}\left(\int_{-\infty}^{\infty} S\frac{\partial q^*}{\partial \nu}\right)dT \tag{9.17}$$

for each parameter ν.

Let us consider a soliton solution given by

$$q(Z, T) = B(Z, T)\exp(-i\theta(Z)), \tag{9.18}$$

where $B(Z, T)$ is the complex amplitude and $\theta(Z)$ gives the phase evolution on propagation. Then, using equation (9.17), we have for a general parameter ν:

$$\frac{\partial L}{\partial \nu} - \frac{d}{dz}\left(\frac{\partial L}{\partial \nu_z}\right) = 2\,\mathrm{Re}\int_{-\infty}^{\infty} \frac{1}{iq^*}\frac{\partial q^*}{\partial \nu}\left(\delta\,|B|^2 + \varepsilon\,|B|^4 + \mu\,|B|^6 + \beta B^*\frac{\partial^2 B}{\partial T^2}\right)dT. \tag{9.19}$$

If $\nu = \theta(Z)$ in equation (9.19), we have

$$\frac{\partial L}{\partial \theta} = 0, \qquad \frac{\partial L}{\partial \theta'} = -Q(Z), \qquad \frac{1}{iq^*}\frac{\partial q^*}{\partial \theta} = 1, \tag{9.20}$$

and

$$\frac{dQ}{dZ} = 2\,\mathrm{Re}\int_{-\infty}^{\infty}\left[\delta\,|B|^2 + \varepsilon\,|B|^4 + \mu\,|B|^6 + \beta B^*\frac{\partial^2 B}{\partial T^2}\right]dT. \tag{9.21}$$

Equation (9.21) corresponds to the balance equation for the energy Q, given by equations (9.4) and (9.5). Obviously, the right-hand side is zero in the case of a stationary soliton solution.

9.3.1 Sech-type ansatz

Let us assume that

$$B(Z, T) = A\,\mathrm{sech}\left(\frac{T}{w(Z)}\right)\exp(iC(Z)T^2), \tag{9.22}$$

where $A(Z)$, $w(Z)$, and $C(Z)$ are the amplitude, width, and chirp parameter, respectively. In this case, we have

$$Q = 2A^2w \tag{9.23}$$

and equation (9.21) becomes

$$\frac{dQ}{dZ} = 2Q\left[\delta - \beta\left(\frac{1}{3w^2} + \frac{\pi^2}{3}C^2w^2\right) + \frac{\varepsilon}{3}\frac{Q}{w} + \frac{2\mu}{15}\frac{Q^2}{w^2}\right]. \tag{9.24}$$

The Lagrangian corresponding to equations (9.18) and (9.22) is given by

$$L = Q\left[\frac{\pi^2C^2w^2}{6} + \frac{1}{6w^2} - \frac{Q}{6w} - \frac{2\nu Q^2}{45w^2} + \frac{\pi^2w^2C'(Z)}{12} - \theta'(Z)\right]. \tag{9.25}$$

If $\nu = C(Z)$ in equation (9.19), we obtain the following equation for the pulse width

$$\frac{dw}{dZ} = 2cw + \beta\left(\frac{8}{\pi^2w} - \frac{16\pi^2}{15}C^2w^3\right) - \frac{2\varepsilon}{\pi^2}Q - \frac{\mu}{\pi^2}\frac{Q^2}{w}. \tag{9.26}$$

The equation for the chirp parameter $C(Z)$ is obtained using the parameter $\nu = w(Z)$ in equation (9.19). We have

$$\frac{dC}{dZ} = 2\left(\frac{1}{\pi^2w^4} - C^2\right) - \frac{1}{\pi^2}\frac{Q}{w^3} - 4\left(\frac{1}{3} + \frac{1}{\pi^2}\right)\beta\frac{C}{w^2} - \frac{8\nu}{15\pi^2}\frac{Q^2}{w^4}. \tag{9.27}$$

Finally, using $\nu = Q(Z)$ in equation (9.19) we obtain the following equation for the phase evolution:

$$\frac{d\theta}{dZ} = \frac{1}{3w^2} - \beta C\left(\frac{1}{3} + \frac{\pi^2}{9}\right) - \frac{5}{12}\frac{Q}{w} - \frac{8\nu}{45}\frac{Q^2}{w^2}. \tag{9.28}$$

9.3.2 Gaussian-type ansatz

As a second example, let us consider a simple Gaussian-type ansatz:

$$B(Z, T) = A(Z)\exp\left(-\frac{T^2}{w^2(Z)} + iC(Z)T^2\right), \tag{9.29}$$

where $A(Z)$, $w(Z)$, and $C(Z)$ are the amplitude, width, and chirp parameter, respectively. In this case, the pulse energy Q is given by

$$Q(Z) = \sqrt{\frac{\pi}{2}}A^2(Z)w(Z), \tag{9.30}$$

while the Lagrangian can be written as

$$L = Q\left[\frac{C^2w^2}{2} + \frac{1}{2w^2} - \frac{Q}{2\sqrt{\pi}w} - \frac{2\nu Q^2}{3\sqrt{3}\pi w^2} + \frac{1}{4}w^2C'(Z) - \theta'(Z)\right]. \tag{9.31}$$

Considering $\nu = \theta, C, w, Q$, in equation (9.19) we obtain the following evolution equations for the pulse parameters:

$$\frac{dQ}{dZ} = 2Q\left(\delta - \frac{\beta}{w^2} - \beta C^2 w^2 + \frac{1}{\sqrt{\pi}}\frac{\varepsilon Q}{w} + \frac{2}{\pi\sqrt{3}}\frac{\mu Q^2}{w^2}\right) \tag{9.32}$$

$$\frac{dw}{dZ} = 2\frac{\beta}{w} + 2Cw - 2\beta C^2 w^3 - \frac{1}{2\sqrt{\pi}}\varepsilon Q - \frac{4}{3\pi\sqrt{3}}\frac{\mu Q^2}{w} \tag{9.33}$$

$$\frac{dC}{dZ} = \frac{2}{w^4} - 2C^2 - 8\frac{\beta C}{w^2} - \frac{1}{\sqrt{\pi}}\frac{Q}{w^3} - \frac{8}{3\pi\sqrt{3}}\frac{\nu Q^2}{w^4} \tag{9.34}$$

$$\frac{d\theta}{dZ} = \frac{1}{w^2} - 2\beta C - \frac{5}{4\sqrt{\pi}}\frac{Q}{w} - \frac{8}{3\pi\sqrt{3}}\frac{\nu Q^2}{w^2}. \tag{9.35}$$

We observe that equations (9.32)–(9.35) and equations (9.24)–(9.28) are similar. The only difference lies in the numerical values of the coefficients of the terms.

A system of evolution equations for the soliton parameters similar to those presented above, considering either the sech or the Gaussian ansatz, can be obtained using the method of moments [27]. We can conclude that both the method of moments and the variational approach provide the possibility of reducing the cubic-quintic CGLE to the same finite-dimensional dynamical system.

An a analysis of the dynamical system given by equations (9.24)–(9.28) has demonstrated a clear correspondence between attractors and various types of localized waves governed by the cubic-quintic CGLE [27]. Both fixed points and limit cycles can be found for such dynamical systems, the former corresponding to stationary solitons and the latter to pulsating solitons. A transformation from a stationary soliton to a pulsating soliton appears as the result of a Hopf bifurcation in the reduced dynamical system [27]. Some examples of stationary soliton solutions were given in section 9.2. Different types of pulsating soliton solutions will be discussed in sections 9.5, 9.6, and 9.7.

Hamiltonian systems do not have pulsating soliton solutions, but some pulsating structures were found in integrable models as a result of the nonlinear superposition of fundamental solitons [38, 39]. However, such structures have nothing in common with pulsating solitons in dissipative systems, which are only observed when the parameters of the CGL equation are far enough from the integrable (NLSE) limit.

9.4 Higher-order effects

Equation (9.1) is adequate to describe the propagation of picosecond optical solitons in fiber systems under the influence of spectral filtering, linear, and nonlinear gains, including the saturation effects of nonlinear gain and of the nonlinear refractive index. However, it becomes inadequate in the case of ultrashort solitons ($t_0 < 1$ ps). The spectral width of these pulses becomes large enough that it becomes necessary to take into account the Raman effect and the frequency dependence of the nonlinear parameter γ. Moreover, it is also useful to include the higher-order terms in the expansion of the propagation constant. This gives rise to three main higher-order effects, namely the third-order dispersion (TOD), self-steepening (SST), and intra-pulse Raman scattering (IRS) effects [24].

Taking into account the above mentioned higher-order effects, the cubic-quintic CGLE given by equation (9.1) is generalized and becomes [24, 51–54]

$$i\frac{\partial q}{\partial Z} + \frac{1}{2}\frac{\partial^2 q}{\partial T^2} + |q|^2 q = i\delta q + i\beta\frac{\partial^2 q}{\partial T^2} + i\varepsilon\,|q|^2 q + i\mu\,|q|^4 q - \nu\,|q|^4 q + \text{HOEs}, \tag{9.36a}$$

where

$$\text{HOE} = i\delta_3\frac{\partial^3 q}{\partial T^3} - is\frac{\partial(|q|^2 q)}{\partial T} + \tau_R q\frac{\partial\,|q|^2}{\partial T}. \tag{9.36b}$$

The parameters δ_3, s, and τ_R, govern, respectively, the effects of TOD, SST, and IRS, and are given by

$$\delta_3 = \frac{\beta_3}{6|\beta_2|t_0}, \qquad s = \frac{1}{\omega_0 t_0}, \qquad \tau_R = \frac{t_R}{t_0}, \tag{9.37}$$

where

$$\beta_j = \left(\frac{d^j\beta}{d\omega^j}\right)_{\omega=\omega_0} \tag{9.38}$$

and β is the propagation constant, ω_0 is the carrier frequency, and t_0 is the initial pulse width. The three parameters given by equation (9.36b) are inversely proportional to the initial pulse width t_0. Assuming a value $t_0 = 30$ fs for pulses propagating at 1550 nm in a standard silica fiber, we have $\delta_3 \approx 0.03$, $s \approx 0.03$, and $\tau_R \approx 0.1$. The IRS effect is often the most relevant among the higher-order effects.

As an asymmetric cubic function with respect to ω, TOD causes the pulse temporal profile to also become asymmetric, moving its peak from the original position. It can be shown that, under the TOD effect only, the soliton peak is shifted from its original position by an amount that increases linearly with the propagation distance. The soliton peak is advanced when $\delta_3 < 0$ and it is delayed when $\delta_3 > 0$. Such temporal shift is negligible for picosecond pulses, but becomes significant for femtosecond pulses [39].

TOD can affect the dynamics of NLSE solitons drastically. In particular, it can determine the break-up of second- and higher-order solitons into multiple fundamental solitons. This phenomenon is called soliton fission [24, 39, 63]. During this process, TOD can also provide the phase matching between the solitons and the linear waves, enabling the resonant transfer of energy between them [63, 64].

Self-steepening arises from the intensity dependence of the group velocity [65–67] and leads to an asymmetry in both the temporal and spectral profiles of ultrashort pulses [39, 68–70]. It can be shown that, as the pulse propagates inside the fiber, its peak falls behind the pedestal, whereas the trailing edge becomes steepened. In the absence of dispersion, the pulse will continue to steepen until a shock front is formed. However, this process is counteracted in the presence of dispersion. Higher frequencies arise to support the steep trailing edge, which leads also to an asymmetric pulse spectrum [39]. The SST effect can also lead to the break-up of higher-order solitons through the phenomenon of soliton fission [71].

Considering only the IRS term in equation (9.36) and applying the soliton perturbation theory, it is found that the soliton frequency varies with distance as [24, 63]

$$\kappa(Z) = -\frac{8}{15}\tau_{\mathrm{R}}\eta^4 Z. \tag{9.39}$$

Equation (9.39) shows that the IRS effect leads to a continuous downshift of the carrier frequency, an effect known as the soliton self-frequency shift (SSFS) [72]. This effect was observed for the first time by Mitschke and Mollenauer in 1986 using 0.5 ps pulses obtained from a passively mode-locked color-center laser [73]. The origin of SSFS can be understood by noting that for ultrashort solitons the pulse spectrum becomes so broad that the high-frequency components of the pulse can transfer energy through Raman amplification to the low-frequency components of the same pulse. Such an energy transfer appears as a red shift of the soliton spectrum, with shift increasing with distance. Since the amplitude and width are inversely related, the frequency shift scales as t_0^4, indicating that it can become quite large for short pulses.

The behavior of the localized fixed-shape solutions of the CGLE in the presence of IRS has been studied using the perturbation theory and a finite-dimensional dynamic system has been derived for the amplitude and frequency of the soliton solution [42, 43]. The analysis of the stationary solutions of such dynamic system has shown that narrow-band filtering and nonlinear gain can control the self-frequency shift due to the IRS of ultrashort optical solitons [42, 43]. An effect of reduction in the time shift due to the IRS in the presence of TOD for ultrashort high-amplitude soliton pulses was found in [56]. It has also been shown that the singularity in the solution of the CGLE when the saturation of the nonlinear gain tends to zero [20] no longer exists in the presence of IRS [56].

9.5 Plain pulsating solitons

A plain pulsating soliton is a localized solution of the CGLE that changes its width, amplitude, and other parameters periodically during propagation. It corresponds to a limit cycle of the infinite-dimensional dynamical system and can be described by a closed loop in the phase space of the system.

Plain pulsating soliton solutions of the cubic-quintic CGLE were first found by Deissler and Brand [74] in the normal dispersion regime. The profile of such solitons shows a periodic variation only in its tails, whereas the energy remains almost constant. In contrast, the plain pulsating solitons found by Akhmediev *et al* [34] were observed in the anomalous dispersion regime and both their shape and energy change quite significantly.

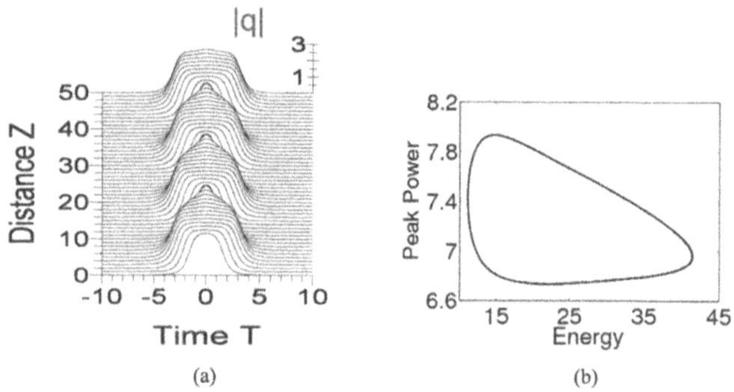

Figure 9.4. (a) Amplitude evolution and (b) peak power versus energy for a plain pulsating soliton, assuming the following parameter values: $\delta = -0.1$, $\beta = 0.08$, $\varepsilon = 0.66$, $\mu = -0.1$, and $\nu = -0.1$. Reproduced from [51], copyright (2011) with permission of Springer.

Figure 9.4(a) shows the amplitude evolution of a plain pulsating soliton assuming the following parameter values: $\delta = -0.1$, $\beta = 0.08$, $\varepsilon = 0.66$, $\mu = -0.1$, and $\nu = -0.1$. The pulse shows a periodic behavior, with a period $Z \approx 14$. It has different shapes at each position Z, but it recovers its initial shape exactly after a period. Figure 9.4(b) shows the soliton peak power, $|q|^2$, against the soliton energy, given by equation (9.2). The trajectory corresponds to a cycle repeating itself indefinitely, which means that the pulse energy and peak power also have a periodic behavior. The pulse energy changes from 10 to 40, while the pulse peak power changes between 6.8 to 8, approximately.

An interesting question concerns the impact of the higher-order effects discussed in section 9.4 on the behavior of the plain pulsating pulse. This is illustrated in figure 9.5, where the upper row shows the amplitude contour plot for the pulsating pulse in the presence of (a) IRS ($\tau_R = 0.01$), (b) negative TOD ($\beta_3 = -0.05$), and (c) IRS plus TOD, respectively. The bottom row of figure 9.5 shows the peak power, $|q|^2$, against the soliton energy, Q, for the same cases, respectively. The other parameter values are the same as for figure 9.4.

Under the IRS effect alone, figure 9.5(a) shows that the pulse moves to the right at a constant velocity, whereas the pulse of figure 9.5(a) evolves with zero velocity. In the presence of negative TOD, the pulse moves to the left at constant velocity, as illustrated in figure 9.5(b). In both cases, the oscillations of the pulse shape are slightly reduced, compared to the case of figure 9.4(a). However, if both effects act together, the pulse shape oscillations are reduced more significantly, as illustrated by figure 9.5(c). This behavior can also be observed from figure 9.5(d)–(f). In particular, when IRS and negative TOD act simultaneously, we observe from figure 9.5(f) that the trajectory slowly spirals inward, converging eventually to an inner limit cycle. The pulse energy oscillates from 18 to 23, while the pulse peak power oscillates from 7 to 7.4, approximately.

Figure 9.6 illustrates the impact of the three higher-order effects acting together on the plain pulsating soliton. The top row shows the amplitude contour plot for the pulsating pulse in the presence of IRS ($\tau_R = 0.01$), SST ($s = 0.005$), and (a) positive TOD ($\beta_3 = 0.05$) or (b) negative TOD ($\beta_3 = -0.05$). In both cases, the periodic

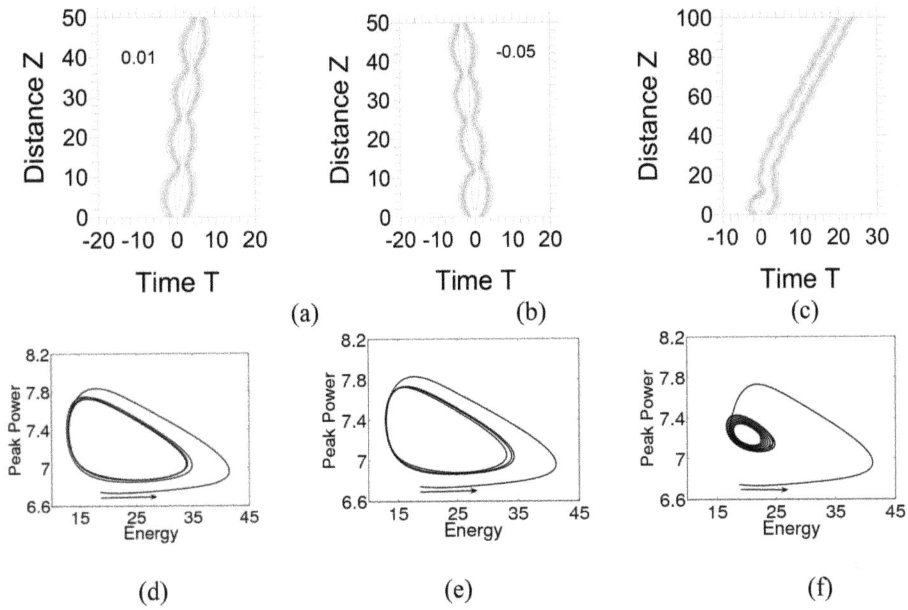

Figure 9.5. Amplitude contour plot (top row) and pulse peak power against pulse energy (bottom row) for a plain pulsating pulse in the presence of (a), (d) IRS ($\tau_R = 0.01$), (b), (e) negative TOD ($\beta_3 = -0.05$), and (c), (f), IRS and negative TOD. The other parameter values are the same as in figure 9.4.

behavior of the pulsating pulse is drastically reduced and the pulse almost achieves a fixed shape in case (b). The negative TOD is more effective than the positive TOD in achieving this goal. The same feature can also be observed from the bottom row of figure 9.6. In the case (d) the cycle trajectory decreases and quickly spirals inward to a fixed point, corresponding to a fixed shape pulse.

9.6 Creeping solitons

A creeping soliton is a pulsating localized solution that changes its shape periodically and shifts a finite distance in the transverse direction after each pulsation. It has a rectangular pulse with two fronts and one sink on the top [35], presenting a shape that resembles that of the composite soliton [11, 75, 76]. Creeping solitons were first observed numerically in [34] and their existence was later confirmed for various dissipative systems [35, 49].

Figure 9.7 shows (a) the amplitude evolution and (b) the soliton peak power versus the energy of a creeping pulsating pulse, assuming the following set of parameters: $\delta = -0.1$, $\beta = 0.101$, $\varepsilon = 1.3$, $\mu = -0.3$, and $\nu = -0.101$. We observe that the two fronts pulsate back and forth relative to the sink asymmetrically at the two sides of the pulse. Due to this asymmetry the position of the center of mass of the whole soliton shifts a finite distance in the transverse direction after each period of oscillation, which generally determines a nonzero constant velocity. The trajectories represented in figure 9.7(b) seem to evolve to a limiting cycle. In this process the pulse energy maximum grows slightly, while the minimum pulse energy remains

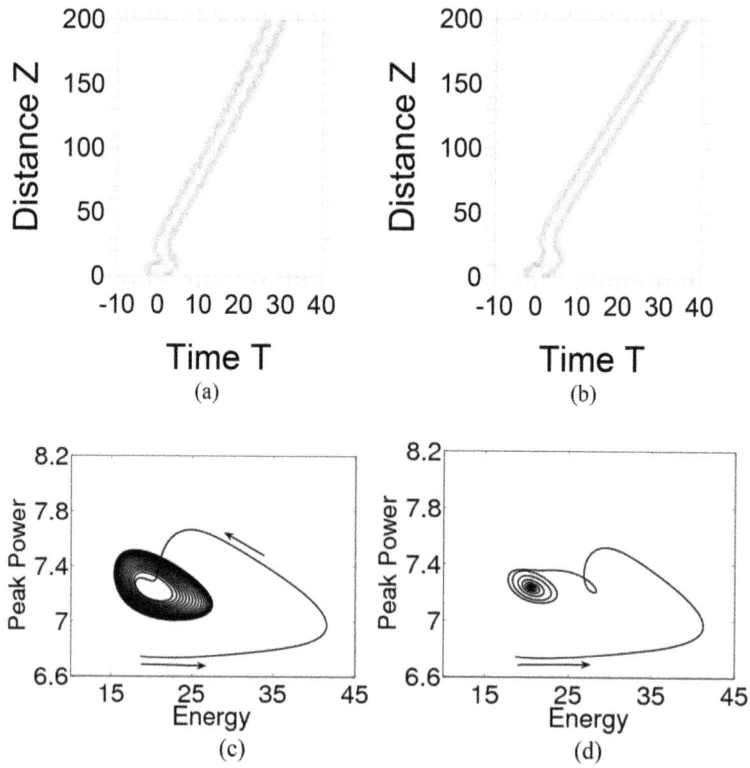

Figure 9.6. Amplitude contour plot (top row) and soliton peak power versus energy (bottom row) for the pulsating pulse in the presence of IRS ($\tau_R = 0.01$), SST ($s = 0.005$), and (a), (c) positive TOD ($\beta_3 = 0.05$), or (b), (d) negative TOD ($\beta_3 = -0.05$). The other parameter values are the same as in figure 9.4.

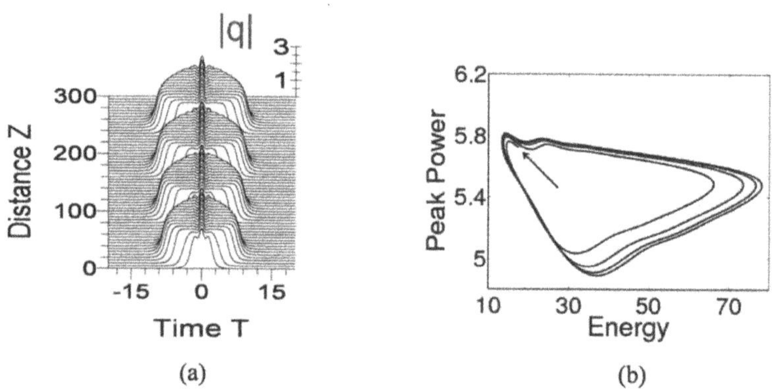

Figure 9.7. (a) Amplitude and (b) soliton peak power versus energy for a creeping soliton, assuming the following set of parameters: $\delta = -0.1$, $\beta = 0.101$, $\varepsilon = 1.3$, $\mu = -0.3$, and $\nu = -0.1$. Reproduced from [51], copyright (2011) with permission of Springer.

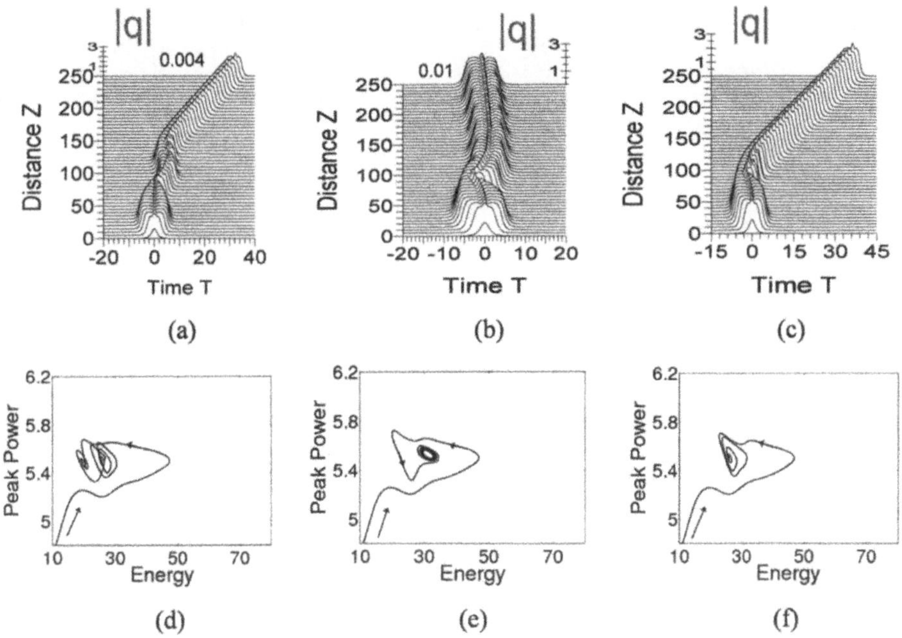

Figure 9.8. Amplitude evolution (top row) and peak power versus energy (bottom row) for a creeping soliton in the presence of (a), (d) IRS ($\tau_R = 0.004$), (b), (e) OD ($\beta_3 = +0.01$), and (c), (f) IRS and TOD. The remaining values of the parameters are the same as for figure 9.7. Reproduced from [51], copyright (2011) with permission of Springer.

almost constant. On the other hand, the minimum value of the pulse peak power tends to decrease, whereas its maximum value remains nearly constant.

Figure 9.8 shows the evolution of the creeping pulse amplitude (upper row) in the presence of (a) IRS ($\tau_R = 0.004$), (b) positive TOD ($\beta_3 = 0.01$), and (c) IRS plus TOD. The values of the remaining parameters are the same as for figure 9.7. We observe from figure 9.8(a) that even for a relatively low value of the IRS coefficient, the creeping soliton is converted into a fixed-shape soliton. This is confirmed by figure 9.8(d), in which it can be seen that the pulse peak power only experiences a small change. After the initial evolution, the values of the pulse peak power oscillate between a minimum of 5.3 and a maximum of 5.7, while the maximum value of the pulse energy suffers a reduction from 80 to 50, approximately. The cycle trajectory tends to spiral inward to a fixed point, the fixed-shape solution, in which the cycle trajectory of the peak power versus pulse energy spirals inward to a fixed point. The resulting pulse moves to the right at a constant velocity.

Figure 9.8(b) shows the impact of positive TOD ($\beta_3 = 0.01$) on the evolution of a creeping pulse propagation. TOD reduces the amplitude and the periodic fluctuations of the pulsating behavior on both sides of the pulse. The cycle trajectory, represented in figure 9.8(e), decreases and slowly spirals inward. This behavior changes if both effects, TOD and IRS, are combined. Figure 9.8(c) shows that if both the IRS and positive TOD effects act simultaneously, the pulse also achieves a fixed shape while moving to the right at constant velocity. Figure 9.8(e) confirms that the pulse quickly evolves to a fixed point in this case.

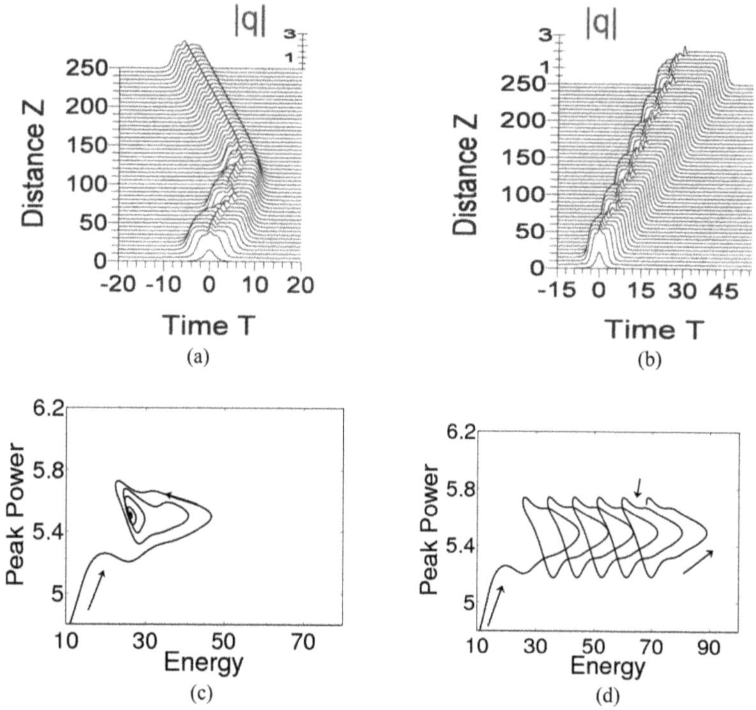

Figure 9.9. Amplitude evolution (top row) and peak power versus energy (bottom row) for a creeping soliton in the presence of IRS ($\tau = 0.004$), SST ($s = 0.002$), and (a), (c) positive TOD ($\beta_3 = 0.001$) and (b), (d) negative TOD ($\beta_3 = -0.01$). The remaining values of the parameters are the same as for figure 9.7. Reproduced from [51], copyright (2011) with permission of Springer.

Finally, figure 9.9 illustrates the propagation of the creeping soliton under the simultaneous action of the three higher-order effects: TOD, IRS, and SST. The top row of figure 9.9 shows the amplitude evolution in the presence of IRS ($\tau_R = 0.004$), SST ($s = 0.002$), and (a) positive TOD ($\beta_3 = 0.01$) and (b) negative TOD ($\beta_3 = -0.01$). The bottom row in figure 9.9 shows the peak power versus energy for the two cases. It can be seen that behavior of the pulse evolution changes drastically with the sign of TOD. For this particular set of parameter values, the pulse achieves a fixed shape in the case of positive TOD. This is confirmed in figure 9.9(c), where the cycle trajectory quickly spirals to a fixed point. However, in the presence of negative TOD the front on the left-hand side pulsates back and forth periodically relative to the sink, while the front on the right-hand side spreads with a fixed velocity far away from the sink. From figure 9.9(d) it can be seen that, in this case, the pulse peak power changes between 5.2 and 5.8, while the pulse mean energy grows, due to the right front expansion.

9.7 Erupting solitons

One of the most striking forms of pulsating solitons is the erupting soliton which was found numerically for the first time in [34], where the soliton eruption manifests itself as a chaotic and quasi-periodic process when the dissipative system is in a meta-stable state. The soliton in the meta-stable dissipative system erupts into pieces abruptly in the temporal domain and gradually recovers its original state after the eruption, which is similar to exploding behavior and thus is regarded as the so-called soliton explosion [34]. The soliton explosion was experimentally observed for the first time by Cundiff *et al* [36] in a solid-state, Kerr-lens mode-locked Ti:sapphire laser. More recently, experimental observation of the spectral and temporal signatures of soliton explosions were experimentally observed in an all-normal dispersion, all-polarization maintaining a passively mode-locked Yb-doped fiber laser [37]. The explosions were observed when the laser was operating in a transition zone between stable mode-locking [77] and noise-like emission [78, 79].

Figure 9.10 shows (a) the amplitude and (b) the peak power versus energy of a CGLE soliton for the following set of parameters: $\delta = -0.1$, $\beta = 0.125$, $\varepsilon = 1.0$, $\mu = -0.1$, and $\nu = -0.6$. The erupting soliton starts from a stationary localized solution. After a while it becomes covered with small ripples which seem to move downwards along the two slopes of the soliton, such that after a short distance the pulse becomes covered with this seemingly chaotic structure. When the ripples increase in size the soliton cracks into pieces, after which it evolves in order to restore its shape. The erupting process repeats forever, but never in exactly equal periods, as can be seen from figure 9.10(b). In fact, the pulse energy and the pulse peak power have a periodic behavior, but each trajectory is different from one explosion to another, having in common only the stationary regime of evolution. In this stationary regime the pulse energy seems to grow as the ripples develop, while the pulse peak power remains almost constant. In figure 9.10(b) the pulse energy changes from about 25 to 98, while the pulse peak power change from about 6.3 to 10.2, approximately.

Linear stability analysis was proposed to explain the origin of the soliton explosion [13, 14]. The behavior of the erupting soliton can be studied using the linear stability analysis when it is in the laminar regime [80]. In order to investigate the stability in this regime we can assume a stationary solution given by $q(Z, T) = q_0 e^{ikZ}$, where

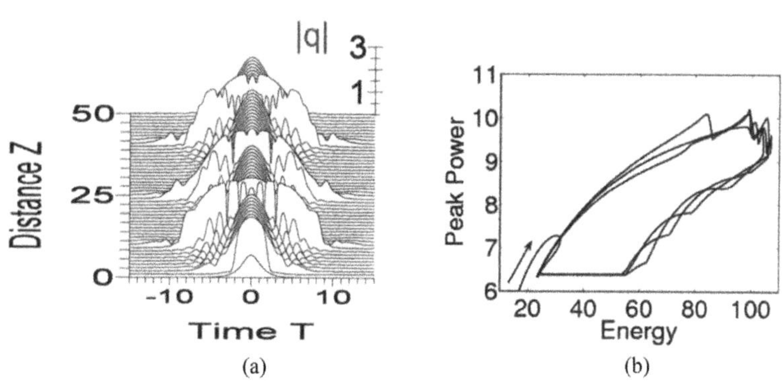

Figure 9.10. (a) Amplitude and (b) peak power against pulse energy of an erupting soliton for the following parameter values: $\delta = -0.1$, $\beta = 0.125$, $\varepsilon = 1.0$, $\mu = -0.1$, and $\nu = -0.6$. Reproduced from [51], copyright (2011) with permission of Springer.

$q_0(T)$ is a complex function of T with exponentially decaying tails and k is the propagation constant, assumed to be real. The evolution of the solution in the vicinity of the stationary solution can be described by an expression of the type

$$q(Z, T) = [q_0(T) + f(T)e^{\lambda Z} + g(T)e^{\lambda^* Z}]e^{ikZ}, \tag{9.40}$$

where $f(T)$ and $g(T)$ are small perturbation functions and λ is the associated perturbation growth rate. When the coefficients in the right-hand side of equation (9.1) are small, the stability analysis can be done analytically. However, when such coefficients are not small we must perform numerical calculations in order to obtain the eigenvalues and the eigenfunctions of the linearized problem. It has been found that the full spectrum of an erupting soliton consists of two complex conjugate eigenvalues with positive real parts and a continuous spectrum of complex conjugate eigenvalues, all with negative real parts [80]. It was verified that the eigenfunctions corresponding to the eigenvalues with positive real parts are nonzero mainly in the wings of the soliton [80].

Considering an initial stationary solution with small perturbations, the positive real parts of the eigenvalues will determine an increase of such perturbations. Since the eigenfunctions are almost zero in the central part of the soliton, its center is not affected by the growing instability. When the amplitude of the perturbations becomes similar to the soliton amplitude, the dynamics becomes strongly nonlinear and chaotic. However, the solution remains localized, both in amplitude and in width. Indeed, a positive value of β guarantees that the total width in the frequency domain remains finite, whereas a negative value of μ ensures that the maximum amplitude is limited. As all radiative waves have eigenvalues with negative real parts, they decay and quickly disappear. As a result, the evolution returns to the state of a stationary soliton with a small perturbation that has an eigenvalue with a positive real part. As a consequence, the instability will develop again later, thus repeating the whole process.

In addition to the erupting solitons, equation (9.36) also has erupting front solutions. An example of an erupting front is shown in figure 9.11. Fronts and solitons can coexist and any of them can be stable or unstable. In general, the transition from solitons to fronts happens along a certain boundary in the parameter space. If explosions existed in the region of solitons they continue to exist after the transition to fronts. Since the perturbation functions are nonzero mainly at the soliton wings, if the soliton splits into fronts the two parts of the perturbation function move away, each attached to the corresponding front.

Figure 9.12 shows the amplitude evolution (upper row) and the peak power versus energy (lower row) for an erupting pulse, in the presence of (a), (d) IRS ($\tau_R = 0.25$), (b), (e) positive TOD ($\beta_3 = +0.1$), and (c), (f) IRS and positive TOD. The other parameter values and the initial condition are the same as for figure 9.10.

Figure 9.12(a) shows that, under the effect of IRS, the pulse moves to the left and only cracks on the left-hand side, compared with the pulse evolution in figure 9.10(a). In the presence of positive TOD the pulse moves to the right and the explosions only occur on the right-hand side, as illustrated in figure 9.12(b).

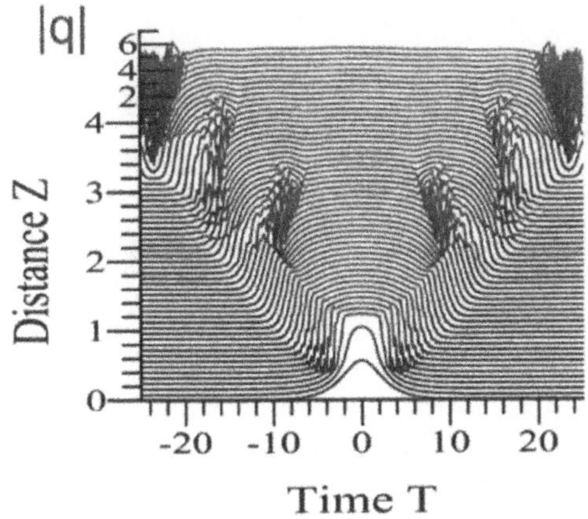

Figure 9.11. Amplitude of an erupting front obtained using the following parameter values: $\delta = -0.1$, $\beta = 0.08$, $\varepsilon = 1.0$, $\mu = -0.03$, and $\nu = -0.08$.

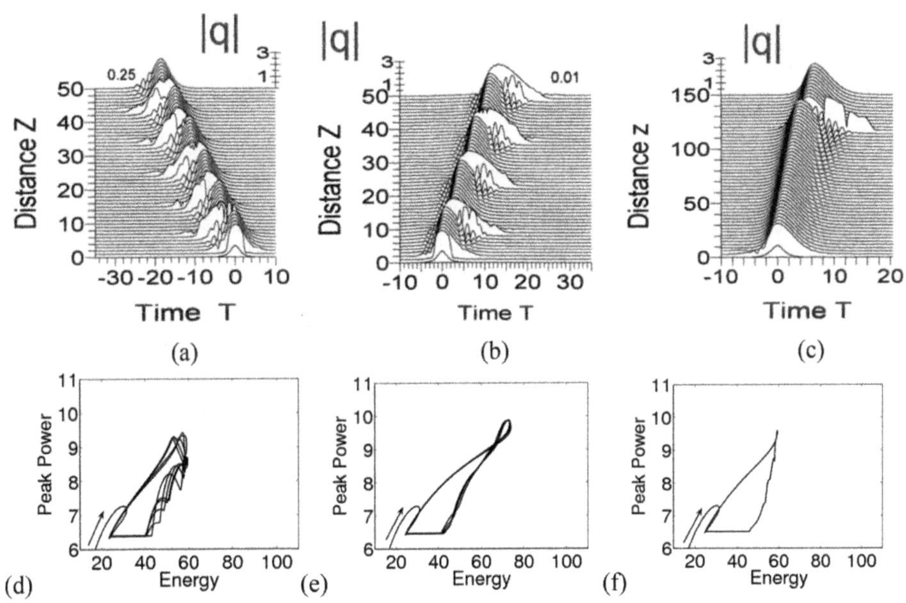

Figure 9.12. Amplitude evolution (upper row) and soliton peak power versus soliton energy (lower row) for an erupting pulse in the presence of (a), (d) IRS ($\tau_R = 0.25$), (b), (e) TOD ($\beta_3 = +0.1$), and (c), (f) IRS plus TOD. The other parameter values are the same as for figure 9.10. Reproduced from [51], copyright (2011) with permission of Springer.

In both cases the pulse moves at constant velocity and the number of explosions increases, compared to the case of figure 9.10. However, if the pulse propagates under the simultaneous presence of both effects, figure 9.12(c) shows that the number of explosions is drastically reduced. Comparing figure 9.12(d)–(f) with figure 9.10(b), it can be seen that the oscillations of the pulse energy are reduced, in

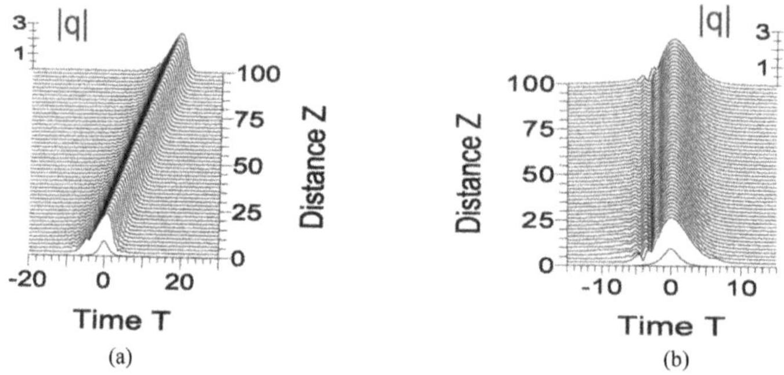

Figure 9.13. Amplitude evolution for an erupting soliton in the presence of IRS ($\tau_R = 0.25$), together with (a) negative TOD ($\beta_3 = -0.125$) and SST ($s = 0.125$), and (b) positive TOD ($\beta_3 = 0.1$) and SST ($s = 0.005$). The other parameter values are the same as in figure 9.10. Reproduced from [51], copyright (2011) with permission of Springer.

both the stationary and chaotic regimes. The oscillations of the pulse peak power also suffer a reduction in the chaotic regime, while the peak power remains constant in the stationary regime.

Figure 9.13 illustrates the propagation of the erupting soliton, in the presence of IRS ($\tau_R = 0.25$), together with (a) negative TOD ($\beta_3 = -0.125$) and SST ($s = 0.125$), and (b) positive TOD ($\beta_3 = +0.1$) and SST ($s = 0.005$). In the case of figure 9.13(a) it can be seen that the pulse achieves a fixed shape and moves to the right at a nonzero velocity. However, the pulse shape is not symmetric, with a trailing edge steeper than the leading edge. In the case of figure 9.13(b) the soliton also achieves a fixed shape, but moves at almost zero velocity. The pulse shape is also asymmetric, with several small peaks in the leading edge.

It has been shown in [51] that, under the effect of positive third other dispersion, the explosions become asymmetric and, in some circumstances, completely eliminated on the left-hand side of the pulse. It was also found that the intrapulse Raman scattering and the positive third-order dispersion can compensate each other, the same occurring for the self-steepening and negative third-order dispersion. Figure 9.13 shows that the soliton explosions can be completely eliminated under the simultaneous influence of the three higher-order effects, by a proper choice of the parameter values.

9.8 Conclusions

We introduce the concept of dissipative solitons, focusing on optical fiber systems which can be described by the cubic-quintic complex Ginzburg–Landau equation. We use the perturbation theory to discuss the conditions to have stable soliton solutions and the variational approach to find some properties of the dissipative solitons. We found that the evolution equations for the soliton parameters derived using such an approach are similar to those obtained by the method of moments. The existence of both stationary and pulsating soliton solutions is indicated by these approaches and confirmed numerically. We described the main properties of three types of pulsating pulses: plain pulsating, erupting, and creeping solitons. We demonstrated the dramatic impact that some higher-order effects, namely the intrapulse Raman, third-order dispersion, and self-steepening effects can have on such pulsating solitons. In particular, we have shown the possibility of converting them into fixed-shape pulses by using a proper combination of the above effects.

Acknowledgments

The authors are grateful for funding from Fundação para a Ciência e a Tecnologia (FCT) (Project UID/CTM/50025/2020).

References

[1] Weiner A M 2009 *Ultrafast Optics* (Hoboken, NJ: Wiley)
[2] Fermann M, Galvanauskas A and Sucha G (ed) 2002 *Ultrafast Lasers: Technology and Applications* (Boca Raton, FL: CRC Press)
[3] Grelu P and Akhmediev N 2012 *Nat. Photonics* **6** 84
[4] Dudley J M, Dias F, Erkintalo M and Genty G 2014 *Nat. Photonics* **8** 755
[5] Haus H A 1975 *J. Phys.* **46** 3049
[6] Haus H A, Fujimoto J G and Ippen E P 1991 *J. Opt. Soc. Am.* B **8** 2068
[7] Moores J D 1993 *Opt. Commun.* **96** 65
[8] Kärtner F X, Ausder Au J and Keller U 1998 *IEEE J. Sel. Top. Quantum Electron.* **4** 159
[9] Matsumoto M, Ikeda H, Uda T and Hasegawa A 1995 *J. Lightw. Technol.* **13** 658
[10] Kodama Y, Romagnoli M and Wabnitz S 1992 *Electron. Lett.* **28** 1981
[11] Akhmediev N and Ankiewicz A 1997 *Solitons: Nonlinear Pulses and Beams* (London: Chapman and Hall)
[12] Akhmediev N and Ankiewicz A 2005 *Dissipative Solitons* (Lecture Notes in Physics vol 661) ed N Akhmediev and A Ankiewicz (Berlin: Springer)
[13] Kalashnikov V L 2012 *Solid State Lasers* ed A Al-Kursan (Rijeka: IntechOpen) pp 145–84
[14] Rozanov N 1997 *Optical Bistability and Hysteresis in Distributed Nonlinear Systems* (Moscow: Nauka)
[15] Taranenko V B, Staliunas K and Weiss C O 1997 *Phys. Rev.* A **56** 1582
[16] Slekys G, Staliunas K and Weis C O 1998 *Opt. Commun.* **149** 113
[17] Staliunas U and Sanchez-Morcillo V J 2003 *Transverse Patterns in Nonlinear Optical Resonators* (Springer Tracts in Modern Physics vol 183) (Berlin: Springer) pp 7–37
[18] Cross M C and Hohenberg P C 1993 *Rev. Mod. Phys.* **65** 851
[19] Aranson I S and Kramer L 2002 *Rev. Mod. Phys.* **74** 99
[20] Akhmediev N, Afanasjev V and Soto-Crespo J 1996 *Phys. Rev.* E **53** 1190
[21] Soto-Crespo J, Akhmediev N and Afanasjev V 1996 *J. Opt. Soc. Am.* B **13** 1439
[22] Pereira N R and Stenflo L 1977 *Phys. Fluids* **20** 1733
[23] Nozaki K and Bekki N 1984 *Phys. Soc. Japan* **53** 1581
[24] Ferreira M F 2011 *Nonlinear Effects in Optical Fibers* (Hoboken, NJ: Wiley)
[25] Maimistov A I 1993 *J. Exp. Theor. Phys.* **77** 727
[26] Zhuravlev M N and Ostrovskaya N V 2004 *J. Exp. Theor. Phys.* **99** 427
[27] Tsoy E, Ankiewicz A and Akhmediev N 2006 *Phys. Rev.* E **73** 036621-1–10
[28] Manousakis M, Papagiannis P, Moshonas N and Hizanidis K 2001 *Opt. Commun.* **198** 351
[29] Ankiewicz A, Akhmediev N and Devine N 2007 *Opt. Fib. Technol.* **13** 91

[30] Mancas S C and Choudhury S R 2007 *Theor. Math. Phys.* **152** 1160
[31] Mancas S C and Choudhury S R 2009 *Chaos, Solitons Fractals* **40** 91
[32] Ferreira M F 2018 *IET Optoelectron* **12** 122
[33] Chang W, Ankiewicz A, Akhmediev N and Soto-Crespo J 2007 *Phys. Rev.* E **76** 016607
[34] Soto-Crespo J, Akhmediev N and Ankiewicz A 2000 *Phys. Rev. Lett.* **85** 2937
[35] Akhmediev N, Soto-Crespo J and Town G 2001 *Phys. Rev.* E **63** 056602
[36] Cundiff S, Soto-Crespo J and Akhmediev N 2002 *Phys. Rev. Lett.* **88** 073903
[37] Runge A, Broderick N and Erkintalo M 2015 *Optica* **2** 36
[38] Hasegawa A and Kodama Y 1995 *Solitons in Optical Communications* (Oxford: Clarendon)
[39] Agrawal G P 2001 *Nonlinear Fiber Optics* 3rd edn (San Diego: Academic)
[40] Horikis T P and Ablowitz M J 2014 *J. Opt. Soc. Am.* B **31** 2748
[41] Song L, Li L, Li Z and Zhou G 2005 *Opt. Commun.* **249** 301
[42] Tian H, Li Z, Hu Z, Tian J and Zhou G 2003 *J. Opt. Soc. Am.* B **20** 59
[43] Latas S C and Ferreira M F 2005 *Opt. Commun.* **251** 415
[44] Facao M V and Carvalho M I 2011 *Phys. Lett.* A **375** 2327
[45] Uzunov I M, Georgiev Z D and Arabadzhiev T N 2014 *Phys. Rev.* E **90** 042906
[46] Cartes C and Descalzi O 2015 *Fiber Integr. Opt.* **34** 14
[47] Cartes C and Descalzi O 2014 *Eur. Phys. J.: Spec. Top.* **223** 91
[48] Li Z, Li L, Tian H, Zhou G and Spatschek K H 2002 *Phys. Rev. Lett.* **89** 263901
[49] Tian H, Li Z, Tian J, Zhou G and Zi J 2004 *Appl. Phys.* B **78** 199
[50] Bélanger P A 2006 *Opt. Express* **14** 12174
[51] Latas S C, Ferreira M F and Facao M V 2011 *Appl. Phys.* B **104** 131
[52] Latas S C and Ferreira M F 2012 *Opt. Lett.* **37** 3897
[53] Facao M V and Carvalho M I 2012 *Phys. Lett.* A **376** 950
[54] Latas S C and Ferreira M F 2018 *J. Nonlinear Opt. Phys. Mat.* **27** 1850008
[55] Uzunov I M, Arabadzhiev T N and Georgiev Z D 2015 *Opt. Fiber Technol.* **24** 15
[56] Latas S C, Ferreira M F and Facao M V 2017 *J. Opt. Soc. Am.* B **34** 1033
[57] Nelson L E, Jones D J, Tamura K, Haus H A and Ippen E P 1997 *Appl. Phys.* B **65** 277
[58] Wise F W, Chong A and Renninger W H 2008 *Laser Phot. Rev.* **2** 58
[59] Renninger W H, Chong A and Wise F W 2011 *Opt. Express* **19** 22496
[60] Komarov A, Leblond H and Sanchez F 2005 *Phys. Rev.* E **72** 025604(R)
[61] Anderson D 1983 *Phys. Rev.* A **27** 3135
[62] Anderson D, Lisak M and Reichel T 1988 *Phys. Rev.* A **38** 1618
[63] Ferreira M F 2020 *Optical Signal Processing in Highly Nonlinear Fibers* (Abingdon: CRC Press)
[64] Akhmediev N and Karlsson M 1995 *Phys. Rev.* A **51** 2602
[65] Jonek R J and Landauer R 1967 *Phys. Lett.* **24A** 228
[66] DeMartini F, Townes C H, Gustafson T K and Kelley P L 1967 *Phys. Rev.* **164** 312
[67] Grischkowsky D, Courtens E and Armstrong J A 1973 *Phys. Rev. Lett.* **31** 422
[68] Ablowitz M J and Clarkson P A 1991 *Solitons, Nonlinear Evolution Equations, and Inverse Scattering* (New York: Cambridge University Press)
[69] Taylor J T (ed) 1992 *Optical Solitons—Theory and Experiment* (New York: Cambridge University Press)
[70] Abdullaev F K 1993 *Optical Solitons* (New York: Springer)
[71] Golovchenko E A, Dianov E M, Prokhorov A M and Serkin V N 1985 *JETP Lett.* **42** 87
[72] Gordon J P 1987 *Opt. Lett.* **11** 659

[73] Mitschke F and Mollenauer L 1986 *Opt. Lett.* **11** 659
[74] Deissler R and Brandt H 1994 *Phys. Rev. Lett.* **72** 478
[75] Latas S and Ferreira M 2005 *Opt. Fiber Technol.* **11** 292
[76] Latas S and Ferreira M 2011 *Appl. Phys.* B **105** 863
[77] Erkintalo M, Aguergaray C, Runge A and Broderick N G R 2012 *Opt. Express* **20** 22669
[78] Aguergaray C, Runge A, Erkintalo M and Broderick N G R 2013 *Opt. Lett.* **38** 2644
[79] Runge A F J, Aguergaray C, Broderick N G R and Erkintalo M 2014 *Opt. Lett.* **39** 319
[80] Akhmediev N and Soto-Crespo J 2003 *Phys. Lett.* A **317** 287

IOP Publishing

Optical Fiber Technology and Applications
Recent advances
Mário F S Ferreira and Mukul Chandra Paul

Chapter 10

Optical fiber networks for 5G environments

Francesco Matera

With the implementation of fifth-generation networks, called 5G, we will see a profound transformation of the entire telecommunications network, in particular in terms of interoperability between heterogeneous wireless networks and wired networks from the access segment to the core. 5G networks will lead to a new series of services that will also affect sectors that until now have not made use of the potential offered by the ICT mode, such as the automotive, energy, and health fields, and so on.

In addition to increasing the transmission capacity of the terminals, 5G will allow very low latencies, a much higher degree of mobility, and above all it will subvert the classic network paradigms by introducing new traffic management methods. This is particularly true in radio access thanks to the introduction of new frequency bands and the integration of wireless networks with wired access and transport networks.

From the architectural perspective, the connection of the radio interfaces will take place through the creation of logical and physical networks that may have very large geographical extensions and this will be achieved through the segmentation of the transport network into many independent subnetworks, each with specific characteristics for the services that they will have to support. The management of these network segments, called *slices*, must take place in the manner typical to software defined networks (SDNs).

In this profound evolution, optical systems and networks will provide fundamental contributions with novel devices dedicated solely to the segments, called xHaul, starting from the antennas to the first point of contact, with specific technologies for backhauling and fronthauling.

This chapter is dedicated to an overview of the evolution of optical networks to support the 5G environment.

10.1 Introduction

The growth of data traffic is pushing towards a revolution in telecommunication networks in order to meet a series of needs ranging from the increasing bandwidth

doi:10.1088/978-0-7503-3243-9ch10

required by users, in particular in the mobile mode [1–5], to the cooperation between different wired and wireless systems with an adequate convergence between fiber optic and radio networks, with particular relevance for the environment that goes by the name of the Internet of things (IoT) and machine to machine (M2M) [6]. Moreover, the evolution of radio access networks (RAN) will increasingly exploit a combination of virtualization, centralization, and coordination techniques, which interact with each other in different ways within a new concept, known as cloud-RAN or centralized-RAN (C-RAN) [7]. All of these issues fall into what is now referred to as the the fifth-generation or 5G network [1–5]. Clearly this revolution will also stimulate great demand for services and applications involving a whole series of subjects (verticals) [8] who will use the telecommunications network in ways that have not been explored until now.

Thus the network will have to meet, in each logical (or physical) segment, specific requirements in terms of bandwidth, reliability, latency, or more generally in terms of quality of service (QoS), which will vary according to the application. To cope with the different performances required by the parties who interface with the network, operators will have to partition of their infrastructures, both physically and logically, in order to ensure the most suitable methods for routing information for each specific service. This way of partitioning the network was recently understood as a new strategy, called network slicing as a service (NsaS) [8]. This approach will help operators to offer ad hoc networks adapted to every need to be offered as a specific service. Alongside the operators there will be other subjects who will be interested in the creation of specific networks, as in the case of the owners of ports, cities, and utilities.

In this evolution, the radio segment will be characterized by the introduction of new interfaces that will also work on new frequency bands and with an increasingly dynamic management of the spectrum and with methods suitable for sharing portions of the band, considering that the number of devices will grow much faster than the frequencies that will gradually become available over time [4]. One of the most important activities is precisely that of regulation of the new frequency bands, which involves carriers at ever higher frequencies (millimeter waves). On these bands, transmission techniques are being developed with increasingly better spectral efficiency and antenna systems capable of managing numerous flows simultaneously [9].

From the point of view of a wired network, 5G will require profound innovations, in particular on the segment that today takes the name of xHaul [3], meaning from the radio interface to the core network, with the already mentioned slicing techniques [8] and with techniques of feeding of antennas with backhauling and fronthauling techniques [10, 11]. In particular, with regard to the integration between wired and mobile networks, it should be noted that the envisaged 5G networks will have to serve millions of connected devices per square kilometer, i.e. thousands of times more than is the case today, and to do this it will be necessary to have smaller and smaller cells, with a consequent increase in the density of installed antennas. In general, a device will have the ability to connect to different types of cells, which may have very different sizes (from femto- to macrocells) as shown in figure 10.1, and all these cells must be connected to the rest of the network via

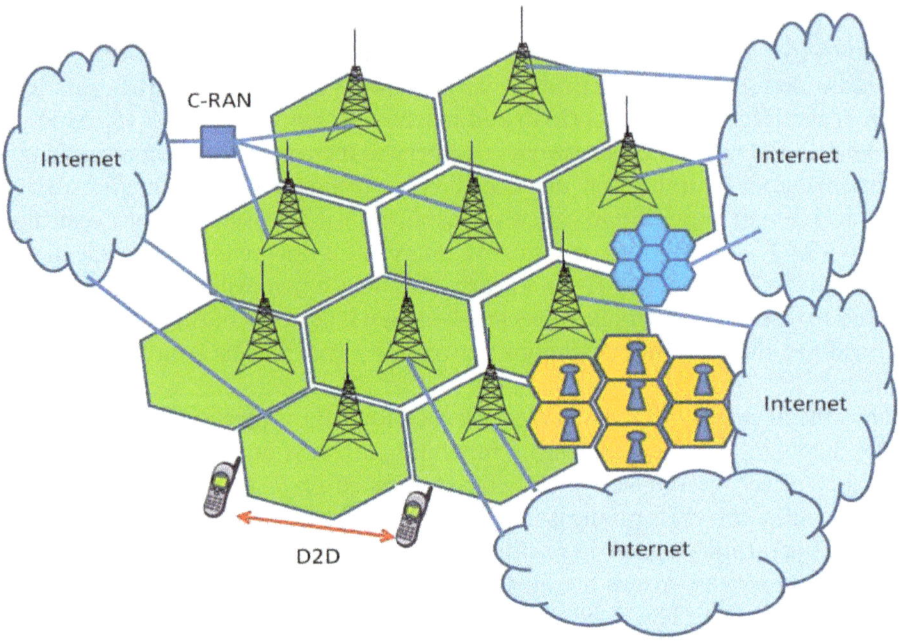

Figure 10.1. A typical 5G network.

high-capacity links since the volume of aggregate traffic can reach many gigabits up to terabits per second. In most cases, these links will be fiber optic connections, but other transmission systems, particularly wireless, could also be used. This evolution will essentially rely on the access network, currently the object of significant investments for the construction of fiber optic cabling, in particular following the plans of European governments to achieve the objectives of the European Digital Agenda 2020, now updated for 2025.

With regard to the routing aspects in 5G networks, it is clear that the connection between wireless networks, operating in even very distant geographic areas, will take place through the creation of logical and physical networks (slices) that will cross different operators. Everything will have to take place with very fast adaptive processes and in particular using the methods typical to software defined networks (SDNs) [12–14] and with the virtualization of the equipment with network function virtualization (NFV) techniques [15].

In this chapter a review is presented of all the technological aspects just mentioned, also reporting the experiences of the Ugo Bordoni Foundation, in particular from an experimental point of view, as described in section 10.4.

10.2 A short overview of 5G networks

The recent development of telecommunication networks has shown a well-defined trend in its transformation. Starting from heterogeneous infrastructures and technologies, dedicated to specific and well-defined services, we are moving towards a convergence of all services on platforms based on IP communication networks. This convergence, with the ability to interconnect heterogeneous radio networks (HetNet), combined with the possibility of having consolidated tools for quality management, is now a reality that all users can use both in the wired and mobile modes.

This evolution is continuing with the development of architectural models, whose transformation is essentially driven by key characteristics. These can be summarized by the concepts of flexibility, scalability, and service-oriented management, with the ability to dynamically configure the network functions necessary to support specific services and use cases. We are therefore moving towards architectural models, including a set of communication paradigms and enabling technologies, aimed at efficiently managing all network segments, from all accesses including radio ones, to the core network.

The goal is to enable the network to control the dynamics of traffic flows, providing the end user with the possibility of using different accesses, in particular of the radio type, in a transparent way, possibly through the same mobile device. This device must consequently being able to support increasingly complex functions, evolving both from the computational point of view and from that of efficiency, in particular from the energy point of view, to increase battery life.

This approach, which we could define as user-centric, involves the need to make a so-called vertical handover between the different access technologies feasible. There is a need interoperate within a heterogeneous context with an increasingly decisive role for technologies such as self-organizing networks (SONs), SDNs [13, 14], and NFV [15], which we will return to later. In this context, it is necessary to distinguish the procedures in which the terminal itself chooses the domain of its interest from that in which the operator assigns specific resources to the terminal, in particular by adopting appropriate classes of service.

Furthermore, the evolution of the RAN infrastructures aims to increase performance through the integration of functionalities in centralized processing, also resorting to the support of procedures supported by cloud computing, with C-RAN type infrastructures where C can mean both centralized and cloud. As shown in figure 10.2, at the basis of C-RAN is the possibility of dividing the radio base station (BS) into two units: the remote radio head (RRH) very simplified in its functionality and in particular with the task of radiating the radio signal, and the baseband unit (BBU) in which all processes take place. These units are connected by an optical fronthaul network [10, 11] capable of preserving the requirements of low latency and high throughput. More BBUs can then be aggregated in the same central office to form a BBU-pool.

The C-RANs bring several advantages to RANs, including the sharing of resources, elasticity, and above all the minimization of costs related to CAPEX and OPEX. The installation and updating phases of the RAN will be faster, more flexible, and can be optimized, allowing the on-demand adaptation of resources to non-uniform traffic patterns. Furthermore, the centralization of processing relating to a cluster of RRHs would also make it possible to make mobility operations between the RRHs themselves more efficient, as well as those for reducing inter-cell

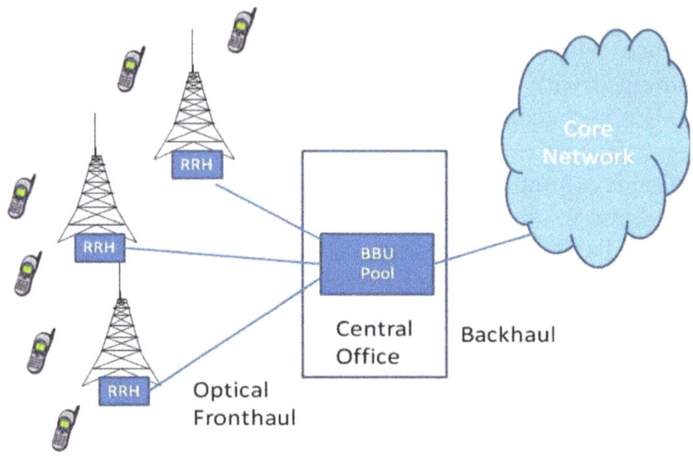

Figure 10.2. Example of a C-RAN network.

interference based on transmission and reception mechanisms such as that of the cooperative multi-point (CoMP).

The architectural evolution described is already present in current 4G networks, but it will become even more evident when radio interfaces evolve with 5G and are able to use new frequency bands, therefore allowing much greater capacities and above all a much higher density of devices. The most important aspects for 5G, in addition to the new radio interfaces, will also concern the whole network, in particular for the management of the many antennas that will be installed in each area, with coverage characteristics that will range from macrocells to femtocells. Furthermore, these new architectures will be able to establish connections with characteristics that are not present today, not only in terms of capacity but above all in terms of reliability (ultra-reliable) and with very low latencies, even below one millisecond [16]. These performance characteristics will allow the management of all those devices that are the basis, for example, of the automatic driving of vehicles.

The requirements for 5G have been defined by the International Telecommunication Union (ITU) in its 'IMT 2020' specification. The Third Generation Partnership Project (3GPP), which is the body that defined the specifications for third- and fourth-generation telephony, is developing the standards for 5G to reflect ITU requirements, which focus on two main aspects: that of the new radio interface (known as 5G new radio (5G NR)) which will cover devices that will operate in a wide range of frequency bands, and that of the network that will have to manage the connections for the antennas, and which in turn is distinguished by the access component (5G new access (5G NA)) and the core component (5G new core (5G NC)). In particular, 3GPP has planned a study divided into two phases with specific technical and temporal objectives: phase 1 which concerns release 15 and phase 2 which focuses on release 16. It should be noted that the first phase of specifications defines a 'non-standalone' solution that has to work in collaboration with existing systems such as the 4G network. Subsequently, the specifications for the 'standalone' version will be completed, which will be independent from the 4G network, allowing the creation of autonomous 5G networks.

5G will therefore allow the network to support very different methods for the dissemination and management of information, allowing a quantity of services not even imaginable today. To better understand what the future offers us, we refer to the grouping of industries, operators, and research centers that have formed the Next Generation Mobile Network (NGMN) Alliance and that has tried to highlight the needs that the networks will have to face and in the NGMN white paper [16], where the main use cases are distinguished. Following the ITU vision, all these use cases can be grouped into three main categories, which are those that are reported most often in the technical and scientific literature:

- *Enhanced Mobile BroadBand (eMBB)*, which concerns services that require more bandwidth, such as video, virtual reality, and augmented reality, and where capacities up to 10 Gb s^{-1} can be achieved.
- *Massive Machine Type Communications (mMTC)*, which concern high concentrations of devices (up to 1 million km^{-1}) and which refer to typical scenarios of smart cities and smart agriculture.
- *Ultra-reliable and low latency communications (URLLC)*, which concerns all those services where very high reliability is required in communication with very low response times (<1 ms), such as in the automatic driving of vehicles.

10.3 From radio to optics

As already extensively described in section 10.2, the radio part of the 5G network will be composed of a vast set of heterogeneous networks that will, in general, operate to meet specific needs. As already shown in figure 10.1, a macrocell will provide general coverage, while specific services, with particular requirements in terms of bandwidth and latency, will be provided by smaller cells. To understand this infrastructure it is first necessary to understand how the radio spectrum will be organized and then we will examine some architectural aspects, also looking at the services defined at the end of section 10.2.

10.3.1 Radio spectrum management

For 5G, first of all new frequency bands are needed and in Europe a number of pioneer bands have been identified by the Radio Spectrum Policy Group (RSPG), in particular the following three bands: 700 MHz, 3.6 GHz, and 26 GHz. These bands offer a mixture of features because they operate both below the 6 GHz and in the millimeter wave spectrum, in order to offer a wide range of 5G services such as very high transmission capacities, URLLC, and mMTC.

The three frequency ranges were identified for the following reasons:

- The spectrum below 1 GHz provides wide area coverage and good in-building penetration with enough bandwidth to provide a mix of services, including smart metering, enhanced mobile broadband, and IoT applications.
- The 1 GHz to 6 GHz spectrum provides extended coverage, additional capacity, and high throughput links.

- The spectrum above 6 GHz provides very wide bandwidths (up to 2 GHz) which are suitable for very high transmission rates (in the order of Gbps) and this part of the spectrum is ideal for hot spots in small areas, both indoors and outdoors.

Other bands that the RSPG considers promising for long-term 5G include the 32 GHz and 42 GHz bands, but it also looks for carriers above 60 GHz.

From the point of view of spectrum management, another important aspect of 5G will concern the evolution of video services and in particular of broadcast techniques, with a comparison between the most popular current methods, such as satellite and digital video broadcasting—terrestrial (DVB-T), compared to the more innovative methods operating with mobile devices such as the evolved Multimedia Broadcast Multicast Service (eMBMS) already adopted in the Long-Term Evolution (LTE) field [17]. Some technical requirements relating to eMBMS technology in the 5G context have already been agreed and included in the 3GPP TS22.261 document dedicated to the requirements for 5G systems (*Service Requirements for Next Generation New Services and Markets*) and in the TR38.913 document dedicated to the requirements of radio access (*Study on Scenarios and Requirements for Next Generation Access Technologies*). Some of these requirements are similar to those that are being defined for LTE-eMBMS technology (Rel.14) while other requirements anticipate new features that will be included in the 5G standard.

10.3.2 Wireless network architectures

The use of portions of the spectrum at higher frequencies and with ever greater bands, together with a greater demand for coverage, result in the so-called densification of wireless infrastructures with the deployment of small cells for offloading the network based on macrocells (figure 10.1). However, this densification of antennas will determine an increase in connections (both in fiber and in radio links) to connect the cells with a consequent increase in infrastructure costs. Following the studies reported in [18], the macrocell operating at a frequency below 3 GHz guarantees coverage in large areas while the smaller cells (operating above 3 GHz) serve several purposes including eMBB ultra-broadband access. The smaller millimeter wave cells (picocells) could also be used in scenarios where the combined use of licensed and unlicensed spectrum is envisaged [19].

A fundamental aspect of the 5G networks will be the connection of the antennas to the transport network which will integrate with the access network that is evolving towards fiber to the building (FTTB) and fiber to the home (FTTH) solutions, even if the necessary capacities for the BS suggest connections logically separate from those specifically made for wired users. After all, this sharing of optical fiber infrastructures is easily obtainable with wavelength division multiplexing (WDM) [3, 10–12] systems. In particular, standard formats such as either Gigabit Ethernet (GbE) or techniques such as radio over fiber (RoF) can be adopted. RoF can operate both in analog form, where the spectrum of the radio signal is entirely translated into the optical domain [20], and in digital where the radio signal is sampled and transmitted in digital format on an optical carrier, as in the case of the Common Public Radio Interface (CPRI) [20]. Furthermore, we must not forget optical solutions such as free spectral optics (FSO) systems, based on the optical transmission of signals in the air, which are very economical and have high performance, although over distances that are not too long (less than 1 km in places where fog and rain can be present) [21].

10.3.3 The xHaul segment: backhauling and fronthauling

As already mentioned above, RANs are evolving towards centralized architectures called C-RAN, as schematized in figure 10.3. The BS of figure 10.3(a) is broken down into the two elements RRH and BBU, which will be installed in different places and must be connected with a very high-capacity channel (typically with optical fiber), as in the case of figure 10.3(b) and figure 10.3(c) in which the distance of the link can also be very long (more than a kilometer). In case (d) for specific

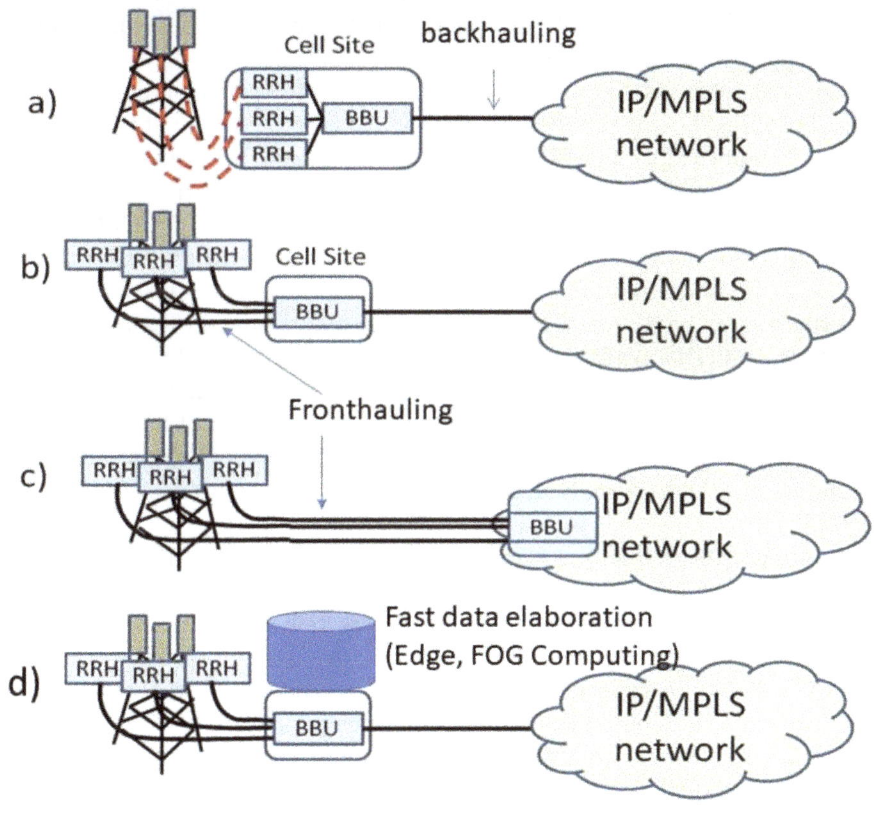

Figure 10.3. The evolution of a base station towards a centralized architecture: (a) traditional, (b) C-RAN with mixed fronthauling/backhauling, (c) C-RAN with pure fronthauling, and (d) edge/fog configuration that requires data elaboration closer and closer to the wireless devices. Solid lines indicate optical fibers while dashed lines show coaxial cables.

services a fast data elaboration is necessary close to the antenna, as in the case of edge and fog computers.

In the (b)–(d) configurations, the RRU has the sole task of receiving the signal already elaborated in a format suitable for being radiated and transmitting it to the antenna. Therefore the baseband modulation format is also created in the BBU and transmitted to the RRU via an optical fiber. However, this BBU–RRU step is not easy because the signal (generally a multi-level orthogonal frequency-division multiplexing (OFDM)) in the double electrical–optical–electrical conversion can undergo strong degradation. The main solution adopted today for the BBU–RRH connection is that of sampling the analog signal and transmitting it on an optical carrier with a digital format, as in the case of CPRI [20]. It should be noted that this process broadens the signal band by up to ten times and therefore the capabilities of CPRI systems are generally very high ($\gg 1$ Gb s^{-1}). For this segment, considering also that the distances of the connections for the fronthauling can also be very extended (figure 10.3(c)), the coarse transmissions WDM and dense WDM have a very important role, also for the topologies of the gigabit passive optical networks (GPONs) [10, 22]. However, it should be noted that, due to the high capacities required by the CPRI, today the analogic RoF transmission techniques are being re-evaluated, as for example shown in [23].

We can therefore say that the most suitable architecture for a complete xHaul network for 5G is the one reported in section 10.4 [18, 21], where at the access level the point to point (P2P) and GPON fiber networks will be managed with slicing techniques designed to create logical and physical circuits with very low latency. The typical configuration used for HetNet will be based on P2P CPRI links for fronthauling from the BBU to RRH and on backhauling with P2P for macrocells, GPON, and Next-Generation Passive Optical Network 2 (NG-PON2) for small cells [24]. Furthermore, the whole network will be able to dynamically manage the CPRI, wavelengths, and GbE available resources according to typical SDN schemes. It has been shown that the architecture of figure 10.4 is also the one that allows the lowest energy consumption [18], also distinguishing the transmission of backhauling from fronthauling [24]. In the same figure twisted-pair transmissions are also considered by adopting very high-capacity systems such as very high-speed digital subscriber line (VDSL2).

Concerning figure 10.3(d) it has to be pointed out that traffic deriving from communications with mission critical devices (such as cars, trains, health apparatus, and so on) leads to the search for architectures that are more suitable for the treatment of some specific data. For example, let us consider the exchange of information between self-guided vehicles that require continuous interactions between cameras, radars, and other sensors. Therefore in this evolution we must also include the new computation and distributed storage models such as fog computing [25] and edge computing. Therefore novel architectures are necessary in order to use computing and analysis equipment located towards the edge of the network in order to physically distribute the computing and data storage capacities within the network itself. In particular, fog computing is a distributed infrastructure in which certain processes or application services are managed at the edge of the

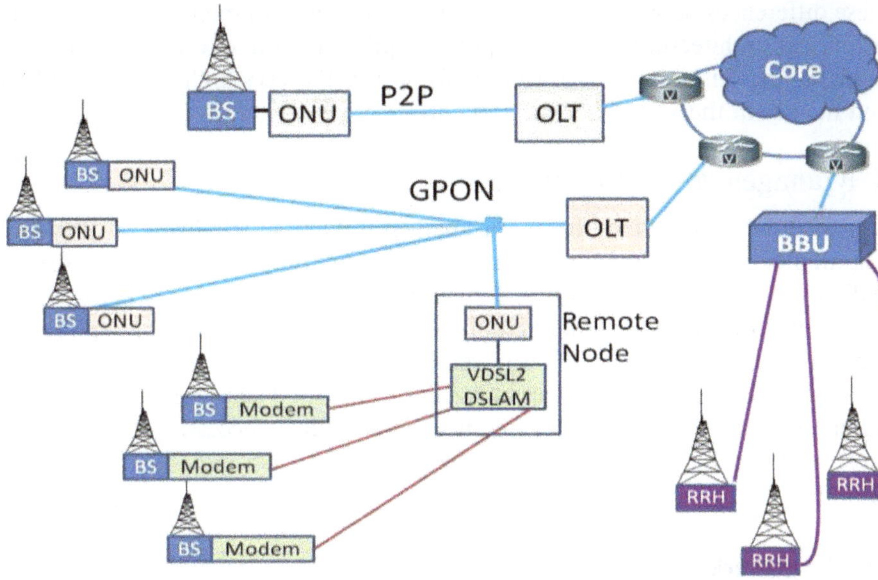

Figure 10.4. Backhauling and fronthauling architectures for macro- and microcells.

network by an intelligent device, while others are still managed in a centralized cloud. Basically, this is a method to allow processing, analysis, and storage of data locally, thus reducing the amount of data that must be sent and managed in a cloud and above all reducing processing times. While fog computing requires the use of an external node or gateway, edge computing processes data directly in the devices themselves.

Fog and the edge are an evolution of cloud computing, in the sense that, in addition to the data storage processes that are typical of the cloud and that take place in sites that can also be very far from the sources, in fog and edge computing the data processing must take place at points as close as possible to the sources. Therefore a fog/edge architecture must guarantee an immediate analysis of the information collected, which must take place as close as possible to the sources, while data analysis that does not require instant decisions can be performed at the cloud level.

A clear example of the use of fog computing is for services related to autonomous driving of vehicles, which represents the sector with the highest safety rate and decision-making factor for data processing. The next 5G networks will connect vehicles that will be equipped with the highest level radar and LiDAR technologies, cameras for image processing, and a whole range of cloud services to support entertainment and maintenance in order to offer remote diagnostics in real time.

All this data must be processed and then transferred, minimizing a series of problems related to latencies. Managing the volume, variety, and speed of IoT data also requires new computational models and data analysis close to the device can make a difference in averting disasters and system failures.

These differences between the cloud and edge/fog environments have consequences in optical architectures since the cloud requires long and very long links with much higher capacities compared to edge/fog where the very high-capacity links are located mainly in the xHaul/access segment.

10.4 Management of 5G networks

Section 10.3 has illustrated the main changes that 5G will bring to the entire telecommunications network, in particular at the architectural level. In this section we will instead address the issue of dynamic resource management in the 5G environment, in particular at the xHaul level. We assume that the modalities that will be adopted are generally already well known in the telecommunications field, but specifically with 5G their use could present a profound acceleration because they would enable a whole series of specific needs that will be necessary for innovative services. Such management is mainly based on two aspects: slicing and SDN procedures.

10.4.1 The network partition for 5G: slicing

To support the immensity of services expected for 5G, it is clear that the entire network must be managed to allow the creation of a vast set of independent subnetworks, each consisting of a set of functions that are selected for a specific service. Such a network partition is called a slice and the operator's physical network is 'sliced' into multiple end-to-end virtual networks (E2E) and each slice is logically isolated [8, 26, 27].

It should be noted immediately that the concept of slicing is not new, and since the birth of the IP network the need to introduce procedures for forwarding traffic with different priorities has been clear. For several years the Multiprotocol Label Switching (MPLS), the Virtual Private LAN Service (VPLS) [28], and more recently the Carrier Ethernet [29] have shown extraordinary properties for the creation of logical networks with different classes of service (CoS), also guaranteeing high reliability. Among the Carrier Ethernet techniques the ones that are currently being considered the most are the Provider Backbone Bridging-Traffic Engineering (PBB-TE) and the MPLS-Transport Profile (MPLS-TP) [26, 30]. As shown in [26, 31], for network slicing different routing techniques can be adopted in the same network, with different encapsulation steps starting from the Virtual LAN (VLAN), up to MPLS and VPLS, with a further advantage offered by the Carrier Ethernet, up to the management of subnetworks on the basis of wavelengths in the framework of WDM architectures. This encapsulation process will, step by step, allow services to be managed that can vary from a few Kb s^{-1}, as in the case of M2M environments, up to the order of Tb s^{-1} per slice that must serve devices for eMBB transmissions. Furthermore, the management of the network must be performed dynamically to follow all the requirements requested by users and which change over time. For this purpose the SDN approach is fundamental [26–33] and also for the management of the activation and deactivation of slicing, following, for example, the procedures defined by 3GPP: preparation, commissioning, operation, and decommissioning.

10.4.2 SDN for 5G: from the centralized up to the distributed model

SDN allows the network to be programmable, easily modulated, and virtualized. These features were initially used primarily for cloud computing environments, where the network infrastructure is shared by a number of competing entities via the NFV mode [15]. Furthermore, extremely fast programmability for most hardware components is now a necessity to make the network dynamic and adaptable to any traffic requirements.

In traditional networks, the data and the control paths occur on the same device (switch, router) and this aspect reduces the flexibility with respect to current and future needs for networks that tend towards a content centric paradigm.

An SDN consists of a centralized control plane with a lower application programming interface (API) for communicating with the hardware infrastructure and an upper API for communicating with network applications (figure 10.5). One of the main protocols that works with the lower APIs is OpenFlow [31], an open standard that has been developed for several years at Stanford University initially to be used in local networks.

It should be emphasized that a certain number of control and management protocols already exist, such as the Simple Network Management Protocol (SNMP), in current routers operating in the field. Therefore this protocol can be considered as a potential lower API and several SDN experiments have been carried out on this aspect by the Fondazione Ugo Bordoni (FUB)-ISCOM Laboratories [33–36].

This SNMP approach falls within the SDN models operating only on the lower layers of the open systems interconnection (OSI) stack (L1–L3) allowing network operators to keep most of the routing devices already installed. Similarly to what happens for slicing processes, the SDN modes for resource management can also use the logical paths based on MPLS, VPLS, and above all on the Carrier Ethernet, through the PBB-TE or the MPLS-TP [30, 33–36]. Clearly the SDN methodology, created above all for the management of the core network, in particular for its connections with cloud elements located in the periphery of the network, has

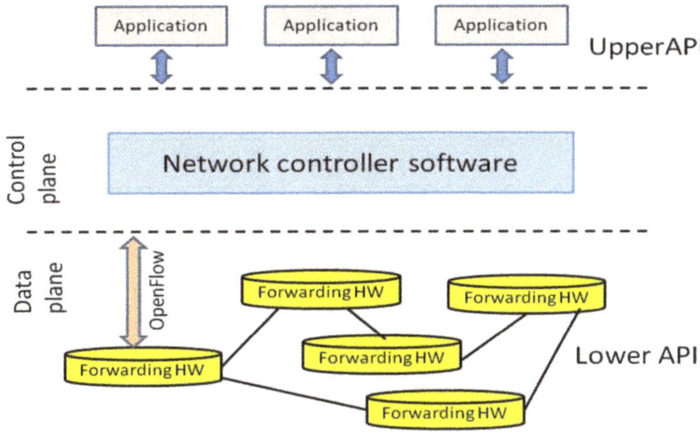

Figure 10.5. Typical SDN architecture.

received a lot of attention from the wireless world. In particular in this area other protocols have been developed that have placed then in competition with OpenFlow, which are more aimed at managing the xHaul network towards radio terminations. This has led to the introduction of a new acronym, SDWN, which stands for software defined wireless network and among the main SDWNs we mention Flow N and Flow Visor [15]. An efficiently functioning SDN needs to know the behavior of the traffic in the network and to monitor the performance, at least in some segments of the network. In this context the model that the FUB has taken into consideration is that of using a measurement plane [35, 36], which allows the state of the network to be known both in terms of traffic and QoS.

Figure 10.6 illustrates this SDN approach based on a measurement plane which is able to analyze through a reasoner [35, 36] all the information arriving from a group of active probes (for example performing QoS tests) located on a set of end-points and passive probes that monitor traffic within the network. For the measurement plane we refer to the mPlane model which was developed within a European project (www.mplane.eu) and for details we refer the reader to the website or to the scientific literature [37, 38]. Figure 10.6 in particular shows the users belonging to three different networks and the measurement plane that collects data both directly from the users (throughput, jitter, etc) and from some measurement points distributed within the networks through the traffic monitoring probes. The reasoner of the measurement plan checks the quality of each end-point and the performance within the network, for example in terms of congestion, and is therefore also able to detect any anomalies that occur in the network. The measurement plane communicates with the SDN and its orchestrator can decide the actions to take on the networks in terms of bandwidth, new connections, and more generally on the allocation of resources.

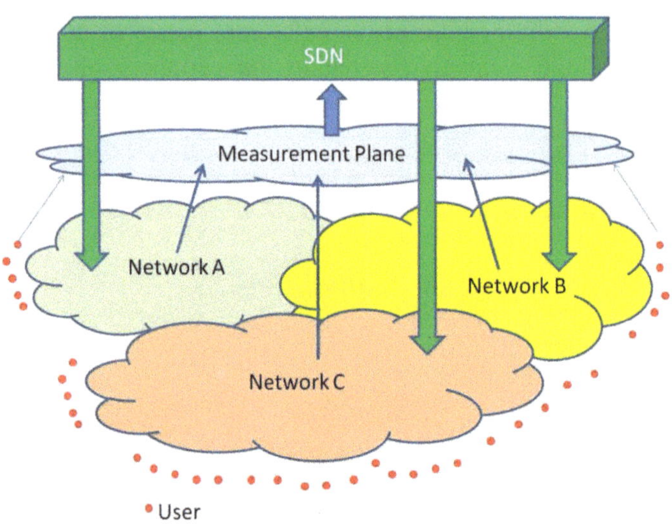

Figure 10.6. Interaction between users, networks, the measurement plane, and SDN in a system with centralized control.

It can be expected that the role of the measurement plane in SDN will become even more important for 5G. In fact, if we refer to the role of the so-called verticals, i.e. entities that require the use of slices, the relationship between users, network operators, measurement planes, and SDNs will be much more complex. To obtain an idea of the different aspects that should be controlled by the measurement plane and managed by the SDN orchestrator in a 5G environment, just think of all the characteristics that must be controlled for each slice, as already described in section 10.4.1.

According to figure 10.6 the SDN is assumed, following its initial introduction, to have a physically centralized controller (orchestrator). However, looking at the evolution towards 5G a centralized control is not able to scale the growth of the network (increase in routing devices, flows, bandwidth, and so on) and moreover it would be very weak in terms of security and reliability. Therefore SDNs are evolving towards distributed controller models following some aspects of well-known routing protocols such as the Border Gateway Protocol (BGP) and Open Shortest Path First (OSPF) [39, 40].

For this purpose it is believed that the architecture of figure 10.6 will have to evolve towards that shown in figure 10.7 which adopts a distributed controller and which would also be very useful in a multi-operator environment, using a single measurement plane that could be useful for complete monitoring of the network. The proposal for a single monitoring plane comes from the experience of the FUB in the mPlane project, with particular regard to the measurement campaigns successfully carried out in the large geographical area of Heidelberg–Milan–Turin–Rome [38].

An example of a distributed SDN controller is shown in figure 10.7 where an architecture with three synchronized SDN controllers is proposed [39]. The cooperation between the SDN controllers is achieved through synchronization protocols

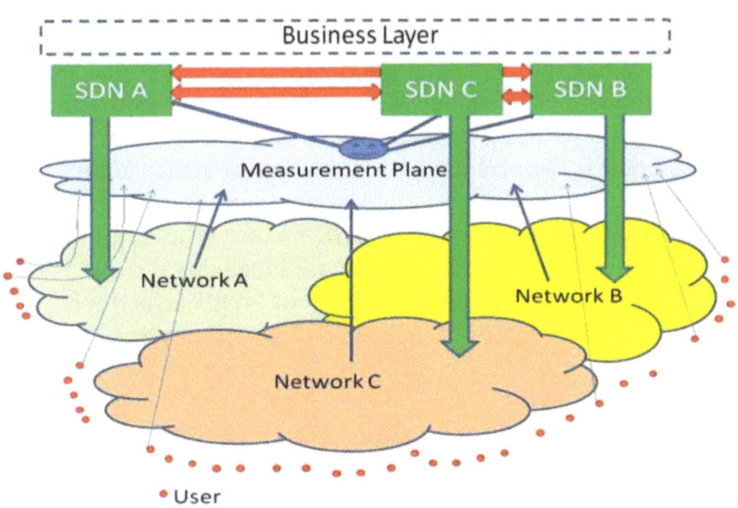

Figure 10.7. Interaction between users, networks, measurement plane, and SDN with distributed controllers.

and each controller directly manages a group of routing devices. Network status information is stored in a distributed database.

In [41] the multi-operator environment is studied in the context of an SDN with distributed controllers, introducing a business layer and a service plan, considering the negotiation procedures between these two levels and the infrastructure levels with the respective orchestrator (an orchestrator for each infrastructure domain).

Returning to the monitoring plan shown in figure 10.7, however, it is necessary to add some consideration of the reasons for adopting a centralized measurement architecture in a distributed SDN approach. In fact, mPlane has been defined with a centralized reasoner and supervisor and this is the reason for the single monitoring plan shown in figure 10.7. However, growing networks can lead to unacceptable processing times for the data collected. As described in [40] for traditional passive flow measurement systems, such as NetFlow and sFlow, which apply packet sampling approaches, the increase in the data to be monitored can induce procedures with low accuracy or with a high cost and consumption of resources. Therefore, following the considerations reported in [40], we can conclude that the architecture of mPlane should be revised in the future to introduce a distributed controller (reasoner).

10.5 Further support of optical technologies for 5G

The description of the 5G network evolution reported in this chapter underlines the fundamental role of optical communications, starting from the xHaul up to the core, with some specific considerations already reported in the previous sections. Certainly, the solutions offered by the optical transmission systems in terms of WDM capacities with coherent and also OFDM formats can satisfy all the Tb s^{-1} necessities requested by some services and adopted in each network segment.

However, some differences have to be emphasized for the xHaul network that shows some peculiarities with respect to the access, metro, and core parts. As described in section 10.3, the main characteristic in the transmission between BBU and RRH is the bandwidth explosion due to the sampling of the radio signal and therefore one of the main challenges will be in finding a methodology to reduce the transmission rate, in particular compared to digital radio over fiber solutions (CPRI). In this contest the exploitation of the analog RoF (ARoF) [23, 25] could be a very interesting solution, even though some signal distortions occurring in the electrical–optical conversion can limit the performance. The ARoF is also important for the generation of radio signal at the millimeter wavelength starting from IF signal, as described for example in figure 3 of [42]. Furthermore, by exploiting the beating between two lasers a sub-millimeter radio signal can be generated, permitting the use of a spectrum of frequency carriers higher than 60 GHz.

Other solutions point to other approaches based on a split of some functions, also introducing a distributed unit (DU) between RRH and BBU [43, 44]. By adopting this architecture some signal processing functions, regarding the first three OSI layers, can be left locally at the antenna site, while others are distributed between the DU and BBU, obtaining so-called functional splits. Several different functional splits are possible in 5G NR [45], even though currently no single optimal splitting solution can be found that meets the trade-off between RAN performance and transport network requirements. Realistic deployments will therefore likely need to adapt to the available transport infrastructure on a case-by-case basis. For this reason 5G networks should ideally support different functional splits. One relevant split specification is the evolved CPRI (eCPRI) [43]. While CPRI is basically a time division multiplexing (TDM) transmission of antenna signal samples, which poses tight requirements in terms of latency, synchronization, and bandwidth, eCPRI transmits the modulation symbols and thus reduces the amount of required bandwidth while making use of statistical multiplexing. This enables scaling up the radio bandwidth by one order of magnitude without a dramatic increase in the bit rate to be transported between the BBU and RRHs compared to CPRI [44].

As illustrated in figure 10.4, a fundamental infrastructure for 5G networks is given by the PON architectures and an important support will be given by the upgrades to WDM-PON and DWDM-PON, operating with channels at 50 Gb s^{-1} and 100 Gb s^{-1} and exploiting all the fiber bandwidth, from the O to the U band [45]. Finally, concerning the role of optical fibers, to cope with the large capacities required by the 5G antennae, particularly in the xHaul segment, multicore and multimode fibers have to be considered, exploiting space division multiplexing software defined networks architectures.

Looking at the core network, most of the issues concerning management have been briefly described in section 10.4. However, other topics should be analyzed regarding the contribution of optical technologies to 5G and one of the most critical relates to the necessity for accurate, stable, and robust frequency, phase, and time synchronization. Therefore, hierarchical synchronization networks are needed that allow frequency, phase, and time coherence of various network components to be controlled. Within the network the required level of stability and accuracy decreases from the core level through aggregation nodes toward the base stations. To ensure synchronization a common time basis, as recommended in the ITU Radio Regulations, optical time transfer (OTT) has been shown as a viable solution [46].

To conclude, mention should be made of the issues around security in 5G networks and the advantages that only optics can provide. The photon with its quantum properties allows an intrinsic management of security that no classical process can allow [47]. Currently, quantum key distribution (QKD) is recognized as one of the most important safe quantum communication approaches [47–50] and it adopts a quantum photon channel to encrypt the information exploiting the polarization state of the light. QKD systems are already operating in several contexts, in different city field tests, and also connecting places over long distances in P2P configurations. Moreover, the coexistence of quantum channels and classical data channels proves the feasibility of adopting QKD in standard WDM telecommunication networks without the need for a fiber surplus [49]. Regulation is also under investigation in the ITU and ETSI organizations and interest is also being shown in the 3GPP for 5G networks [51].

QKD exploits a quantum channel only to produce and distribute keys to encrypt (and decrypt) a message exchanged on a standard classical communication channel [47].

More information on the physical approaches, the protocols, and the performance of this security procedure can be found elsewhere [49] and for reference the most well known protocol is the BB84 protocol.

However, with respect to classical systems some precautions are necessary and some key components of the optical networks cannot be used as optical amplifiers since the amplifier spontaneous emission noise would have a severe degrading effect and the feasibility of quantum repeaters in practical long distance QKD networks has yet to be demonstrated [49]. Fortunately, at the moment the transmission of quantum secure information over long distance can be achieved by means of trusted repeater nodes (TRNs) [48, 49], located in the same places as the optical amplifiers, which have been widely adopted for real-world deployed QKD networks, as is the case between Beijing and Shanghai [48].

QKD will also play a fundamental role in the area of access and examples of QKD approaches are already present, not only in P2P links but also in GPONs [52] and FSO architectures [53]. Furthermore, QKDs have shown interesting applications in multimode and multicore fibers [54].

However, for the complete introduction of QKD in all telecommunication networks, from the access to the core, several technological steps are still necessary. Despite these difficulties there are already many proposals to include QKD techniques in novel wide area networks with methodologies for also managing quantum resources in large geographic networks, adopting SDN approaches and operating in slice contexts for 5G networks [55].

10.6 Conclusions

In this chapter an overview has been presented of the evolution that the entire telecommunications network will undergo in the context of 5G, not only in terms of fiber optic architectures, but also for radio interfaces, information routing, and data processing. We have seen that one of the main characteristics of 5G is that of distributing traffic over different networks of a heterogeneous type, carrying out coverage with a set of small and large cells.

For the development of 5G it will first be necessary to make new frequency bands available for radio transmissions, but several architectural changes will also be required to ensure efficient connection of the antennas, even if appropriate modalities for traffic management, with the implementation of appropriate logical subnetworks and techniques, are also necessary, in particular to allow the management of the dynamics of resources respecting the requests of the users and the necessary operating modes for services.

References

[1] Andrews J, Buzzi S, Choi W, Hanly S V, Lozano A, Soon A and Zhang J C 2014 What will 5G be? *IEEE J. Sel. Area Telecommun.* **32** 1065–82

[2] Sun S, Gong L, Rong B and Lu K 2015 An intelligent framework for 5G heterogeneous networks *IEEE Comm. Mag.* **53** 142–7

[3] Li X, Casellas R, Landi G, de Olivera A, Costa-Perez X, Garcia-Saavedra A, Deiss T, Comardi L and Vilalta R 2017 5G-crosshaul network slicing: enabling multi-tenant in mobile transport networks *IEEE Commun. Mag.* **55** 128–37
[4] European Commission 2016 *5G for Europe: An Action Plan* COM(2016)588
[5] Rost P, Banchs A, Berberania I, Breibach M, Doll M, Droste H, Manweiler C, Puente M A, Samdanis K and Sayadi B 2016 Mobile network architecture evolution towards 5G *IEEE Comm. Mag.* **54** 84–91
[6] Settembre M 2012 Towards a hyperconnected world *15th Int. Telecommunications Network Strategy and Planning Symp., IEEE Networks 2012 (Rome, October 12–15, 2012)*
[7] Fan C, Zhang C F and Yuan X 2016 Advances and challenges towards a scalable cloud radio access network *IEEE Commun. Mag.* **54** 29–35
[8] Zhou L X, Chen R and Zhang H 2016 Network slicing as a service: enabling enterprises' own software-defined cellular networks *IEEE Commun. Mag.* **54** 146–53
[9] Nikolikj V and Janevski T 2014 A comparative cost-capacity modeling of wireless heterogeneous networks, implemented within the 0.7 GHz, 2.6 GHz, 5 GHz and 28 GHz bands *IEEE Int. Conf. on Ultra-WideBand (ICUWB) (Paris, 1–3 September)*
[10] Mitchel J E 2014 Integrated wireless backhaul over optical access networks *J. Lightw. Technol.* **32** 3373–82
[11] Musumeci F *et al* 2016 Optimal BBU placement for 5G C-RAN deployment over WDM aggregation networks *J. Lightw. Technol.* **34** 1963–70
[12] Sezer S 2013 Are we ready for SDN? Implementation challenges for software-defined networks *IEEE Commun. Mag.* **51** 36–43
[13] Foster N 2013 Languages for software-defined networks *IEEE Commun. Mag.* **51** 128–34
[14] 5GPPP 2017 Vision on software networks and 5G, January https://5g-ppp.eu/wp-content/uploads/2014/02/5G-PPP_SoftNets_WG_whitepaper_v20.pdf
[15] Zhou Q, Wang C-X, McLaughlin S and Zhou X 2015 Network virtualization and resource description in software-defined wireless networks *IEEE Commun. Mag.* **51** 110–7
[16] NGMN Alliance 2015 5G White Paper https://www.ngmn.org/wp-content/uploads/NGMN_5G_White_Paper_V1_0.pdf
[17] Hein T *et al* 2016 Integration of broadcast and broadband in LTE/5G (IMB5)—experimental results from the eMBMS testbeds *2016 European Conf. on Networks and Communications (EuCNC)*
[18] Carciofi C, Grazioso P and Matera F 2017 Power consumption in mobile HetNets *AEIT Int. Annual Conf. 2017 (Cagliari)*
[19] Guiducci D *et al* 2017 Sharing under licensed shared access in a live LTE network in the 2.3–2.4 GHz band end-to-end architecture and compliance results *IEEE International Symp. on Dynamic Spectrum Access Networks*
[20] Chen L 2014 Adaptive photonics-aided coordinated multipoint transmissions for next generation mobile fontthauling *J. Lightw. Technol.* **32** 1907–14
[21] Parca G, Shahpari A, Carrozzo V, Tosi Beleffi G M and Teixeira A L J 2013 Optical wireless transmission at 1.6 Tb/s (16 × 100 Gb/s) for next-generation convergent urban infrastructures *J. Opt. Eng.* **52** 102–16
[22] Pizzinat A, Chanclou P, Saliou F and Diallo T 2015 Things you should know about fronthaul *J. Lightw. Technol.* **33** 1077–83
[23] Agus M *et al* 2018 Novel 5G broadband front-haul architectures for ultra-dense and hot-spot areas exploiting optical technologies and photonic integrated circuits *20th Italian National Conf. on Photonic Technologies (Lecce, 23–25 May)*

[24] Matera F *et al* 2018 Fronthauling–backhauling contributions to power consumption in 5G optical links *20th Italian National Conf. on Photonic Technologies (Lecce, 23–25 May)*

[25] Sung M *et al* 2020 RoF-based radio access network for 5G mobile communication systems in 28 GHz millimeter-wave *J. Lightw. Technol.* **38** 409–20

[26] Katsalis K, Nikaein N, Schiller E, Ksentini A and Braun T 2017 Network slices towards 5G communications: slicing the LTE network *IEEE Commun. Mag.* **55** 146–54

[27] 5GPPP 2017 Vision on software networks and 5G https://5g-ppp.eu/wp-content/uploads/2014/02/5G-PPP_SoftNets_WG_whitepaper_v20.pdf

[28] Valenti A, Bolletta P, Pompei S and Matera F 2009 Experimental investigations on restoration techniques in a wide area Gigabit Ethernet optical test-bed based on virtual private LAN service *Proc. of ICTON 09 (Ponta Delgada (PO), 29 June–2 July)*

[29] Coiro A, Valenti A, Pompei S, Matera F, Testa P and Settembre M 2009 Network evolution towards carrier-grade ethernet transport network *Fiber Integr. Opt.* **28** 393–411

[30] Matera F *et al* 2017 Optical network slicing approaches with carrier ethernet tests *Fotonica 2017 (Padova, 3–5 May)*

[31] McKeown N, Anderson T, Balakrishnan H, Parulkar G, Peterson L, Rexford J, Shenker S and Turner J 2008 Openflow: enabling innovation in campus networks *SIGCOMM CCR* **38** 69–74

[32] Persia S, Carciofi C and Faccioli M 2017 NB-IoT and LoRA connectivity analysis for M2M/IoT smart grids applications *AEIT Int. Annual Conf. 2017 (Cagliari, 20–22 September)*

[33] Tego E *et al* 2014 Facing the reality: validation of energy saving mechanisms on a testbed *J. Elect. Comp. Engin.* **2014** 806960

[34] Tego E, Matera F, Del Buono D and Attanasio V 2014 Quality of service management based on software defined networking approach in wide GbE networks *EuMed Telco (Naples)*

[35] Tego E, Carciofi C, Grazioso P, Petrini V, Pompei S, Matera F, Attanasio V, Nastri E and Restuccia E 2017 A measurement plane for optical networks to manage emergency events *Fiber Integr. Opt.* **36** 227–41

[36] Tego E, Matera F, Attanasio V and Nastri E 2017 A measurement plane to monitor and manage QoS in optical access networks *IEEE Fiber Optics in Access Networks 2017 (FOAN 2017) Conf. (Munich, 7 November)*

[37] Casas P, Fiadino P, Wassermann S, Traverso S, D'Alconzo A, Tego E, Matera F and Mellia M 2016 Unveiling network and service performance degradation in the wild with mPlane *IEEE Commun. Mag.* **54** 71–6

[38] Mellia M, Traverso S, Tego E, Rufini A, Valenti A, Matera F, Scarpino M, Kahveci A and Invernizzi F 2016 Quality of service monitoring adopting correlation among active and passive measurements *17th Int. Network Strategy and Planning Symp.—Network 2016 (Montreal, 26–28 September)*

[39] Blial O, Mamoun M B and Banaini R 2016 An overview on SDN architectures with multiple controllers *J. Comp. Commun.* **2016** 9396525

[40] Tahaei H, Salleh R, Razak M, Ko K and Anuar N B 2018 Cost effective network flow measurement for software defined networks: a distributed controller scenario *IEEE Access* **6** 5182

[41] Flavio B *et al* 2012 Fog computing and its role in the internet of things *Proc. of the First Edition of the MCC Workshop on Mobile Cloud Computing in SIGCOMM '12: ACM SIGCOMM 2012 Conf. (Helsinki, 17 August 2012)*

[42] Rommel E *et al* 2020 Towards a scaleable 5G fronthaul: analog radio-over-fiber and space division multiplexing *J. Lightw. Technol.* **31** 5412–22

[43] Jovanna P, Cavaliere F, Stracca S, Giorgi L and Ubaldi F 2020 5G xHaul and service convergence: transmission, switching and automation enabling technologies *J. Lightw. Technol* **38** 2799–806
[44] Sabella R, Iovanna P, Cavalieri F and Bottari G 2020 Optical transport for industry 4.0 *J. Opt. Commun. Netw.* **12** 264–72
[45] Wey J 2020 The outlook for PON standardization: a tutorial *J. Lightw. Technol.* **38** 31–42
[46] Śliwczyński Ł, Krehlik P, Imlau H, Ender H, Schnatz H, Piester D and Bauch A 2020 Fiber-based UTC dissemination supporting 5G telecommunications networks *IEEE Commun. Mag.* **58** 66–73
[47] Diamanti E *et al* 2016 Practical challenges in quantum key distribution *Quantum Inf.* **2** 16025
[48] Zhang Q, Xu F, Chen Y, Peng C and Pan J 2018 Large scale quantum key distribution: challenges and solutions *Opt. Express* **26** 24260–73
[49] Cao Y *et al* 2019 Cost-efficient quantum key distribution (QKD) over WDM networks *J. Opt. Comm. Netw.* **11** 285–98
[50] Eraerds P *et al* 2010 Quantum key distribution and 1 Gbps data encryption over a single fibre *New J. Phys.* **12** 063027
[51] Jin C *et al* 2020 A survey on security aspects for 3GPP 5G networks *IEEE Commun. Tutor.* **22** 170–95
[52] Trinh P V *et al* 2018 Design and security analysis of quantum key distribution protocol over free-space optics using dual-threshold direct-detection receiver *IEEE Access* **6** 4159–75
[53] Hugues-Salas E *et al* 2019 Monitoring and physical-layer attack mitigation in SDN-controlled quantum key distribution networks *J. Opt. Commun. Netw.* **11** A209–18
[54] Cao Y *et al* 2017 Resource assignment strategy in optical networks integrated with quantum key distribution *J. Opt. Commun. Netw.* **9** 995–1004
[55] Wang R *et al* 2020 End-to-end quantum secured inter-domain 5G service orchestration over dynamically switched flex-grid optical networks enabled by a q-ROADM *J. Lightw. Technol.* **38** 139–40

www.ingramcontent.com/pod-product-compliance
Ingram Content Group UK Ltd.
Pitfield, Milton Keynes, MK11 3LW, UK
UKHW051041120926
8556IPUK00030B/678

9 780750 332415